AF901147

Mothe
85,- €
'20

Birkhäuser

Mängelexemplar

Trends in Mathematics

Trends in Mathematics is a series devoted to the publication of volumes arising from conferences and lecture series focusing on a particular topic from any area of mathematics. Its aim is to make current developments available to the community as rapidly as possible without compromise to quality and to archive these for reference.

Proposals for volumes can be submitted using the Online Book Project Submission Form at our website www.birkhauser-science.com.

Material submitted for publication must be screened and prepared as follows:

All contributions should undergo a reviewing process similar to that carried out by journals and be checked for correct use of language which, as a rule, is English. Articles without proofs, or which do not contain any significantly new results, should be rejected. High quality survey papers, however, are welcome.

We expect the organizers to deliver manuscripts in a form that is essentially ready for direct reproduction. Any version of TeX is acceptable, but the entire collection of files must be in one particular dialect of TeX and unified according to simple instructions available from Birkhäuser.

Furthermore, in order to guarantee the timely appearance of the proceedings it is essential that the final version of the entire material be submitted no later than one year after the conference.

More information about this series at http://www.springer.com/series/4961

Vladimir Georgiev • Tohru Ozawa •
Michael Ruzhansky • Jens Wirth
Editors

Advances in Harmonic Analysis and Partial Differential Equations

Birkhäuser

Editors
Vladimir Georgiev
Department of Mathematics
University of Pisa
Pisa, Italy

Tohru Ozawa
Department of Applied Physics
Waseda University
Tokyo, Japan

Michael Ruzhansky
Department of Mathematics
Ghent University
Gent, Belgium

Jens Wirth
Department of Mathematics
University of Stuttgart
Stuttgart, Germany

ISSN 2297-0215 ISSN 2297-024X (electronic)
Trends in Mathematics
ISBN 978-3-030-58214-2 ISBN 978-3-030-58215-9 (eBook)
https://doi.org/10.1007/978-3-030-58215-9

Mathematics Subject Classification: 35-XX, 42Bxx, 42Cxx, 43Axx, 47Gxx, 35Q55, 35L70, 42B37

© Springer Nature Switzerland AG 2020
This work is subject to copyright. All rights are reserved by the Publisher, whether the whole or part of the material is concerned, specifically the rights of translation, reprinting, reuse of illustrations, recitation, broadcasting, reproduction on microfilms or in any other physical way, and transmission or information storage and retrieval, electronic adaptation, computer software, or by similar or dissimilar methodology now known or hereafter developed.
The use of general descriptive names, registered names, trademarks, service marks, etc. in this publication does not imply, even in the absence of a specific statement, that such names are exempt from the relevant protective laws and regulations and therefore free for general use.
The publisher, the authors, and the editors are safe to assume that the advice and information in this book are believed to be true and accurate at the date of publication. Neither the publisher nor the authors or the editors give a warranty, expressed or implied, with respect to the material contained herein or for any errors or omissions that may have been made. The publisher remains neutral with regard to jurisdictional claims in published maps and institutional affiliations.

This book is published under the imprint Birkhäuser, www.birkhauser-science.com, by the registered company Springer Nature Switzerland AG.
The registered company address is: Gewerbestrasse 11, 6330 Cham, Switzerland

Preface

Harmonic analysis and the analysis of partial differential equations are two active fields of mathematical research with many close interactions and emerging new ideas. The aim of this volume is to bring together a diverse range of current research topics, providing insights in novel approaches and highlighting some of the inspiring connections between seemingly different areas of pure and applied mathematics. We also hope to set new impulses for future interactions.

Most of the contributions in this volume are related to talks given and results presented at the 12th ISAAC Congress in Aveiro in July/August 2019. The special session *Harmonic Analysis and PDEs* attracted 43 participants from 15 countries.

Pisa, Italy	Vladimir Georgiev
Tokyo, Japan	Tohru Ozawa
Gent, Belgium	Michael Ruzhansky
Stuttgart, Germany	Jens Wirth
June 2020	

Contents

Contributors

Jonas Brinker Instut für Analysis, Dynamik und Modellierung, Universität Stuttgart, Stuttgart, Germany

Duván Cardona Ghent University, Department of Analysis, Logic and Discrete Mathematics, Ghent, Belgium

Marianna Chatzakou Imperial College London, London, UK

César Del Corral Universidad de Los Andes, Department of Mathematics, Bogotá, Colombia

Jishan Fan Department of Applied Mathematics, Nanjing Forestry University, Nanjing, P.R. China

Vladimir Georgiev Dipartimento di Matematica, Università di Pisa, Pisa, Italy
Faculty of Science and Engineering, Waseda University, Tokyo, Japan
Institute of Mathematics and Informatics at BAS, Sofia, Bulgaria

Masaru Hamano Department of Mathematics, Graduate School of Science and Engineering, Saitama University, Saitama, Japan

Takahisa Inui Department of Mathematics, Graduate School of Science, Osaka University, Toyonaka, Osaka, Japan

Masahiro Ikeda Center for Advanced Intelligence Project, RIKEN, Saitama, Japan
Department of Mathematics, Faculty of Science and Technology, Keio University, Yokohama, Japan

Keiichi Kato Department of Mathematics, Tokyo University of Science, Tokyo, Japan

Annunziata Loiudice Department of Mathematics, University of Bari, Bari, Italy

Ramesh Manna Harish-Chandra Research Institute, Jhusi, Allahabad, India

Hayato Miyazaki Teacher Training Courses, Faculty of Education, Kagawa University, Takamatsu, Kagawa, Japan

Kiyoshi Mochizuki Emeritus, Department of Mathematics, Chuo University, Kasuga, Tokyo, Japan

Tohru Ozawa Department of Applied Physics, Waseda University, Tokyo, Japan

P. K. Ratnakumar Harish-Chandra Research Institute, HBNI, Jhunsi, Allahabad, India

Motohiro Sobajima Department of Mathematics, Faculty of Science and Technology, Tokyo University of Science, Noda-shi, Chiba, Japan

George Venkov Faculty of Applied Mathematics and Informatics, Technical University of Sofia, Sofia, Bulgaria

Jens Wirth Instut für Analysis, Dynamik und Modellierung, Universität Stuttgart, Stuttgart, Germany

Local Smoothing of Fourier Integral Operators and Hermite Functions

Ramesh Manna and P. K. Ratnakumar

(Dedicated to Prof. S. Thangavelu)

Abstract We prove local smoothing estimates for the Fourier integral operators of the form

$$\mathcal{F}f(x,t) = \int_{\mathbb{R}^2} e^{i(x\cdot\xi + t|\xi|)}\, a(x,t,\xi)\, \hat{f}(\xi)\, d\xi$$

for a wide class of symbols $a \in S^m(\mathbb{R}^2 \times \mathbb{R} \times \mathbb{R}^2)$, $m \leq 0$. Our result generalises the local smoothing estimate of Mockenhaupt, Seeger and Sogge, and is a global result with respect to the space variable. The novelty in our approach is the use of harmonic analysis of Hermite functions in the study of Fourier integral operators.

Keywords Fourier integral operator · Wave front set · Local smoothing · Littlewood-Paley square function · Hermite expansion

1 Introduction

A Fourier integral operator is an integral operator of the form

$$\mathcal{F}f(x) = \int_{\xi\in\mathbb{R}^M} \int_{y\in\mathbb{R}^n} e^{i\phi(x,y,\xi)}\, a(x,y,\xi)\, f(y)\, dy\, d\xi, \quad f \in \mathcal{S}(\mathbb{R}^n) \tag{1}$$

R. Manna
Harish-Chandra Research Institute, Jhunsi, Allahabad, India
e-mail: rameshmanna@hri.res.in

P. K. Ratnakumar (✉)
Harish-Chandra Research Institute, HBNI, Jhunsi, Allahabad, India
e-mail: ratnapk@hri.res.in

© Springer Nature Switzerland AG 2020

V. Georgiev et al. (eds.), *Advances in Harmonic Analysis and Partial Differential Equations*, Trends in Mathematics, https://doi.org/10.1007/978-3-030-58215-9_1

$x \in \mathbb{R}^N$, where the phase function ϕ is real valued and positively homogeneous of degree 1 with respect to the ξ variable and smooth for $\xi \neq 0$. The amplitude function $a \in S^m(\mathbb{R}^N \times \mathbb{R}^n \times \mathbb{R}^M)$, the symbol class of order $m \in \mathbb{R}$: i.e. a is a smooth function on $\mathbb{R}^N \times \mathbb{R}^n \times \mathbb{R}^M$ satisfying the estimate $|\partial_{x,y}^{\beta} \partial_{\xi}^{\alpha} a(x, y, \xi)| \leq C_{\alpha,\beta}(1 + |\xi|)^{m-|\alpha|}$ for some constant $C_{\alpha,\beta}$, for all multi-indices α, β.

The basic theory of Fourier integral operators was developed by Hörmander in [13] in 1971, soon after the work of Eskin [9], who has actually studied such operators as degenerate elliptic pseudo differential operators. Hörmander established the local L^2 regularity estimate for such operators in terms of certain geometric conditions on the phase function, when the amplitude function $a \in S^0$. The global L^2 regularity estimate has been obtained by K. Asada and D. Fujiwara under certain assumptions on amplitude and phase functions, see [1]. This result has been extended to a larger class of Fourier integral operators by Ruzhansky and Sugimoto in [20].

The local L^p regularity estimates has been proved by Seeger, Sogge and Stein [21] for Fourier integral operators with amplitude function $a \in S^m$ for the range $m \leq -(n-1)|1/p - 1/2|$, $1 < p < \infty$. When $p = 2$ this shows that a can be in S^0. Also letting p tend to 1 or ∞ in the above inequality, we see that the local L^p boundedness for $1 < p < \infty$, requires that $a \in S^m$ with $-\frac{n-1}{2} < m \leq 0$, as $S^\mu \subset S^\nu$ if $\mu < \nu$. The above range of p is also optimal in view of the result of Peral [16] and Miyachi [17] on wave equation, which corresponds to the phase function $\phi(x, t, \xi) = x \cdot \xi + t|\xi|$, $x, \xi \in \mathbb{R}^n, t > 0$ and $a \equiv 1$. The global L^p estimate has been obtained by Coriasco and Ruzhansky, under additional assumptions, see [5].

In the context of wave equation, C. D. Sogge observed that averaging over $t \in [1, 2]$, results in a gain of regularity in L^p for $2 < p < \infty$, see [23]. In fact, Sogge showed that there is an $\epsilon(p) > 0$ such that the following estimate

$$\left(\int_{t=1}^{2} \int_{\mathbb{R}^2} |(I - \Delta)^{\frac{\sigma}{2}} \mathcal{F}f(x, t)|^p \, dxdt \right)^{\frac{1}{p}} \leq c_{\sigma,p} \|f\|_{L^p(\mathbb{R}^2)} \tag{2}$$

holds for all $\sigma \in \mathbb{C}$ with $\mathrm{Re}(\sigma) < \left(\frac{1}{p} - \frac{1}{2}\right) + \epsilon(p)$, $2 < p < \infty$. Compared to the result of [16, 17, 21] this amounts to a gain in regularity for $\mathcal{F}f$ by $\epsilon(p)$. The estimate (2) is called the local smoothing estimate of order $\epsilon(p)$. Local smoothing estimates were previously obtained for Schrödinger equation in [22], see also [29],[6].

Mockenhaupt, Seeger and Sogge later gave an estimate for $\epsilon(p)$ for the Fourier integral operators with amplitude function of the form $a(x, t, \xi) = \rho_1(x, t)\, a_1(\xi)$ with $\rho_1 \in C_c^\infty(\mathbb{R}^2 \times \mathbb{R})$. They showed that $\epsilon(p) \leq \frac{1}{2p}$ for $p \geq 4$ and $\epsilon(p) \leq \frac{1}{2}(\frac{1}{2} - \frac{1}{p})$ for $2 < p \leq 4$, see [18]. It is conjectured that $\epsilon(p) = \frac{1}{p}$ for $p \geq 4$, see [23]. They have also extended their results to manifolds in [19]. Recently some improvement on regularity has been obtained by Lee and Vargas [15], as a consequence of their study on cone multipliers associated to wave equation in $\mathbb{R}^3$. For more on cone multipliers and its connection with the local smoothing estimates, we refer to Wolff [30] and the references therein.

The main result in this paper concerns the local smoothing estimate for Fourier integral operators of the form

$$\mathcal{G}f(x,t) = \int_{\mathbb{R}^2} e^{i(x\cdot\xi + t|\xi|)}\, a(x,t,\xi)\, \hat{f}(\xi)\, d\xi, \ \ f \in \mathcal{S}(\mathbb{R}^2) \tag{3}$$

with a general amplitude function $a \in S^m(\mathbb{R}^2 \times \mathbb{R} \times \mathbb{R}^2)$, $m \leq 0$, depending on (x,t) and ξ variables with some mild decay assumption with respect to the (x,t) variables. Here $\hat{f}$ denotes the Fourier transform of f. Since we are interested in the local (in t-variable) smoothing estimate, it is enough to work with the Fourier integral operators of the form

$$\mathcal{F}f(x,t) = \rho_1(t) \int_{\mathbb{R}^2} e^{i(x\cdot\xi + t|\xi|)}\, a(x,t,\xi)\, \hat{f}(\xi)\, d\xi, \ \ f \in \mathcal{S}(\mathbb{R}^2) \tag{4}$$

for $\rho_1 \in C_c^\infty(\mathbb{R})$. The local smoothing estimate for $\mathcal{G}f$ then follows from the estimate for $\mathcal{F}f$, by a suitable choice of $0 \leq \rho_1 \in C_c^\infty(\mathbb{R})$.

The local smoothing estimate is given in terms of the fractional L^p Sobolev spaces. Recall that the L^p Sobolev space of order $\alpha \geq 0$ is defined by $L^p_\alpha(\mathbb{R}^n) = (-\Delta + I)^{-\alpha/2} L^p(\mathbb{R}^n)$, the Sobolev space of L^p functions on $\mathbb{R}^n$ with 'α derivatives' in L^p, see [25]. L^p_α becomes a Banach space with norm $\|f\|_{L^p_\alpha} = \|(-\Delta + I)^{\alpha/2} f\|_p$. Note that L^p_α is also defined for complex α, and are spaces of tempered distributions when $\mathrm{Re}(\alpha) < 0$.

We also need to consider the Hermite Sobolev spaces $W_H^{k,p}(\mathbb{R}^3)$, $k \geq 0$, $1 \leq p \leq \infty$. Let $H = -\Delta_{x,t} + |x|^2 + t^2$ denote the Hermite operator on $\mathbb{R}^3$, where $\Delta_{x,t}$ is the Laplacian in the (x,t) variable. Then $W_H^{k,p}(\mathbb{R}^3)$ consists of tempered distributions u on $\mathbb{R}^3$, for which $\|H^k u\|_{L^p(\mathbb{R}^3)} < \infty$. In fact the only assumption we make on the amplitude function $a(x,t,\xi) \in S^m$, $m \leq 0$ is that it satisfies the following Hermite-Sobolev estimate:

$$\|\partial_\xi^\alpha a(\cdot,\cdot,\xi)\|_{W_H^{4,\infty}(\mathbb{R}^2\times\mathbb{R})} := \|H^4 \partial_\xi^\alpha a\|_{L^\infty(\mathbb{R}^2\times\mathbb{R})} \leq C_\alpha (1 + |\xi|)^{m-|\alpha|} \tag{5}$$

for all multi indices α, with a constant C_α independent of ξ. Our main result is the following local smoothing estimate, which extends the result obtained in [18] to much more general class of amplitude functions $a(x,t,\xi)$, in particular leading to global result with respect to the space variable.

Theorem 1.1 *Let $\mathcal{G}f$ be the Fourier integral operator given by* (3) *with amplitude function $a \in S^m$, $m \leq 0$ satisfying* (5). *Then for any compact t-interval I, $\mathcal{G}f$ satisfies the estimate:*

$$\|\mathcal{G}f\|_{L^p(\mathbb{R}^2\times I)} \leq C_\sigma \|f\|_{L^p_{m-\sigma}(\mathbb{R}^2)} \tag{6}$$

for all $f \in L^p_{m-\sigma}(\mathbb{R}^2)$ *with a constant* C_σ *depending on the length of I, where*

$$\begin{cases} Re(\sigma) < \frac{1}{2}(\frac{1}{p} - \frac{1}{2}), & \textit{for } 2 < p \leq 4, \\ Re(\sigma) < \frac{3}{2p} - \frac{1}{2}, & \textit{for } 4 \leq p < \infty. \end{cases} \tag{7}$$

In fact, Theorem 1.1 gives the local smoothing of order upto $\epsilon(p) = \frac{1}{2p}$ for $4 \leq p < \infty$. This is the same order of regularity established in [18] for Fourier integral operators with amplitude functions of the form $\rho_1(x,t)\, a(\xi)$. Note that since $\rho_1(x,t)$ is compactly supported, the result in [18] is local in space variable. In contrast, our result given by Theorem 1.1 is a global result with respect to the space variable.

Note that the arguments in [18] relies on the Fourier multipliers arguments, valid only for amplitude function depending on ξ variable alone, hence do not carry over to the general amplitude function of the form $a(x,t,\xi)$. However, we show that the Fourier integral operators with general amplitude function could be dealt with, considering a family of Fourier integral operators $\{\mathcal{F}^n\}_n$ given by

$$\mathcal{F}^n f(x,t) = \rho_1(t) \int_{\mathbb{R}^2} e^{i(x\cdot\xi + t|\xi|)}\, a_n(\xi)\, \hat{f}(\xi)\, d\xi, \quad f \in \mathcal{S}(\mathbb{R}^2) \tag{8}$$

with $a_n \in S^0(\mathbb{R}^2)$, $n \in \mathbb{N}_0^3$ where $\mathbb{N}_0 = \mathbb{N} \cup \{0\}$.

Unfortunately, we cannot appeal to the local smoothing estimate proved in [18] for $\mathcal{F}^n$ in our situation, as we need more refined estimate in terms of n. These estimates are established in Sects. 6 and 7. We also adopt some methods used in [18] like the wave front set analysis, to isolate the region where the Fourier transform of $\mathcal{F}^n f$ has rapid decay, and the maximal function estimate etc. However, our approach is more direct, using a duality argument via an improved estimate for square function based purely on angular decomposition. In the process we also obtain a square function estimate that is new for the range $4/3 \leq p < 2$, see Theorem 5.1.

Now we define the wave front set of a distribution, which is a refinement of the notion of singular support of a distribution, that also carries the information on the direction along which the singularity exists.

Definition 1.1 Let u be a distribution defined on an open set $X \subset \mathbb{R}^2$. The wave front set of u, denoted by $WF(u)$ is defined as the complement of the largest open cone in $X \times (\mathbb{R}^2 \setminus \{0\})$ of $(x_0, \xi_0) \in X \times (\mathbb{R}^2 \setminus \{0\})$ such that the Fourier transform inequality

$$|\widehat{\varphi u}(\xi)| \leq C_N (1 + |\xi|)^{-N} \tag{9}$$

holds for all φ compactly supported in a neighborhood of x_0 and for all ξ in a conic neighborhood V of ξ_0, for some constant C_N, for all $N \in \mathbb{N}$.

We refer the reader to [13] for more on the wave front set of a distribution and also for Fourier integral operators, see also the books [7, 24]. It is possible to locate the wave front set of a distribution given by oscillatory integrals. In fact by Proposition 2.5.7 in [13, page 123] we have the following

Remark 1.1 The wave front set of the distribution given by the oscillatory integral (8) on $\mathbb{R}^3$, is contained in $\{(x, t, \xi, |\xi|) : |x| = t,\ \xi \in \mathbb{R}^2 \setminus \{0\}\}$, provided $a_n \equiv 0$ near the zero section $\xi \equiv 0$. Thus if $a_n \equiv 0$ near the zero section, then the Fourier transform of $\mathcal{F}^n f(x, t)$ has rapid decay away from the light cone $(\xi, |\xi|)$.

2 Decompositions of Fourier Integral Operators

Now we analyse the Fourier integral operators $\mathcal{F}^n$ given by (8). We first express $\mathcal{F}^n$ as an infinite sum of dyadic Fourier integral operators $\{\mathcal{F}^n_j\}_{j\geq 0}$ as follows: Choose $\rho_0 \in C_c^\infty([\frac{1}{2}, 2])$ such that $\sum\limits_{j\in\mathbb{Z}} \rho_0(2^{-j}|\xi|) = 1$. See [8, page 162], for the construction of such a $\rho_0 \geq 0$. For technical reasons, we take ρ_0 to be of the form $\rho_0 = \rho^2$ with $0 \leq \rho \in C_c^\infty([\frac{1}{2}, 2])$. Setting $\varphi_0 = \sum_{j\leq 0} \rho_0(2^{-j}|\xi|)$, we can write $1 = \varphi_0 + \sum_{j\in\mathbb{N}} \rho_0(2^{-j}|\xi|)$, where φ_0 is a smooth function supported in the ball $|\xi| \leq 2$. Note that since $\rho_0(2^{-j}|\xi|)$ is supported on the annulus $2^{j-1} \leq |\xi| \leq 2^{j+1}$, the functions $\xi \to \rho_0(2^{-j}|\xi|)\, a_n(\xi)$ are vanishing on $|\xi| < 1$ for all $j \in \mathbb{N}$. Thus for each $j \in \mathbb{N}$, we set

$$\mathcal{F}^n_j f(x, t) = \rho_1(t) \int_{\mathbb{R}^2} e^{i(x\cdot\xi + t|\xi|)}\, \rho_0(2^{-j}|\xi|)\, a_n(\xi)\, \hat{f}(\xi)\, d\xi \tag{10}$$

so that

$$\mathcal{F}^n f(x, t) = \mathcal{F}^n_0 f(x, t) + \sum_{j\in\mathbb{N}} \mathcal{F}^n_j f(x, t) \tag{11}$$

as a tempered distribution. Note that $\mathcal{F}^n_0 f$ is a Fourier integral operator with compactly supported amplitude function $b_n(\xi) := a_n(\xi)\varphi_0(\xi)$ with support $\{|\xi| \leq 2\}$. It turns out that $f \to \mathcal{F}^n_0 f(\cdot, t)$ is a smoothing operator, see Theorem 3.1 in Sect. 3.

For $\mathcal{F}^n_j f,\ j \in \mathbb{N}$, we need more precise regularity estimates in terms of j. In fact, the regularity estimate for $\mathcal{F}^n_j f$ will be deduced from the corresponding L^p estimate. The L^p estimate for $\mathcal{F}^n_j f$ for $4 \leq p \leq \infty$, follows by Riesz–Thorin interpolation theorem [27], once we have the following estimates

$$\|\mathcal{F}^n_j f\|_{L^4(\mathbb{R}^2\times\mathbb{R})} \leq C_\epsilon \frac{2^{j(3\epsilon+1/8)}}{(1+|n|)^{3+1/4}} \|f\|_{L^4(\mathbb{R}^2)},\ \epsilon > 0 \tag{12}$$

$$\|\mathcal{F}^n_j f\|_{L^\infty(\mathbb{R}^2\times\mathbb{R})} \leq C \frac{2^{j/2}}{(1+|n|)^{3+1/4}} \|f\|_{L^\infty(\mathbb{R}^2)} \tag{13}$$

for all $j \in \mathbb{N}, n \in \mathbb{N}_0^3$, with constants C, C_ϵ independent of j and n. The estimate (13) follows by standard arguments expressing $\mathcal{F}_j^n$ as a convolution operator with an L^1 kernel, see Proposition 3.1 in Sect. 3. The estimate (12) is subtle and requires more sophisticated arguments. First, by a wavefront set analysis, we can identify the region where $\mathcal{F}_j^n f$ has rapid decay. In fact, by Remark 1.1, we see that the wave front set of the distributions $\mathcal{F}_j^n f,\ j \in \mathbb{N}$ given by (10) is actually contained in the conic set

$$C = \{(x, t, \xi, |\xi|) : |x| = t,\ \xi \in \mathbb{R}^2 \setminus \{0\}\}. \tag{14}$$

Note that each $\mathcal{F}_j^n, j \geq 1, n \in \mathbb{N}_0^3$ are integral operators with a distribution kernel having singularities "along the direction" $\tau = |\xi|$ in the frequency domain. So it is natural to split the kernel localising around the wave front set and away from it and analyze separately. This leads to the two operators $\mathcal{Q}_\delta$ and $\mathcal{R}_\delta$ defined as follows.

Choose an even function $\psi \in C_c^\infty(-2, 2)$ such that $\psi = 1$ in $[-1, 1]$. This gives a cutoff function ψ^δ supported near the cone $|\xi| = \tau$ in $\mathbb{R}^3$ by

$$\psi^\delta(\xi, \tau) = \psi\left(\frac{|\xi| - \tau}{\delta}\right), \quad (\xi, \tau) \in \mathbb{R}^2 \times \mathbb{R},\ \delta \geq 1. \tag{15}$$

Let $\mathcal{Q}_\delta$ and $\mathcal{R}_\delta$ denote the multiplier operators on R^3 with multipliers ψ^δ and $1 - \psi^\delta$ respectively:

$$\begin{aligned} \widehat{\mathcal{Q}_\delta(\mathcal{F}_j^n f)}(\xi, \tau) &= \psi^\delta(\xi, \tau)\, \widehat{\mathcal{F}_j^n f}(\xi, \tau), \\ \widehat{\mathcal{R}_\delta(\mathcal{F}_j^n f)}(\xi, \tau) &= [1 - \psi^\delta(\xi, \tau)]\, \widehat{\mathcal{F}_j^n f}(\xi, \tau). \end{aligned} \tag{16}$$

Since $\mathcal{F}_j^n f = \mathcal{Q}_\delta(\mathcal{F}_j^n f) + \mathcal{R}_\delta(\mathcal{F}_j^n f)$, the L^p estimate for $\mathcal{F}_j^n f$ follows from the corresponding estimates for $\mathcal{Q}_\delta(\mathcal{F}_j^n f)$ and $\mathcal{R}_\delta(\mathcal{F}_j^n f)$. We require these estimates with appropriate decay in $n \in \mathbb{N}_0^3$. The estimate for $\mathcal{R}_\delta(\mathcal{F}_j^n f)$ is easy and follows via standard kernel estimate, see Proposition 4.1 in Sect. 4.

The estimate for $\mathcal{Q}_\delta(\mathcal{F}_j^n f)$ is more delicate and uses Littlewood–Paley type argument as in [18]. Here we give a more direct proof based on a duality argument and a refined estimate for square function based purely on angular decomposition.

3 Some Auxiliary Estimates

In this section we prove the estimates for various oscillatory integrals that we require. We start with showing that the Fourier integral operators $\mathcal{F}f$ with compactly supported amplitude functions are infinitely smoothing. For $\rho_1 \in C_c^\infty(I)$ and

$b \in C_c^\infty(\mathbb{R}^2)$ set

$$\mathcal{F}_0 f(x,t) = \rho_1(t) \int_{\mathbb{R}^2} e^{i(x\cdot\xi + t|\xi|)}\, b(\xi)\, \hat{f}(\xi)\, d\xi, \;\; f \in \mathcal{S}(\mathbb{R}^2). \tag{17}$$

Theorem 3.1 *Let $\mathcal{F}_0 f(x,t)$ be as in (17), with $\operatorname{supp} b \subset \{\xi \in \mathbb{R}^2 : |\xi| \le 2\}$. Then the operator $f \to \mathcal{F}_0 f(\cdot,t)$ is a smoothing operator. In fact, for each $\sigma \in \mathbb{C}$ the estimate*

$$\|(I - \Delta_x)^\sigma \mathcal{F}_0 f\|_{L^p(\mathbb{R}^2\times\mathbb{R})} \le C_\sigma\, \|f\|_{L^p(\mathbb{R}^2)}, \;\; 1 < p < \infty \tag{18}$$

holds for all $f \in L^p(\mathbb{R}^2)$ with $C_\sigma = \sup_{|\alpha|\le 2} \|\partial^\alpha b(\xi)\|_\infty$.

Proof For each $t \in \mathbb{R}$ and $\sigma \in \mathbb{C}$, the operator $T_t^\sigma : f(x) \to (I - \Delta_x)^\sigma \mathcal{F}_0 f(\cdot,t)$ is a multiplier operator on $L^2(\mathbb{R}^2)$ with multiplier function

$$M_t^\sigma(\xi) = \rho_1(t)\, e^{it|\xi|}\, (1 + |\xi|^2)^\sigma\, b(\xi), \;\; \xi \in \mathbb{R}^2 \tag{19}$$

as $b \in C_c^\infty(\mathbb{R}^2)$. A simple computation shows that

$$|\xi|^{|\alpha|} |\partial_\xi^\alpha M_t^\sigma(\xi)| \le C\, \rho_1(t) \sup_{|\alpha|\le 2} \|\partial^\alpha b(\xi)\|_\infty \tag{20}$$

for $|\alpha| \le 2$, with a constant C independent of t. Hence by the Mihlin multiplier theorem (see [2]), applied to T_t^σ for each t, we get the estimate

$$\|(I - \Delta_x)^\sigma \mathcal{F}_0 f(\cdot,t)\|_{L^p(\mathbb{R}^2)} \le C\, \rho_1(t) \sup_{|\alpha|\le 2} \|\partial^\alpha b(\xi)\|_\infty\, \|f\|_{L^p(\mathbb{R}^2)}, \tag{21}$$

for $1 < p < \infty$. A further integration over $t \in \mathbb{R}$ after taking pth power will give the required estimate. □

Next we discuss a further decomposition of the operators $\mathcal{F}_j^n$ in terms of the angular variable as in [10, 21], which is useful to deal with $Q^\delta \mathcal{F}_j^n$. For fixed $j \ge 1$, let $N = N(j)$ be the largest integer less than or equal to $2^{j/2}$, so that $2^{j/2} - 1 < N \le 2^{j/2}$. We now choose N equally spaced points $\xi_0, \xi_1, \ldots, \xi_{N-1}$ on the unit circle $S^1 = \{\xi \in \mathbb{R}^2 : |\xi| = 1\}$ with $\xi_0 = e_1 := (1,0)$. In fact, we take $\xi_\nu = O\xi_0$ for $1 \le \nu \le N-1$, where O is the rotation in the counter clockwise direction by an angle $2\pi\nu/N$.

With $N = N(j)$ as above, let $\{\chi_\nu\}_{\nu=0}^{N-1}$ be a partition of unity on $\mathbb{R}^2 \setminus \{0\}$ with respect to the angular variable with the following properties:

$$\chi_\nu(\xi) = \chi_0(O^{-1}\xi), \;\; 1 \le \nu \le N-1 \tag{22}$$

where O is the rotation by an angle $2\pi\nu/N$, and

$$|\partial_{\xi_1}^k \chi_0(\xi)| \le C_k, \; |\partial_{\xi_2}^k \chi_0(\xi)| \le C_k N^k \approx C_k\, 2^{\frac{jk}{2}} \text{ for } |\xi| = 1 \tag{23}$$

for all $k \in \mathbb{N}$, with a constant C_k independent of ν (hence independent of j). Here and elsewhere, by $A(s) \approx B(s)$ we mean there exist positive constants c_1, c_2 such that $c_1 A(s) \le B(s) \le c_2 A(s)$.

Note that the functions χ_ν are homogeneous functions of degree zero on $\mathbb{R}^2 \setminus \{0\}$. An explicit construction of such a partition of unity is carried out in Appendix. Using the homogeneous partitions of unity $\{\chi_\nu\}_\nu$, we define the operators

$$\mathcal{F}_{j,\nu}^n f(x,t) = \rho_1(t) \int_{\mathbb{R}^2} e^{i(x\cdot\xi + t|\xi|)}\, \rho_0(2^{-j}|\xi|)\, a_n(\xi)\, \chi_\nu(\xi)\, \hat{f}(\xi)\, d\xi \tag{24}$$

for $0 \le \nu \le N-1$, $j \in \mathbb{N}$ and $n \in \mathbb{N}_0^3$. Note that $\mathcal{F}_j^n f = \sum_{\nu=0}^{N-1} \mathcal{F}_{j,\nu}^n f$. We also need to consider the Fourier integral operators $\tilde{\mathcal{F}}_{j,\nu}^n$ given by

$$\tilde{\mathcal{F}}_{j,\nu}^n f(x,t) = \rho_1(t) \int_{\mathbb{R}^2} e^{i(x\cdot\xi + t|\xi|)}\, \rho(2^{-j}|\xi|)\, a_n(\xi)\, \chi_\nu(\xi)\, \hat{f}(\xi)\, d\xi, \; j \in \mathbb{N} \tag{25}$$

where ρ is a smooth non negative function such that $\rho_0 = \rho^2$. We have

$$\tilde{\mathcal{F}}_{j,\nu}^n f(x,t) = \int_{y\in\mathbb{R}^2} \tilde{K}_{j,\nu}^n(x-y,t)\, f(y)\, dy \tag{26}$$

where

$$\tilde{K}_{j,\nu}^n(x,t) := \tilde{K}_{j,\nu}^{a_n}(x,t) = \rho_1(t) \int_\xi e^{i(x\cdot\xi + t|\xi|)}\, \rho(2^{-j}|\xi|)\, a_n(\xi)\, \chi_\nu(\xi)\; d\xi. \tag{27}$$

The following kernel estimate is a refinement of the estimate obtained in [21], and is crucial in our argument for dealing with general amplitude functions depending on (x,t). Let $\{a_n(\xi)\}_{n\in\mathbb{N}_0^3}$ be a family of symbols satisfying

$$|\partial^\alpha a_n(\xi)| \le C_\alpha \frac{(1+|\xi|)^{-|\alpha|}}{(1+|n|)^{3+1/4}}, \tag{28}$$

for all multi-indices $\alpha \in \mathbb{N}_0^2$, $n \in \mathbb{N}_0^3$ with constant C_α independent of n.

Lemma 3.1 *Let $\{a_n(\xi)\}_{n\in\mathbb{N}_0^3}$ be a family of symbols in $S^0(\mathbb{R}^2)$ satisfying (28). Then the Kernels $\tilde{K}_{j,\nu}^n$ given by (27) satisfy the estimates*

$$|\tilde{K}_{j,\nu}^n(x,t)| \le \frac{C\, 2^{3j/2}}{(1+|n|)^{3+1/4}} \rho_1(t)\, \Psi_j(Tx + te_1), \tag{29}$$

where $\Psi_j(x) := \Psi_{j,k}(x) = \left[1 + 2^{2j}|x_1|^2\right]^{-k} \left[1 + 2^j|x_2|^2\right]^{-k}$, $k \geq 1$, $j \in \mathbb{N}$ *and the constant* $C = C_k$ *is independent of* j, ν *and* n. *Here* $T \in SO(2)$ *is such that* $T\xi_\nu = e_1$, $0 \leq \nu \leq N - 1$.

Proof We first consider the case $\xi_\nu = \xi_0 = e_1$ and estimate $\tilde{K}^n_{j,0}(x, t)$ by oscillatory integral techniques. By (27)

$$\tilde{K}^n_{j,0}(x, t) = \rho_1(t) \int_{\mathbb{R}^2} e^{i(x\cdot\xi+t|\xi|)} \rho(2^{-j}|\xi|)\, a_n(\xi)\, \chi_0(\xi)\, d\xi. \tag{30}$$

Consider the differential operator $L = \left(I - 2^{2j}\partial^2_{\xi_1}\right)\left(I - 2^j\partial^2_{\xi_2}\right)$ as in [21], so that

$$L^k e^{i(x+te_1)\cdot\xi} = \left[1 + 2^{2j}|x_1 + t|^2\right]^k \left[1 + 2^j|x_2|^2\right]^k e^{i(x+te_1)\cdot\xi}, k \in \mathbb{N}. \tag{31}$$

Re writing $e^{i(x\cdot\xi+t|\xi|)}$ as $e^{it(|\xi|-\xi_1)}\, e^{i(x+te_1)\cdot\xi}$ and using the above formula, we get

$$e^{i(x\cdot\xi+t|\xi|)} = \left[1 + 2^{2j}|x_1 + t|^2\right]^{-k} \left[1 + 2^j|x_2|^2\right]^{-k} e^{it(|\xi|-\xi_1)}\, L^k e^{i(x+te_1)\cdot\xi}. \tag{32}$$

Using this formula in (30) and an integration by parts shows that

$$\tilde{K}^n_{j,0}(x, t) = \Psi_j(x + te_1)\, A^{a_n}_j(x, t) \tag{33}$$

where $\Psi_j(x) = \left[1 + 2^{2j}|x_1|^2\right]^{-k} \left[1 + 2^j|x_2|^2\right]^{-k}$ and

$$A^a_j(x, t) = \int_{\mathbb{R}^2} e^{i[x\cdot\xi+t\xi_1]}\, L^k \left[e^{it(|\xi|-\xi_1)}\, \rho(2^{-j}|\xi|)\, a(\xi)\, \chi_0(\xi)\right] d\xi \tag{34}$$

for $a \in S^0(\mathbb{R}^2)$. Note that the integrand in (34) is supported on the set

$$E = \operatorname{supp} \chi_0 \cap \{\xi : 2^{j-1} \leq |\xi| \leq 2^{j+1}\}. \tag{35}$$

To complete the proof for $\nu = 0$, we need to show that

$$|A^{a_n}_j(x, t)| \leq \frac{C_k\, 2^{3j/2}}{(1 + |n|)^{3+1/4}}. \tag{36}$$

This will follow from (34) with $a = a_n$, once we verify the following;

- The measure of E is bounded by a constant times $2^{3j/2}$
- $L^k\left[e^{it(|\xi|-\xi_1)}\, \rho(2^{-j}|\xi|)\, a_n(\xi)\, \chi_0(\xi)\right] \leq C_k/(1+|n|)^{3+1/4}$ for some constant C_k independent of j and n.

The first statement is clear. In fact for all $\xi \in E$, we have $|\xi_2| \leq \xi_1 \sin(2\pi/N) \lesssim 2^{j/2}$ as $\xi_1 \leq 2^{j+1}$. To verify the second statement, we observe that L^k is a linear combination of various derivatives $(2^{2j}\partial^2_{\xi_1})^{k_1}(2^j\partial^2_{\xi_2})^{k_2}$ with $k_1+k_2 \leq 2k$. Note that each of the above derivative of the functions $\rho(2^{-j}|\xi|)$ are uniformly bounded in j. In view of (23) and the fact that $\chi_0(\xi)$ is homogeneous of degree zero, the above derivatives of $\chi_0(\xi)$ are uniformly bounded in $j \in \mathbb{N}$. Also in view of (28), the above derivatives falling on a_n gives the bound $C_k/(1+|n|)^{3+1/4}$, with C_k independent of j and n. All the above derivatives applied to $e^{it(|\xi|-\xi_1)}$ also give functions bounded uniformly in j on E, in view of Lemma 3.2 below.

To estimate $\tilde{K}^n_{j,\nu}$ for general ν, first note that $\chi_\nu(\xi) = \chi_0(O^{-1}\xi)$ by (22) where $O \in SO(2)$ is such that $\xi_\nu = Oe_1$. Thus using the change of variable $\xi \to O\xi$ in (27), we see that $\tilde{K}^n_{j,\nu}(x,t) := \tilde{K}^{a_n}_{j,\nu} = \tilde{K}^{a^\nu_n}_{j,0}(O^{-1}x,t)$ where $a^\nu_n(\xi) = a_n(O\xi)$. Thus in view of (33) and (34) we have

$$\tilde{K}^n_{j,\nu}(x,t) = \Psi_j(O^{-1}x+te_1)\, A^{a^\nu_n}_j(O^{-1}x,t). \tag{37}$$

Notice that the estimate for $|A^{a^\nu_n}_{j,0}(y,t)|$ depends on the derivatives of $a^\nu_n = a_n \circ O$, and these derivatives have the bound twice as that of a_n given by (28). Hence the proof follows with $T = O^{-1}$, once we prove the following lemma. □

Lemma 3.2 *Let $h(\xi) = |\xi| - \xi_1$ for $\xi = (\xi_1,\xi_2) \in \mathbb{R}^2$, then we have*

$$|\partial^k_{\xi_1}h(\xi)| \leq A_k\, 2^{-kj},\ |\partial^k_{\xi_2}h(\xi)| \leq B_k\, 2^{-\frac{kj}{2}},\ \textit{for } k \in \mathbb{N}, \tag{38}$$

on the set $E = supp\, \chi_0 \cap \{\xi = (\xi_1,\xi_2) : 2^{j-1} \leq |\xi| \leq 2^{j+1}\},\ j \geq 1$.

Proof This lemma has already been proved in [26], for more general function $h(\xi) = \Phi(x,\xi) - \Phi_\xi(x,\bar{\xi})\cdot\xi$ corresponding to a general phase function Φ. In our special case, we give a proof based on geometric reasoning.

Writing $\xi = |\xi|(\cos\theta,\sin\theta)$, we see that $\partial_{\xi_1}h(\xi) = \frac{\xi_1}{|\xi|} - 1 = -2\sin^2(\theta/2)$. Since $|\sin\theta| \leq |\theta|$ and $|\theta| \leq 2\pi/N$ on the support of χ_0, we have

$$|\partial_{\xi_1}h(\xi)| \leq \frac{2\pi^2}{N^2} \leq \frac{2\pi^2}{2^j},\ \xi \in E \tag{39}$$

as $N \geq 2^{j/2}$, which proves the case $k=1$. To deal with the case $k>1$, we write $\partial^k_{\xi_1}h(\xi) = \partial^{k-1}_{\xi_1}g(\xi)$, where $g = \partial_{\xi_1}h$ which is a function homogeneous of degree zero on $\mathbb{R}^2$, hence $\partial^{k-1}_{\xi_1}g$ is homogeneous of degree $1-k$. It follows that

$$\partial^k_{\xi_1}h(\xi) = |\xi|^{1-k}(\partial^{k-1}_{\xi_1}g)(\xi/|\xi|). \tag{40}$$

Now note that $g(\xi) = -2\sin^2(\theta/2) := \tilde{g}(\theta)$ as computed above and also $\partial_{\xi_1} = -\sin\theta\,\partial_\theta$ on functions homogeneous of degree zero. An easy induction argument shows that $(-\sin\theta\,\partial_\theta)^{k-1}\tilde{g}(\theta) = P_k(\cos\theta)\sin^2\theta$ where P_k is a polynomial of degree $k-1$. Now for $\xi = (r\cos\theta, r\sin\theta) \in E$, we have $|\theta| \le 2\pi/N$ and hence $|P_k(\cos\theta)\,\sin^2\theta| \le c_k|\sin^2\theta| \le c_k 4\pi^2 N^{-2} \approx C_k 2^{-j}$ for some constant C_k independent of j. It follows from (40) that, for $k > 1$

$$\left|\partial^k_{\xi_1} h(\xi)\right| \le C_k 2^{-kj} \tag{41}$$

as $|\xi| \approx 2^j$ on E.

Since $\partial_{\xi_2} h(\xi) = \frac{\xi_2}{|\xi|} = \sin\theta$, the required inequality follows as above on E, for $k=1$. For $k \ge 2$, note that $\partial^k_{\xi_2} h(\xi) = \partial^k_{\xi_2}|\xi|$. Since the function $g_1(\xi) = |\xi|$ is homogeneous of degree 1, these derivatives are homogeneous functions of degree $1-k$. It follows that $|\partial^k_{\xi_2} h(\xi)| \le C_k|\xi|^{1-k} \le C_k|\xi|^{-k/2}$ on E, for $k \ge 2$ and hence the required inequality holds on E. □

Remark 3.1 We note that the estimate for the kernel $\tilde{K}^n_{j,\nu}$ can also be expressed in the form.

$$|\tilde{K}^n_{j,\nu}(x,t)| \le \frac{C2^{3j/2}}{(1+|n|)^{3+1/4}}\left[1+2^{2j}|\langle x,\xi_\nu\rangle + t|^2\right]^{-k}\left[1+2^j|\langle x,\xi_\nu^\perp\rangle|^2\right]^{-k} \tag{42}$$

where ξ_ν is as in Lemma 3.1 and $\xi_\nu^\perp$ is a unit vector perpendicular to ξ_ν. In fact, the function $\Psi_j(x)$ defining $\tilde{K}^n_{j,\nu}$ in Lemma 3.1 is a function of $2^j|x_1|$ and $2^{j/2}|x_2|$. Now for $T \in SO(2)$, $(Tx)_1 := \langle Tx, e_1\rangle = \langle x, \xi_\nu\rangle$ if $T\xi_\nu = e_1$. Similarly $(Tx)_2 = \langle x, \xi_\nu^\perp\rangle$, where $T\xi_\nu^\perp = e_2$. It follows that

$$\Psi_j(Tx+te_1) = \left[1+2^{2j}|\langle x,\xi_\nu\rangle + t|^2\right]^{-k}\left[1+2^j|\langle x,\xi_\nu^\perp\rangle|^2\right]^{-k}. \tag{43}$$

Hence (42) follows from Lemma 3.1 and the fact that $|\rho_1(t)| \le 1$.

Now we prove the L^∞ estimate (13) for $\mathcal{F}^n_j$ mentioned in Sect. 2.

Proposition 3.1 *Let $\mathcal{F}^n_j$ be the operator given by* (10) *with amplitude function a_n as in* (28)*. Then $\mathcal{F}^n_j$ satisfies the inequality*

$$\|\mathcal{F}^n_j f\|_{L^\infty(\mathbb{R}^3)} \le \frac{C\,2^{j/2}}{(1+|n|)^{3+1/4}}\|f\|_{L^\infty(\mathbb{R}^2)} \tag{44}$$

with a constant C independent of $j \in \mathbb{N}$ and $n \in \mathbb{N}_0^3$.

Proof We have $\mathcal{F}^n_j = \sum_{\nu=0}^{N-1}\mathcal{F}^n_{j,\nu}$, where $\mathcal{F}^n_{j,\nu}$ is the operator given by (24) which is a convolution operator (in the x-variable, for each t) with kernel

$$K^n_{j,\nu}(x,t) = \rho_1(t)\int_{\mathbb{R}^2} e^{i(x\cdot\xi + t|\xi|)}\,\rho_0(2^{-j}|\xi|)\,a_n(\xi)\,\chi_\nu(\xi)\,d\xi. \tag{45}$$

Note that $K^n_{j,\nu}$ differs from $\tilde{K}^n_{j,\nu}$ defined by (27), only in the power of ρ: $\rho_0 = \rho^2$. Hence $K^n_{j,\nu}(\cdot, t)$ also satisfies the pointwise estimate as in Lemma 3.1. Hence we have

$$\|K^n_{j,\nu}(\cdot, t)\|_{L^1(\mathbb{R}^2)} \leq \frac{C_1\,|\rho_1(t)|}{(1+|n|)^{3+1/4}} \tag{46}$$

with C_1 independent of j, ν, n and t. It follows that

$$\|\mathcal{F}_{j,\nu} f(\cdot, t)\|_{L^\infty(\mathbb{R}^2)} \leq \frac{C\,|\rho_1(t)|}{(1+|n|)^{3+1/4}}\|f\|_{L^\infty}, \tag{47}$$

for $0 \leq \nu \leq N-1$. Summing over ν, this gives the required estimate after a t integration, observing that there are $N = N(j) \approx 2^{j/2}$ terms in the sum. □

4 L^p-Estimates for $\mathcal{R}_\delta(\mathcal{F}^n_j f)$

Recall that $\mathcal{R}_\delta$ was defined in (16) as a multiplier operator on $\mathbb{R}^3$ with multiplier $1-\psi^\delta(\xi, \tau)$. The estimate for $\mathcal{R}_\delta(\mathcal{F}^n_j f)$ relies on the rapid decay of its Fourier transform. The following lemma is the key ingredient for the same.

Lemma 4.1 *For $j \in \mathbb{N}$ and $1 \leq \delta < 2^{j/2}$ consider the set*

$$B^\delta_j = \{(\xi, \tau) \in \mathbb{R}^2 \times \mathbb{R} : 2^{j-1} < |\xi| \leq 2^{j+1}, |\tau - |\xi|| > \delta\}. \tag{48}$$

Then for each $0 < \epsilon \leq \frac{1}{2}$, the estimate

$$|\tau - |\xi|| > C_{j,\epsilon,\delta}(|\tau| + |\xi|)^\epsilon \tag{49}$$

holds for all $(\xi, \tau) \in B^\delta_j$ with $C_{j,\epsilon,\delta} = \frac{1}{2}\left(\frac{2}{3}\right)^\epsilon \frac{\delta}{2^{j\epsilon}}$.

Proof The required inequality clearly holds when $\tau \leq 0$, as $\tau - |\xi| = -(|\tau| + |\xi|)$ and $\delta < 2^{j/2}$. So we assume $\tau > 0$. We write $B^\delta_j = B_1 \cup B_2$ where

$$\begin{aligned} B_1 &= \{(\xi, \tau) \in B^\delta_j : \tau > 2|\xi|\}, \\ B_2 &= \{(\xi, \tau) \in B^\delta_j : \tau \leq 2|\xi|\} \end{aligned} \tag{50}$$

We show that $\inf_{(\xi,\tau)\in B_i} \frac{|\tau-|\xi||}{(\tau+|\xi|)^\epsilon} \geq C_{j,\epsilon,\delta}$ for $i = 1, 2$.

Since $\tau > 2|\xi|$ on B_1, we have

$$\frac{|\tau - |\xi||}{(\tau + |\xi|)^\epsilon} = \frac{\tau - |\xi|}{(\tau + |\xi|)^\epsilon} = \tau^{1-\epsilon}\frac{1-\theta}{(1+\theta)^\epsilon} \tag{51}$$

where $\theta = \frac{|\xi|}{\tau} < \frac{1}{2}$ on B_1. Since $1 - \theta > \frac{1}{2}$ and $1 + \theta < \frac{3}{2}$, we see that for $(\xi, \tau) \in B_1$,

$$\frac{|\tau - |\xi||}{(\tau + |\xi|)^{\epsilon}} > \frac{1}{2}\left(\frac{2}{3}\right)^{\epsilon} 2^{j(1-\epsilon)} > \frac{1}{2}\left(\frac{2}{3}\right)^{\epsilon} \frac{\delta}{2^{j\epsilon}}. \tag{52}$$

On the other hand on B_2, we have

$$\frac{|\tau - |\xi||}{(\tau + |\xi|)^{\epsilon}} > \frac{\delta}{(\tau + |\xi|)^{\epsilon}} \geq \frac{\delta}{(3|\xi|)^{\epsilon}} \geq \frac{\delta}{2^{j\epsilon}} 6^{-\epsilon} \tag{53}$$

which dominates $\frac{\delta}{2^{j\epsilon}} \frac{1}{2} \left(\frac{2}{3}\right)^{\epsilon}$ since $\epsilon \leq 1/2$. □

Proposition 4.1 *For $j \in \mathbb{N}$, let $\mathcal{R}_\delta$ be as in* (16) *with $\delta \geq 1$, and $\mathcal{F}_j^n f$ be as in* (10) *with amplitude function a_n satisfying* (28). *Then for each $\epsilon > 0$, there exist a constant C_ϵ independent of j and n, such that the inequality*

$$\|\mathcal{R}_\delta(\mathcal{F}_j^n f)\|_{L^p(\mathbb{R}^2 \times \mathbb{R})} \leq \frac{C_\epsilon \, [C_{j,\epsilon,\delta}]^{-3/\epsilon}}{(1 + |n|)^{3+1/4}} \|f\|_{L^p}, \tag{54}$$

holds for all $f \in L^p(\mathbb{R}^2)$, $1 \leq p < \infty$ with $C_{j,\epsilon,\delta}$ as in Lemma 4.1.

Proof Since $f \to \mathcal{R}_\delta(\mathcal{F}_j^n f)$ is a linear map, it is enough to estimate (54) for $f \in \mathcal{S}(\mathbb{R}^2)$ by density of $\mathcal{S}(\mathbb{R}^2)$ in $L^p(\mathbb{R}^2)$, $1 \leq p < \infty$. In view of (16) and (10), we have

$$\widehat{\mathcal{R}_\delta(\mathcal{F}_j^n f)}(\xi, \tau) = [1 - \psi^\delta(\xi, \tau)] \, \hat{f}(\xi) \, a_n(\xi) \, \rho_0(2^{-j}|\xi|) \, \hat{\rho}_1(\tau - |\xi|). \tag{55}$$

Thus by the Fourier inversion formula and expanding $\hat{f}$ we see that

$$\begin{aligned} \mathcal{R}_\delta(\mathcal{F}_j^n f)(x, t) &= \int_{\mathbb{R}^2 \times \mathbb{R}} e^{i(x\cdot\xi + t\tau)} \, \widehat{\mathcal{R}_\delta(\mathcal{F}_j^n f)}(\xi, \tau) \, d\xi d\tau \\ &= \int_{\mathbb{R}^2} f(y) \, \mathcal{K}_j^\delta(x - y, t) \, dy \end{aligned} \tag{56}$$

where,

$$\mathcal{K}_j^\delta(x, t) = \int_{\mathbb{R}^2 \times \mathbb{R}} e^{i(x\cdot\xi + t\tau)} \, [1 - \psi^\delta(\xi, \tau)] \, a_n(\xi) \, \rho_0(2^{-j}|\xi|) \, \hat{\rho}_1(\tau - |\xi|) \, d\xi d\tau. \tag{57}$$

Thus in view of Young's inequality (see [11]), it is enough to prove the estimate

$$\|\mathcal{K}_j^\delta(\cdot,t)\|_{L^1(\mathbb{R}^2)} \lesssim \frac{C_\epsilon\,(C_{j,\epsilon,\delta})^{-3/\epsilon}}{(1+|n|)^{3+1/4}(1+|t|^2)} \tag{58}$$

for each $t \in \mathbb{R}$, with C_ϵ independent of j, n and t. Now observe that for all $N \in \mathbb{N}$,

$$(1+|x|^2)^N(1+|t|^2)^N\, e^{i(x\cdot\xi+t\tau)} = (I-\Delta_\xi)^N(I-\partial_\tau^2)^N\, e^{i(x\cdot\xi+t\tau)}. \tag{59}$$

Hence an integration by parts in (57) shows that

$$\begin{aligned}(1+|x|^2)^N(1+|t|^2)^N\,\mathcal{K}_j^\delta(x,t)\\ = \int_{\xi,\tau} e^{i(x\cdot\xi+t\tau)}\,(I-\Delta_\xi)^N(I-\partial_\tau^2)^N\, b_j(\xi,\tau)\,d\xi d\tau,\end{aligned} \tag{60}$$

where $b_{j,n}(\xi,\tau) = [1-\psi^\delta(\xi,\tau)]\,a_n(\xi)\,\rho_0(2^{-j}|\xi|)\,\hat{\rho_1}(\tau-|\xi|)$. Note that $(I-\Delta_\xi)^N(I-\partial_\tau^2)^N\, b_{j,n}(\xi,\tau)$ is a sum of terms that involves various partial derivatives of order up to $4N$, of the functions $\psi^\delta(\xi,\tau), a_n(\xi), \rho_0(2^{-j}|\xi|)$ and $\hat{\rho_1}(\tau-|\xi|)$. Each derivative on ψ^δ brings in a negative power of $\delta \geq 1$ and hence all these derivatives are bounded uniformly in δ. In the same way, all the above derivatives acting on $\rho_0(2^{-j}|\xi|)$ are bounded uniformly in $j \in \mathbb{N}$. By assumption a_n and all its derivatives are bounded by a constant times $\frac{1}{(1+|n|)^{3+1/4}}$. Also, since $\hat{\rho_1}$ is a Schwartz class function we have $|\partial^\alpha\hat{\rho_1}(y)| \leq C_{M,N}(1+|y|)^{-M}$ for $|\alpha| \leq N$, for some constant $C_{M,N}$.

It follows that for each $N, M \in \mathbb{N}$, there is a constant $C_{M,N}$ independent of ϵ, j, δ and n such that

$$\begin{aligned}|(I-\Delta_\xi)^N(I-\partial_\tau^2)^N\, b_j(\xi,\tau)| &\leq \frac{C_{M,N}}{(1+|n|)^{3+1/4}}(1+|\tau-|\xi||)^{-M}\\ &\leq \frac{C_{M,N}}{(1+|n|)^{3+1/4}}(1+C_{j,\epsilon,\delta}(\tau+|\xi|)^\epsilon)^{-M}\end{aligned} \tag{61}$$

for $|\tau-|\xi|| > \delta$, by Lemma 4.1. Note that the integral in (57) and hence in (60) is actually over the set $|\tau-|\xi|| > \delta$, as $1-\psi^\delta(\xi,\tau) = 0$ on $|\tau-|\xi|| \leq \delta$, hence Lemma 4.1 is applicable. Using (61) in (60), the right hand side of (60) is bounded by

$$\begin{aligned}&\frac{C_{M,N}}{(1+|n|)^{3+1/4}}\int_{\mathbb{R}^2\times\mathbb{R}}[1+C_{j,\epsilon,\delta}\,(|\tau|+|\xi|)^\epsilon\,]^{-M}\,d\xi d\tau\\ &\quad = \frac{C_{M,N}}{(1+|n|)^{3+1/4}}(C_{j,\epsilon,\delta})^{-3/\epsilon}\int_{\mathbb{R}^2\times\mathbb{R}}[1+(|\tau|+|\xi|)^\epsilon]^{-M}\,d\xi d\tau\end{aligned} \tag{62}$$

by a change of scale in the variables ξ and τ. Choosing $M > M_\epsilon := [3/\epsilon + 1]$, the last integral is finite, and (60) translates to the inequality

$$\mathcal{K}_j^\delta(x,t) \le \frac{C_{M(\epsilon),N}}{(1+|n|)^{3+1/4}} \frac{(C_{j,\epsilon,\delta})^{-3/\epsilon}}{(1+|x|^2)^N(1+|t|^2)^N}. \tag{63}$$

Choosing $N = 2$, we see that $\mathcal{K}_j(\cdot, t) \in L^1(\mathbb{R}^2)$ and the estimate (58) holds, which completes the proof. □

5 A Square Function Estimate

In this section, we prove the L^p estimate for a square function based on angular decomposition. This estimate for $4/3 \le p < 2$ is new to the best of our knowledge, and the estimate for $p = 4/3$ is crucial in our proof of L^4 estimate for $Q_\delta(\mathcal{F}_j^n f)$. Given $j \in \mathbb{N}$ and $\delta \ge 1$, consider the family of Fourier multiplier operators $\{T_{\nu,j}^\delta\}_{\nu=0}^{N-1}$ on $\mathbb{R}^3$ given by

$$\widehat{T_{\nu,j}^\delta g}\,(\xi,\tau) = \rho(2^{-j}|\xi|)\,\tilde{\chi}_\nu(\xi)\,\psi\left(\frac{|\xi|-\tau}{\delta}\right)\hat{g}(\xi,\tau),\ g \in \mathcal{S}(\mathbb{R}^3) \tag{64}$$

where ρ is as in (25), $\tilde{\chi}_\nu$ is a homogeneous function (smooth and compactly supported as a function on S^1), such that $\tilde{\chi}_\nu \chi_\nu = \chi_\nu$ for χ_ν given by (22) and ψ is as in (15). Consider the square function

$$S(g)(x,t) = \left(\sum_{\nu=0}^{N-1} \left|T_{\nu,j}^\delta g\right|^2\right)^{\frac{1}{2}}. \tag{65}$$

For the square function $S(g)$, we establish the following L^p boundedness result.

Theorem 5.1 *Let Sg be as in* (65)*. Then the inequality*

$$\|Sg\|_{L^p(\mathbb{R}^3)} \le C\,2^{j/8}\,j^b\,\delta^{1/4}\,\|g\|_{L^p(\mathbb{R}^3)},\ 4/3 \le p \le 4 \tag{66}$$

holds for all $g \in \mathcal{S}(\mathbb{R}^3)$ with constants C and b independent of $j \in \mathbb{N}$ and $\delta \ge 1$.

Note that the result was previously known for the range $2 \le p \le 4$, see [18]. However, our approach for estimating $Q_\delta(\mathcal{F}_j^n f)$ requires the square function estimate for $p = 4/3$. In fact, we first establish the case $p = 4$, which will be used to establish the case $4/3 \le p < 4$. The proof of the following square function estimate is essentially based on the approach of Cordoba [3]. In fact we use a reduction to a Kakeya type maximal operator as in [18]. Such control by Kakeya type maximal operator was originally used by C. Fefferman in [10], as far as we know.

Proposition 5.1 *Let $T^\delta_{\nu,j}$ be as in (64). Then the following square function estimate holds*

$$\left\| \left(\sum_{\nu=0}^{N-1} |T^\delta_{\nu,j} g|^2 \right)^{\frac{1}{2}} \right\|_{L^4(\mathbb{R}^3)} \leq C\,\delta^{1/4} j^b\, \|g\|_{L^4(\mathbb{R}^3)} \tag{67}$$

for all $g \in \mathcal{S}(\mathbb{R}^3)$ with constants C and b independent of j.

Proof In view of (64) we have

$$T^\delta_{\nu,j} g(x,t) = \int_{\mathbb{R}^3} \tilde{k}^\delta_{j,\nu}(x-y, t-s)\, g(y,s)\, dyds, \tag{68}$$

where

$$\begin{aligned} \tilde{k}^\delta_{j,\nu}(x,t) &= \int_{\mathbb{R}^3} e^{i(x\cdot\xi+t\tau)} \left[\tilde{\chi}_\nu(\xi)\, \rho(2^{-j}|\xi|)\, \psi\left(\frac{|\xi|-\tau}{\delta} \right) \right] d\xi\, d\tau \\ &= \delta \psi^\vee(\delta t)\, K_{j,\nu}(x,t), \end{aligned} \tag{69}$$

with $K_{j,\nu}(x,t) = \int_{\mathbb{R}^2} e^{i(x\cdot\xi+t|\xi|)} \left[\tilde{\chi}_\nu(\xi)\, \rho(2^{-j}|\xi|) \right] d\xi$. Note that $K_{j,\nu}$ is same as $\tilde{K}^n_{j,\nu}$ in (27) with $a_n \equiv 1$. Hence by the same argument as in Lemma 3.1, we see that $K_{j,\nu}$ and hence $\tilde{k}^\delta_{j,\nu}$ satisfies the pointwise estimate as in (29). It follows that

$$\int_{\mathbb{R}^2\times\mathbb{R}} |\tilde{k}^\delta_{j,\nu}(x,t)|\, dxdt \leq C \tag{70}$$

with a constant C independent of j and δ. Thus an application of Cauchy–Schwarz inequality in (68) yields

$$|T^\delta_{\nu,j} g(x,t)|^2 \leq C \int_{\mathbb{R}^3} |g(y,s)|^2 |\tilde{k}^\delta_{j,\nu}(x-y,t-s)| dy\, ds, \tag{71}$$

with C independent of j and δ. Note that $T^\delta_{\nu,j} g = T^\delta_{\nu,j} g_\nu$, where g_ν is given by the Fourier transform:

$$\widehat{g_\nu}(\xi, s) = \tilde{\Phi}_\nu(\xi)\, \widehat{g}(\xi, s), \tag{72}$$

with $\tilde{\Phi}_\nu$ denoting the characteristic function of the support of $\tilde{\chi}_\nu$. Thus summing over ν in (71) with g replaced by g_ν, squaring, integrating and taking square root leads to the inequality

$$\begin{aligned}
&\left\| \left(\sum_{\nu=0}^{N-1} |T^{\delta}_{\nu,j} g|^2 \right)^{\frac{1}{2}} \right\|^2_{L^4(\mathbb{R}^3)} \\
&\leq C \left(\int_{\mathbb{R}^3} \left[\int_{\mathbb{R}^3} \sum_{\nu=0}^{N-1} |g_\nu(y,s)|^2 |\tilde{k}^{\delta}_{j,\nu}(x-y,t-s)|\, dy\, ds \right]^2 dx dt \right)^{1/2} \\
&= C \sup_{\|h\|_{L^2}=1} \left| \int_{\mathbb{R}^3} \int_{\mathbb{R}^3} \sum_{\nu=0}^{N-1} |g_\nu(y,s)|^2 |\tilde{k}^{\delta}_{j,\nu}(x-y,t-s)|\, dy\, ds\, h(x,t)\, dx dt \right| .
\end{aligned} \tag{73}$$

Since $\tilde{k}^{\delta}_{j,\nu}(x,t) = K_{j,\nu}(x)\, \delta\psi^{\vee}(\delta t)$, using the change of variable $t \to t+s$, we see that

$$\begin{aligned}
&\int_{\mathbb{R}^3} |\tilde{k}^{\delta}_{j,\nu}(x-y,t-s) h(x,t)|\, dx\, dt \\
&\qquad\qquad \leq \sup_{\nu} \int_{\mathbb{R}^3} |K_{j,\nu}(x-y,t)|\, |h_s(x,t)|\, dx\, dt
\end{aligned} \tag{74}$$

where $h_s(x,t) = \delta\psi^{\vee}(\delta t) h(x, t+s)$. Thus an interchange of integrals in (73) shows that the right hand side of (73) is at most a constant times

$$\begin{aligned}
&\sup_{\|h\|_{L^2}=1} \left| \int_{\mathbb{R}^3} \sum_{\nu} |g_\nu(y,s)|^2 \left[\sup_{\nu} \int_{\mathbb{R}^3} |K_{j,\nu}(x-y,t)|\, |h_s(x,t)|\, dx dt \right] dy\, ds \right| \\
&\quad \leq C \left[\int_{\mathbb{R}^3} \left(\sum_{\nu} |g_\nu(y,s)|^2 \right)^2 dy\, ds \right]^{1/2} \times \\
&\quad \sup_{\|h\|_{L^2}=1} \left[\int_{\mathbb{R}^3} \sup_{\nu} \left| \int_{\mathbb{R}^3} |K_{j,\nu}(x-y,t)|\, |h_s(x,t)|\, dx dt \right|^2 dy\, ds \right]^{\frac{1}{2}}
\end{aligned} \tag{75}$$

by Cauchy–Schwarz inequality in the variables (y, s). We first show that the second term on the right hand side of (75) is bounded by $C\,\delta^{1/2}\, j^{3/2}$. Note that the kernel $K_{j,\nu}$ satisfies the pointwise estimate (29) as mentioned above and hence by Remark 3.1, $K_{j,\nu}$ also satisfies the estimate (42). Hence we have the maximal inequality

$$\int_{\mathbb{R}^2} \sup_{\nu} \left| \int_{\mathbb{R}^3} |K_{j,\nu}(x-y,t)|\, |h_s(x,t)|\, dxdt \right|^2 dy \leq Cj^3 \, \|h_s\|^2_{L^2(\mathbb{R}^3)} \tag{76}$$

with C independent of j. This is a restatement of the maximal inequality (1.11) in [18], in view of Lemma 1.4 with $\delta = 2^{-j/2}$ as per the notation in the same paper. Since $h_s(x,t) = \delta\psi^{\vee}(\delta t)h(x, t+s)$, integrating with respect to the s-variable over $\mathbb{R}$ on both sides gives

$$\begin{aligned} \int_{\mathbb{R}^3} \sup_{\nu} &\left| \int_{\mathbb{R}^3} |K_{j,\nu}(x-y,t)|\, |h_s(x,t)|\, dxdt \right|^2 dy\, ds \\ &\leq C\, j^3 \int_{s\in\mathbb{R}} \int_{\mathbb{R}^3} \left|\delta\psi^{\vee}(\delta t)\, h(x, t+s)\right|^2 dxdt\, ds \\ &= C\, j^3\, \delta \int_{t\in\mathbb{R}} |\psi^{\vee}(t)|^2 \left[\int_{\mathbb{R}^3} |h(x,s)|^2\, dxds \right] dt \\ &\leq C\, \delta\, j^3 \, \|h\|^2_{L^2(\mathbb{R}^3)} \end{aligned} \tag{77}$$

for each $h \in L^2(\mathbb{R}^3)$. Taking square root and supremum over $\|h\|_{L^2(\mathbb{R}^3)} = 1$ leads to the required estimate for the second term. Also the first term on the right hand side of (75) is bounded by $C\, j^{2b_1} \|g\|^2_{L^4(\mathbb{R}^3)}$ with some constants C and b_1 independent of j. In fact, since the multiplier function $\tilde{\Phi}_\nu$ in (72) is independent of s, we can appeal to the two dimensional square function estimate of Cordoba [4] for each s to get

$$\int_{\mathbb{R}^2} \left| \left(\sum_{\nu} |g_\nu(y,s)|^2 \right)^2 \right| dy \leq Cj^{4b_1} \|g(\cdot,s)\|^4_{L^4(\mathbb{R}^2)} \tag{78}$$

with C and b_1 independent of s. From this the assertion follows by integration over $s \in \mathbb{R}$. It follows that the right hand side of (75) is bounded by $\delta^{1/2} j^{3/2+2b_1} \|g\|^2_{L^4(\mathbb{R}^3)}$. Using this in (73) and taking square root, the required estimate follows with $b = 3/4 + b_1$. □

Now we prove the square function estimate given by Theorem 5.1.

Proof of Theorem 5.1 We use the Rademacher function argument as in Stein [25] (page 106), to reduce the square function estimate to a multiplier problem. Recall that the Rademacher functions $\{r_k\}_{k\geq 0}$ are functions on $\mathbb{R}$ defined as follows. Let r_0 be the periodic function on $\mathbb{R}$ with period 1 defined by

$$r_0(s) = \chi_{[0,1/2]}(s) - \chi_{(1/2,1)}(s), \text{ for } 0 \leq s < 1. \tag{79}$$

For $k \in \mathbb{N}$, define $r_k(s) = r_0(2^k s)$. The Rademacher functions have the following interesting property: if $F(s) = \sum_\nu a_\nu r_\nu(s) \in L^2([0,1])$ then $F \in L^p([0,1])$ for all $p \in (1,\infty)$. In fact, we have

$$c_1\|F\|_p \le \|F\|_2 \le c_2\|F\|_p \tag{80}$$

for positive constants c_1, c_2 depending only on p (and not on the particular function F), see [25, page 277]. For each $s \in [0,1)$, set

$$P(s,x,t) = \sum_{\nu=0}^{N-1} r_\nu(s)\, T^{\delta}_{\nu,j} g(x,t). \tag{81}$$

By the orthonormality of the Rademacher functions r_ν, we have

$$\int_0^1 |P(s,x,t)|^2 ds = |Sg(x,t)|^2. \tag{82}$$

Hence in view of (80), we have

$$|Sg(x,t)| \le C_p \left(\int_0^1 |P(s,x,t)|^p ds\right)^{1/p} \tag{83}$$

for $1 < p < \infty$, with a constant C_p independent of $(x,t) \in \mathbb{R}^3$. It follows that

$$\int_{\mathbb{R}^3} |Sg(x,t)|^p dx\, dt \le C_p^p \int_{\mathbb{R}^3}\int_{[0,1)} |P(s,x,t)|^p ds\, dxdt. \tag{84}$$

In view of (81) and (64), we have $P(s,x,t) = T_s g(x,t)$, where T_s is the multiplier operator on $\mathbb{R}^3$, defined by

$$\widehat{T_s g}(\xi,\tau) = \tilde{m}_j^{\delta,s}(\xi,\tau)\widehat{g}(\xi,\tau) \tag{85}$$

with $\tilde{m}_j^{\delta,s}(\xi,\tau) = \sum_{\nu=0}^{N-1} r_\nu(s)\, \tilde{\chi}_\nu(\xi)\, \rho(2^{-j}|\xi|)\, \psi\left(\frac{|\xi|-\tau}{\delta}\right)$. Thus (84) reads as

$$\int_{\mathbb{R}^3} |Sg(x,t)|^p dx\, dt \le c_2^p \int_{[0,1)}\int_{\mathbb{R}^3} |T_s g(x,t)|^p dx\, dt\, ds. \tag{86}$$

Hence the L^p boundedness of S for $4/3 \le p \le 4$, will follow once we show that the operators T_s are uniformly bounded in L^p for the above range of p. This will follow from the next proposition. □

Proposition 5.2 *Let T_s be the multiplier operator given by* (85)*. Then T_s satisfy the estimate*

$$\|T_s g\|_{L^p} \leq C\, 2^{j/8}\delta^{1/4} j^b \|g\|_p \tag{87}$$

for all $g \in L^p(\mathbb{R}^3)$, $4/3 \leq p \leq 4$, with constants C and b independent of δ, j and s.

Proof We have $T_s g = \sum_{\nu=0}^{N-1} r_\nu(s) T^\delta_{\nu,j} g$, where $T^\delta_{\nu,j}$ is as in (64). We first prove the estimate for a related operator, acting on vector valued functions. For $s \in [0, 1]$, consider the operator $\tilde{T}_s : L^2(\mathbb{R}^3 : \mathbb{R}^N) \to L^2(\mathbb{R}^3)$ given by

$$\tilde{T}_s(h) = \sum_{\nu=0}^{N-1} r_\nu(s)\, T^\delta_{\nu,j}(h_\nu),\ h = (h_0, h_1, \ldots, h_{N-1}) \tag{88}$$

where $T^\delta_{\nu,j}$ is as in (64), $h_\nu \in \mathcal{S}(\mathbb{R}^3)$ and r_ν, $\nu = 0, 1, \cdots, N-1$ are Rademacher functions as above. Then the following estimates hold

$$\|\tilde{T}_s h\|_2 \leq \sqrt{5} \left\| \left(\sum_\nu |T^\delta_{\nu,j} h_\nu|^2 \right)^{1/2} \right\|_2 \tag{89}$$

$$\|\tilde{T}_s h\|_\infty \leq C 2^{j/4} \left\| \left(\sum_\nu |T^\delta_{\nu,j} h_\nu|^2 \right)^{1/2} \right\|_\infty . \tag{90}$$

The estimate (90) follows from (88) by Cauchy–Schwarz inequality with respect to the sum over ν as $|r_\nu| = 1$ and the fact that $N \leq 2^{j/2}$. The estimate (89) uses Plancheral theorem. Note that

$$\begin{aligned} |\widehat{\tilde{T}_s h}(\xi, \tau)|^2 &= \sum_{\nu,\nu'=0}^{N-1} r_\nu(s) r_{\nu'}(s)\, \widehat{T^\delta_{\nu,j} h_\nu}(\xi, \tau)\, \overline{\widehat{T^\delta_{\nu',j} h_{\nu'}}(\xi, \tau)} \\ &\leq \sum_{\nu=0}^{N-1} \sum_{|\nu-\nu'|\leq 2} \left| \widehat{T^\delta_{\nu,j} h_\nu}(\xi, \tau)\, \overline{\widehat{T^\delta_{\nu',j} h_{\nu'}}(\xi, \tau)} \right| . \end{aligned} \tag{91}$$

This follows from the fact that for any given ν, the support of χ_ν intersects the supports of $\chi_{\nu'}$ only for $|\nu - \nu'| \leq 2$ by the choice of χ_ν, and hence $\widehat{T^\delta_{\nu,j} g}\widehat{T^\delta_{\nu',j} g} \not\equiv 0$ only for $|\nu - \nu'| \leq 2$. It follows that the last expression is a sum of five terms of the form

$$\sum_{\nu=0}^{N-1} \left| \widehat{T^\delta_{\nu,j} h_\nu}(\xi, \tau)\, \overline{\widehat{T^\delta_{\nu',j} h_{\nu'}}(\xi, \tau)} \right|, \text{ with } \nu - 2 \leq \nu' \leq \nu + 2. \tag{92}$$

Setting $\nu' = \nu + l$ where $|l| \leq 2$, by Cauchy–Schwarz inequality we see that

$$\sum_{\nu=0}^{N-1} \left| \widehat{T^{\delta}_{\nu,j} h_\nu}(\xi,\tau) \, \overline{\widehat{T^{\delta}_{\nu+l,j} h_{\nu+l}}(\xi,\tau)} \right| \leq \left(\sum_{\nu=0}^{N-1} \left| \widehat{T^{\delta}_{\nu,j} h_\nu}(\xi,\tau) \right|^2 \right)^{1/2} \left(\sum_{\nu=0}^{N-1} \left| \widehat{T^{\delta}_{\nu+l,j} h_{\nu+l}}(\xi,\tau) \right|^2 \right)^{1/2} \tag{93}$$

Observe from (91) that, when $\nu = N-1$ the terms $\widehat{T^{\delta}_{\nu',j} h_{\nu'}}$ that matters are only those with $\nu' = N-2, N-1, 0, 1, 2$. Similar interpretation holds for the cases when $\nu + l \notin \{0, 1, \ldots, N-1\}$. Hence we see that the second term on the right hand side of the above inequality is same as $\left(\sum_{\nu=0}^{N-1} \left| \widehat{T^{\delta}_{\nu,j} h_\nu}(\xi,\tau) \right|^2 \right)^{1/2}$. It follows that

$$\sum_{\nu=0}^{N-1} \left| \widehat{T^{\delta}_{\nu,j} h_\nu}(\xi,\tau) \, \overline{\widehat{T^{\delta}_{\nu+l,j} h_{\nu+l}}(\xi,\tau)} \right| \leq \sum_{\nu=0}^{N-1} \left| \widehat{T^{\delta}_{\nu,j} h_\nu}(\xi,\tau) \right|^2. \tag{94}$$

Now using (94) in (91) and integrating both sides yields the estimate (89). Interpolating (89) and (90) (see [12, Theorem 1.19]), we get the estimate

$$\|\tilde{T}_s h\|_p \leq 5^{1/p} \, 2^{\frac{j}{2}(\frac{1}{2}-\frac{1}{p})} \left\| \left(\sum_{\nu} |T^{\delta}_{\nu,j} h_\nu|^2 \right)^{1/2} \right\|_p, \quad 2 \leq p \leq \infty. \tag{95}$$

Note that for $g \in \mathcal{S}(\mathbb{R}^3)$, we have $T_s(g) = \tilde{T}_s(h)$ with $h = (g, g, \ldots, g)$, in view of (85) and (88). Hence the inequality (95) with $p = 4$ gives

$$\begin{aligned} \|T_s g\|_4 &\leq C 2^{j/8} \left\| \left(\sum_{\nu} |T^{\delta}_{\nu,j} g|^2 \right)^{1/2} \right\|_4 \\ &\leq C 2^{j/8} \delta^{1/4} \, j^b \|g\|_4 \end{aligned} \tag{96}$$

by Proposition 5.1. Since T_s given by (85) is a multiplier operator, the above estimate holds for the dual index $p = 4/3$ as well. Hence the required estimate follows by Riesz–Thorin interpolation theorem between these two estimates. □

6 L^p-Estimates for $\mathcal{F}_j^n f$

The main difficulty in proving L^p estimate for $\mathcal{F}_j^n f$ lies in the L^4 estimate for $Q_\delta(\mathcal{F}_j^n f)$. We estimate $\|Q_\delta(\mathcal{F}_j^n f)\|_4$ by duality, which requires the L^p estimate for the square function discussed in the previous section for $p = 4/3$. We start with the following

Proposition 6.1 *Let $Q_\delta(\mathcal{F}_j^n f)$ and $\tilde{\mathcal{F}}_{j,\nu}^n f$ be as in* (16) *and* (25) *respectively. Then the inequality*

$$\|Q_\delta(\mathcal{F}_j^n f)\|_{L^4(\mathbb{R}^3)} \leq C\,\delta^{1/4}\,j^b\,2^{j/8} \left\| \left(\sum_{\nu=0}^{N-1} |\tilde{\mathcal{F}}_{j,\nu}^n f|^2 \right)^{\frac{1}{2}} \right\|_{L^4(\mathbb{R}^3)}, \tag{97}$$

holds for all $f \in \mathcal{S}(\mathbb{R}^2)$ with constants C and b independent of j and n.

Proof We use duality to estimate the L^4 norm. For $H \in L^{4/3}(\mathbb{R}^3)$, writing $Q_\delta(\mathcal{F}_j^n f) = \sum_{\nu=0}^{N-1} Q_\delta(\mathcal{F}_{j,\nu}^n f)$, we have

$$\langle Q_\delta(\mathcal{F}_j^n f), H\rangle = \int_{\mathbb{R}^3} \sum_\nu Q_\delta(\mathcal{F}_{j,\nu}^n f)(x,t)\,\overline{H}(x,t)\,dxdt. \tag{98}$$

By Parseval's theorem for the Fourier transform, and in view of (16) we see that

$$\begin{aligned} \int_{\mathbb{R}^3} Q_\delta(\mathcal{F}_{j,\nu}^n f)(x,t)\,\overline{H}(x,t)\,dxdt &= \int_{\mathbb{R}^3} \widehat{\mathcal{F}_{j,\nu}^n f}(\xi,\tau)\,\overline{\widehat{Q_\delta(H)}}(\xi,\tau)\,d\xi d\tau \\ &= \int_{\mathbb{R}^3} (\tilde{\mathcal{F}}_{j,\nu}^n f)(x,t)\,\overline{T_{\nu,j}^\delta H(x,t)}\,dxdt \end{aligned} \tag{99}$$

where $\tilde{\mathcal{F}}_{j,\nu}^n$ is as in (25) and $T_{\nu,j}^\delta$ is the multiplier operator given by (64). Now summing over ν and using Cauchy–Schwarz inequality with respect to sum over ν on the right hand side of (99), followed by an application of Hölder's inequality yields

$$|\langle Q_\delta(\mathcal{F}_j^n f), H\rangle| \leq \left\| \left(\sum_\nu |\tilde{\mathcal{F}}_{j,\nu}^n f|^2 \right)^{\frac{1}{2}} \right\|_4 \left\| \left(\sum_\nu \left| T_{\nu,j}^\delta H \right|^2 \right)^{\frac{1}{2}} \right\|_{4/3}. \tag{100}$$

By Theorem 5.1, the second term on the right hand side of (100) is bounded by $C\,\delta^{1/4}\,j^b\,2^{j/8}\,\|H\|_{4/3}$. Taking supremum over $\|H\|_{4/3} \leq 1$, the required estimate follows. □

Next we proceed to estimate the square function appearing in Proposition 6.1, following similar arguments as in the proof of Proposition 5.1.

Proposition 6.2 *Let* $\tilde{\mathcal{F}}^n_{j,\nu} f$ *be as in* (25) *with amplitude function* a_n *satisfying* (28). *Then the square function estimate*

$$\left\| \left(\sum_{\nu=0}^{N-1} |\tilde{\mathcal{F}}^n_{j,\nu} f|^2 \right)^{\frac{1}{2}} \right\|_{L^4(\mathbb{R}^3)} \leq C \, \frac{j^{b+3/4}}{(1+|n|)^{3+1/4}} \, \|f\|_{L^4(\mathbb{R}^2)} \tag{101}$$

holds for all $f \in \mathcal{S}(\mathbb{R}^2)$ *with constants* C, b *independent of* j *and* n.

Proof Let Φ_ν denote the characteristic function of the support of χ_ν defined in (22) so that $\chi_\nu = \Phi_\nu \chi_\nu$. Then $\tilde{\mathcal{F}}^n_{j,\nu} f = \tilde{\mathcal{F}}^n_{j,\nu} f_\nu$ where $\widehat{f_\nu}(\xi) = \Phi_\nu(\xi)\, \hat{f}(\xi)$. It follows that

$$\tilde{\mathcal{F}}^n_{j,\nu} f(x,t) = \int_{\mathbb{R}^2} \tilde{K}^n_{j,\nu}(x-y,t)\, f_\nu(y)\, dy, \tag{102}$$

with $\tilde{K}^n_{j,\nu}$ as in (27). Using Cauchy–Schwarz inequality in (102) and summing over ν, we get

$$\begin{aligned} &\sum_{\nu=0}^{N-1} |\tilde{\mathcal{F}}^n_{j,\nu} f_\nu(x,t)|^2 \\ &\quad \leq \sum_\nu \left(\int_{\mathbb{R}^2} |f_\nu(y)|^2 |\tilde{K}^n_{j,\nu}(x-y,t)|\, dy \right) \left(\int_{\mathbb{R}^2} |\tilde{K}^n_{j,\nu}(x-y,t)|\, dy \right) \\ &\quad \leq \frac{C}{(1+|n|)^{3+1/4}} \int_{\mathbb{R}^2} \sum_\nu |f_\nu(y)|^2 |\tilde{K}^n_{j,\nu}(x-y,t)|\, dy \end{aligned} \tag{103}$$

since $\left\| \tilde{K}^n_{j,\nu}(\cdot,t) \right\|_{L^1(\mathbb{R}^2)} \leq \frac{C}{(1+|n|)^{3+1/4}}$ for some constant C independent of t, j and n, in view of Lemma 3.1. Squaring, integrating and taking square root in (103) leads to the inequality

$$\begin{aligned} &\left\| \left(\sum_{\nu=0}^{N-1} |\tilde{\mathcal{F}}^n_{j,\nu} f_\nu|^2 \right)^{\frac{1}{2}} \right\|^2_{L^4(\mathbb{R}^3)} \\ &\leq \frac{C}{(1+|n|)^{3+1/4}} \sup_{\|g\|_{L^2}=1} \left| \int_{\mathbb{R}^3} \left[\int_{\mathbb{R}^2} \sum_\nu |f_\nu(y)|^2 |\tilde{K}^n_{j,\nu}(x-y,t)|\, dy \right] g(x,t)\, dx dt \right| \\ &\leq \frac{C}{(1+|n|)^{3+1/4}} \sup_{\|g\|_{L^2}=1} \left| \int_{\mathbb{R}^2} \sum_\nu |f_\nu(y)|^2 \left[\int_{\mathbb{R}^3} |\tilde{K}^n_{j,\nu}(x-y,t)|\, |g(x,t)|\, dx dt \right] dy \right| \end{aligned} \tag{104}$$

where we used duality in the above equality for the $L^2(dxdt)$ norm as in (73), and Fubini's theorem in the last step. By Cauchy–Schwarz inequality in y variable, the term inside the modulus sign is at most

$$\left[\int_{\mathbb{R}^2}\left(\sum_{\nu=0}^{N-1}|f_\nu(y)|^2\right)^2 dy\right]^{1/2}\left[\int_{\mathbb{R}^2}\sup_\nu\left|\int_{\mathbb{R}^3}|\tilde{K}^n_{j,\nu}(x-y,t)|\,|g(x,t)|\,dxdt\right|^2 dy\right]^{\frac{1}{2}}. \tag{105}$$

Note that the first term above is

$$\left\|\left(\sum_{\nu=0}^{N-1}|f_\nu|^2\right)^{1/2}\right\|^2_{L^4(\mathbb{R}^2)} \le C[\log N]^{2b}\,\|f\|^2_{L^4(\mathbb{R}^2)} \tag{106}$$

for constants $b > 0$ and C independent of N, as shown by A. Cordoba in [4]. Since $N \approx 2^{j/2}$, we see that the first term above is at most $Cj^{2b}\,\|f\|^2_{L^4}$, with C and b independent of j. In view of the pointwise estimate for $\tilde{K}^n_{j,\nu}$ given by (42), we have the maximal inequality

$$\left[\int_{\mathbb{R}^2}\sup_\nu\left|\int_{\mathbb{R}^3}|\tilde{K}^n_{j,\nu}(x-y,t)|\,|g(x,t)|\,dxdt\right|^2 dy\right]^{\frac{1}{2}} \le \frac{C\,j^{3/2}}{(1+|n|)^{3+1/4}}\|g\|_{L^2(\mathbb{R}^3)} \tag{107}$$

as in (76), with C independent of j and n. Using these estimates in (104) and taking the square root, the proof follows. □

Proposition 6.3 *Let $Q_\delta\mathcal{F}^n_j f$ be as in* (16) *with $\delta \ge 1$. Then the estimate*

$$\|Q_\delta\mathcal{F}^n_j f\|_4 \le C\,\frac{\delta^{1/4}\,2^{j/8}\,j^{2b+3/4}}{(1+|n|)^{3+1/4}}\|f\|_{L^4(\mathbb{R}^2)} \tag{108}$$

holds for all $f \in L^4(\mathbb{R}^2)$, with constants $C, b > 0$ independent of j, n and δ.

Proof The proof follows from Propositions 6.1 and 6.2, by the density of $\mathcal{S}(\mathbb{R}^2)$ in $L^4(\mathbb{R}^2)$. □

Proposition 6.4 *Let $\mathcal{F}^n_j f$ be as in* (10) *with amplitude function satisfying* (28)*. Then for each $0 < \epsilon \le 1/2$, there exists a constant C_ϵ independent of j and n such that* *the estimate*

$$\|\mathcal{F}_j^n f\|_4 \leq C_\epsilon \frac{2^{j(3\epsilon+1/8)}}{(1+|n|)^{3+1/4}} \|f\|_{L^4(\mathbb{R}^2)} \tag{109}$$

holds for all $f \in L^4(\mathbb{R}^2)$.

Proof In view of (16), we have $\mathcal{F}_j^n f = \mathcal{R}_\delta(\mathcal{F}_j^n f) + Q_\delta(\mathcal{F}_j^n f)$. For each $0 < \epsilon \leq 1/2$, by Proposition 6.3 with $\delta = 2^{\epsilon j}$, we have

$$\begin{aligned} \|Q_\delta \mathcal{F}_j^n f\|_4 &\leq C \frac{2^{j(\epsilon/4+1/8)}\, j^{2b+3/4}}{(1+|n|)^{3+1/4}} \|f\|_{L^4(\mathbb{R}^2)} \\ &\leq C'_{b,\epsilon} \frac{2^{j(3\epsilon+1/8)}}{(1+|n|)^{3+1/4}} \|f\|_{L^4(\mathbb{R}^2)} \end{aligned} \tag{110}$$

by writing $j^{2b+3/4} \leq C_{b,\epsilon} 2^{11\epsilon j/4}$. For the same choice of ϵ and δ, Proposition 4.1 gives

$$\|\mathcal{R}_\delta(\mathcal{F}_j^n f)\|_{L^4(\mathbb{R}^2\times\mathbb{R})} \leq \frac{C'_\epsilon}{(1+|n|)^{3+1/4}} \|f\|_{L^4}. \tag{111}$$

The required estimate follows from these two estimates. □

The above results lead to the following L^p estimate for $\mathcal{F}_j^n f$.

Theorem 6.1 *Let* $\mathcal{F}_j^n f$ *be as in* (10) *with amplitude function satisfying* (28)*. Then for each* $0 < \epsilon \leq 1/2$*, the estimate*

$$\|\mathcal{F}_j^n f\|_p \leq C_\epsilon \frac{2^{(1/2-3/2p)j}\, 2^{12\epsilon j/p}}{(1+|n|)^{3+1/4}} \|f\|_{L^p(\mathbb{R}^2)} \tag{112}$$

holds for all $f \in L^p(\mathbb{R}^2)$, $4 \leq p \leq \infty$ *with a constant* C_ϵ *independent of* j *and* n.

Proof By Riesz–Thorin interpolation, Propositions 6.4 and 3.1 yields

$$\|\mathcal{F}_j^n f\|_{L^p(\mathbb{R}^2\times\mathbb{R})} \leq \frac{C_\epsilon}{(1+|n|)^{3+1/4}} 2^{j(3\epsilon+1/8)(1-t)}\, 2^{tj/2} \|f\|_{L^p(\mathbb{R}^2)}, \tag{113}$$

for $4 \leq p \leq \infty$, where $\frac{1}{p} = \frac{1-t}{4}$. Substituting $t = 1 - \frac{4}{p}$ on the right hand side gives the required estimate. □

7 Local Smoothing Estimate

In this section we give the proof of Theorem 1.1. We will show that the local smoothing estimate for the Fourier integral operator $\mathcal{F}$ will follow from the L^p Sobolev estimate for $\mathcal{F}_j^n f$ via a summability argument. This L^p Sobolev estimate in turn follows from the L^p estimates for $\mathcal{F}_j^n f$ given by Theorem 6.1. A key step in this reduction is the following Lemma.

Lemma 7.1 *For $\sigma \in \mathbb{C}$, $j \in \mathbb{N}$ and $\rho \in C_c^\infty([\frac{1}{2}, 2])$, define $f_{\sigma,j}$ by*

$$\widehat{f_{\sigma,j}}(\xi) = \hat{f}(\xi)\, \rho(2^{-j}|\xi|)\, (1+|\xi|^2)^{\sigma/2}, \;\; f \in \mathcal{S}(\mathbb{R}^2). \tag{114}$$

Then the estimate

$$\| f_{\sigma,j} \|_{L^p(\mathbb{R}^2)} \le C_\sigma\, 2^{jRe(\sigma)}\, \| f \|_{L^p(\mathbb{R}^2)}, \;\; 1 \le p \le \infty \tag{115}$$

holds for all $f \in \mathcal{S}(\mathbb{R}^2)$ for $Re(\sigma) \le 0$, with a constant C_σ independent of j.

Proof We have,

$$f_{\sigma,j}(x) = f * k_\sigma(x), \tag{116}$$

where

$$\begin{aligned} k_\sigma(x) &= \int_{\mathbb{R}^2} e^{ix\cdot\xi} \rho(2^{-j}|\xi|)\, (1+|\xi|^2)^{\sigma/2}\, d\xi \\ &= 2^{2j} \int_{\mathbb{R}^2} e^{i2^j x\cdot\xi} \rho(|\xi|)\, (1+2^{2j}|\xi|^2)^{\sigma/2}\, d\xi. \end{aligned} \tag{117}$$

Note that the above integral is actually over the annulus $1/2 \le |\xi| \le 2$ by the support property of ρ. Hence an integration by parts and the fact that $\mathrm{Re}(\sigma) \le 0$ shows that

$$|k_\sigma(x)| \le C_\sigma\, 2^{j\mathrm{Re}(\sigma)} \frac{2^{2j}}{(1+|2^j x|^2)^2}. \tag{118}$$

Hence $\|k_\sigma\|_1 \le C_\sigma 2^{j\mathrm{Re}(\sigma)}$ and the proof follows by Young's inequality. □

We also consider the Fourier integral operator $\tilde{\mathcal{F}}_j^n$ given by

$$\tilde{\mathcal{F}}_j^n f(x,t) = \rho_1(t) \int_{\mathbb{R}^2} e^{i(x\cdot\xi + t|\xi|)}\, \rho(2^{-j}|\xi|)\, a_n(\xi)\, \hat{f}(\xi)\, d\xi \tag{119}$$

where $\rho^2 = \rho_0$. Note that $\tilde{\mathcal{F}}_j^n$ and $\mathcal{F}_j^n$ given by (10) differs only in the power of ρ, and we have $\tilde{\mathcal{F}}_j^n f = \sum_{\nu=0}^{N-1} \tilde{\mathcal{F}}_{j,\nu}^n f$ where $\tilde{\mathcal{F}}_{j,\nu}^n f$ is as in (25). It follows that $\tilde{\mathcal{F}}_j^n f$ also satisfies the same norm estimates as in Theorem 6.1:

$$\|\tilde{\mathcal{F}}_j^n f\|_p \leq C_\epsilon \frac{2^{(1/2-3/2p)j}\, 2^{12\epsilon j/p}}{(1+|n|)^{3+1/4}} \|f\|_{L^p(\mathbb{R}^2)},\ 4 \leq p \leq \infty \tag{120}$$

with constant C_ϵ independent of j and n. In fact, the L^4 and L^∞ estimates for $\mathcal{F}_j^n f$ involve the bound for ρ_0 and its derivatives, which in turn depend only on the bound for ρ and its derivatives since $\rho_0 = \rho^2$, as seen in Lemma 3.1, Propositions 4.1, 6.1, and 6.2. Hence the assertion.

Theorem 7.1 *Let $\mathcal{F}_j^n f$ be as in* (10) *with the amplitude function $a_n \in S^m$, $m \leq 0$ satisfying $|\partial_\xi^\alpha a_n(\xi)| \leq C_\alpha \frac{(1+|\xi|)^{m-|\alpha|}}{(1+|n|)^{3+1/4}}$ for all multi-indices α, with C_α independent of $n \in \mathbb{N}_0^3$. Then for each $\epsilon > 0$, there exist a constant $C_\epsilon > 0$ independent of j and n such that the estimate*

$$\|(I-\Delta_x)^{(\sigma-m)/2}\mathcal{F}_j^n f\|_{L^p(\mathbb{R}^2\times\mathbb{R})} \leq \frac{C_\epsilon 2^{j\theta}}{(1+|n|)^{3+1/4}} \|f\|_{L^p(\mathbb{R}^2)},\ j \in \mathbb{N} \tag{121}$$

holds for all $f \in L^p(\mathbb{R}^2)$, $4 \leq p \leq \infty$, whenever $Re(\sigma) \leq 0$, where $\theta = 12\epsilon/p + (1/2 - 3/2p) + Re(\sigma)$.

Proof First we consider the case $m = 0$. Set $\mathcal{L} = (I - \Delta_x)^{1/2}$. Then we have

$$\mathcal{L}^\sigma(\mathcal{F}_j^n f) = \tilde{\mathcal{F}}_j^n(f_{\sigma,j}) \tag{122}$$

where $\tilde{\mathcal{F}}_j^n$ and $f_{\sigma,j}$ are as in (119) and Lemma 7.1 respectively. This follows by taking Fourier transform in the x-variable and using the fact that $\rho_0 = \rho^2$. Thus in view of (122), and (120) with $f_{\sigma,j}$ together with Lemma 7.1, we see that

$$\begin{aligned}\|\mathcal{L}^\sigma(\mathcal{F}_j^n f)\|_{L^p(\mathbb{R}^2\times\mathbb{R})} &\leq \frac{C_\epsilon}{(1+|n|)^{3+1/4}} 2^{(1/2-3/2p)j}\, 2^{12\epsilon j/p}\, 2^{j\mathrm{Re}(\sigma)} \|f\|_{L^p(\mathbb{R}^2)} \\ &= \frac{C_\epsilon}{(1+|n|)^{3+1/4}} 2^{j\theta} \|f\|_{L^p(\mathbb{R}^2)}\end{aligned} \tag{123}$$

where $\theta = 12\epsilon/p + (1/2 - 3/2p) + \mathrm{Re}(\sigma)$. This settles the case $m = 0$. Now if $a_n \in S^m$, $m < 0$, then $\mathcal{L}^{-m}\mathcal{F}_j^n$ is a Fourier integral operator with amplitude function $(1+|\xi|^2)^{-m/2} a_n(\xi) \in S^0$. Since $\mathcal{L}^{\sigma-m}\mathcal{F}_j^n = \mathcal{L}^\sigma(\mathcal{L}^{-m}\mathcal{F}_j^n)$, the required estimate follows from the case $m = 0$, hence the proof. □

Remark 7.1 Note that $(I-\Delta_x)^{(\sigma-m)/2}\mathcal{F}_j^n f = \mathcal{F}_j^n((I-\Delta_x)^{(\sigma-m)/2} f)$ as in (122). Hence the inequality in Proposition 7.1 can be re written as

$$\|\mathcal{F}_j^n f\|_{L^p(\mathbb{R}^2\times\mathbb{R})} \leq \frac{C_\epsilon 2^{j\theta}}{(1+|n|)^{3+1/4}} \|f\|_{L^p_{m-\sigma}(\mathbb{R}^2)}. \tag{124}$$

Now we proceed to prove the local smoothing estimate for general Fourier integral operators of the form

$$\mathcal{G}f(x,t) = \int e^{i(x\cdot\xi+t|\xi|)} a(x,t,\xi)\, \hat{f}(\xi)\, d\xi \tag{125}$$

where $a \in S^m(\mathbb{R}^2\times\mathbb{R}\times\mathbb{R}^2)$, $m \leq 0$. First note that the L^p estimate for $\mathcal{G}f$ on $\mathbb{R}^2\times I$ will follow from the L^p estimate for the Fourier integral operator given by (4):

$$\mathcal{F}f(x,t) = \rho_1(t)\int e^{i(x\cdot\xi+t|\xi|)} a(x,t,\xi)\, \hat{f}(\xi)\, d\xi \tag{126}$$

by choosing $\rho_1 \in C_c^\infty(\mathbb{R})$ which is non negative and identically equal to 1 on the compact time interval I.

The key idea in the proof is to expand the general amplitude function $a(x,t,\xi)$ in terms of Hermite functions in (x,t) variables for each fixed ξ, to write it as a sum of Fourier integral operators as in (8) with amplitude function independent of x and t. Recall that the one dimensional Hermite functions h_n are given by

$$h_n(x) = \frac{(-1)^n}{\sqrt{2^n n!\sqrt{\pi}}}\left(\frac{d^n}{dx^n}e^{-x^2}\right)e^{\frac{x^2}{2}},\ x\in\mathbb{R}. \tag{127}$$

The k-dimensional Hermite functions are defined by the tensor product of the one dimensional Hermite functions: i.e., for $x \in \mathbb{R}^k$ and $n = (n_1, n_2, \dots, n_k)$ we have $h_n(x) = \Pi_{i=1}^k h_{n_i}(x_i)$, $x = (x_1, x_2, \dots, x_k)$. Then the functions $h_n \in L^2(\mathbb{R}^k)$ and we have $\|h_n\|_2 = 1$ for each multi-index $n \in \mathbb{N}_0^k$.

The functions h_n are eigenfunctions of the Hermite operator $H = -\Delta_x + |x|^2$ with eigenvalue $2|n| + k$, for the multi index $n = (n_1, n_2, \dots, n_k)$. The family $\{h_n\}_{n\in\mathbb{N}_0^k}$ form an orthonormal basis for $L^2(\mathbb{R}^k)$, hence every L^2 function has an L^2 convergent Hermite expansion. For more details on the Hermite functions and Hermite polynomials, we refer to [14, 28].

Proof of Theorem 1.1 We expand $a(x,t,\xi)$ in terms of Hermite functions in (x,t) variables, for each fixed $\xi \in \mathbb{R}^2$:

$$a(x,t,\xi) = \sum_{n\in\mathbb{N}_0^3} a_n(\xi)\, h_n(x,t),\ \ (x,t)\in\mathbb{R}^2\times\mathbb{R} \tag{128}$$

where a_n denotes the Fourier–Hermite coefficient given by

$$a_n(\xi) = \int_{\mathbb{R}^3} a(x,t,\xi)\, h_n(x,t)\, dxdt \tag{129}$$

and h_n stands for the normalised Hermite function on $\mathbb{R}^3$ discussed above. It is easy to see that the series given by (128) converges absolutely in (x,t,ξ) if $a(\cdot,\cdot,\xi)$ belongs to the Hermite–Sobolev space $W_H^{4,\infty}(\mathbb{R}^3)$ for each ξ, with the Hermite–Sobolev norm $\|a\|_{W_H^{4,\infty}} := \|H^4 a\|_{L^\infty(dx,dt)}$ bounded uniformly in ξ. In fact, using

$$H^4 h_n = (2|n|+3)^4 h_n \tag{130}$$

in (129) and an integration by parts together with the fact that $a \in S^m(\mathbb{R}^2 \times \mathbb{R} \times \mathbb{R}^2)$, $m \leq 0$ shows that

$$|\partial_\xi^\alpha a_n(\xi)| \leq \|H^4 \partial_\xi^\alpha a\|_{L^\infty(\mathbb{R}^2,\, dxdt)} \frac{\|h_n\|_{L^1(\mathbb{R}^3,\, dxdt)}}{(2|n|+3)^4}. \tag{131}$$

In view of Lemma 1.5.2 in [28], we have $\|h_n\|_{L^1(\mathbb{R}^3)} \approx (n_1 n_2 n_3)^{1/4} \leq |n|^{3/4}$ for $n = (n_1, n_2, n_3) \in \mathbb{N}_0^3$. Thus we get

$$|\partial_\xi^\alpha a_n(\xi)| \leq \frac{C\, \|H^4 \partial_\xi^\alpha a(\cdot,\cdot,\xi)\|_{L^\infty(\mathbb{R}^2,\, dxdt)}}{(1+|n|)^{3+1/4}} \tag{132}$$

for all $\alpha \in \mathbb{N}_0^2$. In particular, when $\alpha = 0$ we see that

$$|a_n(\xi)| \leq \frac{C}{(1+|n|)^{3+1/4}}, \quad n \in \mathbb{N}_0^3 \tag{133}$$

with a constant C independent of n and ξ, provided $\|H^4 a(\cdot,\cdot,\xi)\|_{L^\infty(\mathbb{R}^2,\, dxdt)}$ is uniformly bounded in ξ. Since $\|h_n\|_\infty$ is bounded uniformly in $n \in \mathbb{N}_0^3$, (see Lemma 1.5.2 in [28]), it follows that the Hermite expansion (128) converges absolutely and uniformly in (x,t) and ξ variables.

In view of (128) and the decomposition (11), we have for $f \in \mathcal{S}(\mathbb{R}^2)$

$$\mathcal{F}f(x,t) = \sum_{n \in \mathbb{N}_0^3} \sum_{j=0}^{\infty} h_n(x,t)\, \mathcal{F}_j^n f(x,t) \tag{134}$$

where $\mathcal{F}_j^n f$ is as in (10). The above step involves an interchange of integral and sum, which is justified by dominated convergence theorem whenever $f \in \mathcal{S}(\mathbb{R}^2)$, and the fact that $\sum_{n,j} |a_n(\xi)|\, \rho(2^{-j}|\xi|)$ is bounded uniformly in ξ, in view of (133). Since

$\|h_n\|_\infty$ is bounded uniformly in $n \in \mathbb{N}_0^3$ as observed above, taking L^p norm on both sides of (134) yields

$$\|\mathcal{F}f\|_{L^p(\mathbb{R}^2\times\mathbb{R})} \le \sum_{n\in\mathbb{N}_0^3}\sum_{j=0}^{\infty} \|\mathcal{F}_j^n f\|_{L^p(\mathbb{R}^2\times\mathbb{R})}. \tag{135}$$

Note that $H^4\partial_\xi^\alpha a$ is a linear combination of terms of the form $(x,t)^\gamma \partial_{x,t}^\beta \partial_\xi^\alpha a$ for $|\beta|+|\gamma| \le 8$. This does not satisfy the decay condition of the symbol. However by assumption (5) we have $\|H^4\partial_\xi^\alpha a\|_{L^\infty dxdt} \le C_\alpha(1+|\xi|)^{m-|\alpha|}$ for all α. Thus from (132) we see that a_n satisfies the estimate

$$|\partial_\xi^\alpha a_n(\xi)| \le C_\alpha \frac{(1+|\xi|)^{m-|\alpha|}}{(1+|n|)^{3+1/4}} \tag{136}$$

for each α, with C_α independent of n. Hence by Theorem 7.1 and Remark 7.1, we get for each $n \in \mathbb{N}_0^3$ and $j \in \mathbb{N}$

$$\|\mathcal{F}_j^n f\|_{L^p(\mathbb{R}^3)} \le \frac{C_\epsilon 2^{j\theta}}{(1+|n|)^{3+1/4}} \|f\|_{L^p_{m-\sigma}(\mathbb{R}^2)} \tag{137}$$

for $4 \le p \le \infty$, where $\theta = 12\epsilon/p + (1/2 - 3/2p) + \mathrm{Re}(\sigma)$. The same estimates holds for $j = 0$ as well, in view of Theorem 3.1 with $b(\xi) = \varphi_0(\xi)\, a_n(\xi)$, and (136), together with the commutativity of $(I-\Delta_x)$ with $\mathcal{F}_0^n$, where $\varphi_0 = \sum_{j\le 0} \rho_0(2^{-j}|\xi|)$.

Since $\theta < 0$ whenever $\mathrm{Re}(\sigma) < \frac{3}{2p} - \frac{1}{2} - 12\epsilon/p$ and $\sum_{n\in\mathbb{N}_0^3} \frac{1}{(1+|n|)^{3+1/4}} < \infty$, we see that $\sum_{n,j} \mathcal{F}_j^n f$ is absolutely summable in $L^p(\mathbb{R}^3)$ and

$$\|\mathcal{F}f\|_{L^p(\mathbb{R}^2\times\mathbb{R})} \le \sum_{n\in\mathbb{N}_0^3}\sum_{j=0}^{\infty} \|\mathcal{F}_j^n f\|_{L^p(\mathbb{R}^2\times\mathbb{R})} \le C_{\epsilon,\sigma}\|f\|_{L^p_{m-\sigma}(\mathbb{R}^2)} \tag{138}$$

for $\mathrm{Re}(\sigma) < \sigma_\epsilon = \frac{3}{2p} - \frac{1}{2} - 12\epsilon/p$, where

$$C_{\epsilon,\sigma} = C_\epsilon \sum_{n\in\mathbb{N}_0^3} \frac{1}{(1+|n|)^{3+1/4}} \sum_{j=0}^{\infty} 2^{j\theta} < \infty. \tag{139}$$

Note that $\epsilon > 0$ is arbitrary, and $\sigma_\epsilon \to \frac{3}{2p} - \frac{1}{2}$ as $\epsilon \to 0$. Thus for any given σ with $\mathrm{Re}(\sigma) < \frac{3}{2p} - \frac{1}{2}$, we have $\mathrm{Re}(\sigma) < \sigma_\epsilon$ for some small $\epsilon = \epsilon(\sigma) > 0$. It follows that for $\mathrm{Re}(\sigma) < \frac{3}{2p} - \frac{1}{2}$, there exists a constant C_σ such that the inequality

$$\|\mathcal{F}f\|_{L^p(\mathbb{R}^2\times\mathbb{R})} \le C_\sigma\|f\|_{L^p_{m-\sigma}(\mathbb{R}^2)} \tag{140}$$

holds for all $f \in L^p(\mathbb{R}^2)$, $4 \leq p < \infty$. The case $2 < p < 4$ follows by an analytic interpolation argument on the Fourier integral operators $\mathcal{F}^n$ given by (8). In fact, re writing (137) for $p = 4$ as

$$\|(I - \Delta_x)^{(\sigma-m)/2}\mathcal{F}_j^n f\|_{L^4(\mathbb{R}^3)} \leq \frac{C_\epsilon 2^{j(3\epsilon+1/8+\mathrm{Re}(\sigma))}}{(1+|n|)^{3+1/4}} \|f\|_{L^4(\mathbb{R}^2)} \tag{141}$$

and summing over j, the analysis carried out above leads to the following L^4 estimate for each $\mathcal{F}^n f$:

$$\|(I - \Delta_x)^{(\sigma-m)/2}\mathcal{F}^n f\|_{L^4(\mathbb{R}^2\times\mathbb{R})} \leq \frac{C}{(1+|n|)^{3+1/4}} \|f\|_{L^p(\mathbb{R}^2)} \tag{142}$$

for all $f \in L^4(\mathbb{R}^2)$, $\mathrm{Re}(\sigma) < -\frac{1}{8}$. From (8) we also see that

$$\begin{aligned}(I - \Delta_x)^{(\sigma-m)/2}&\mathcal{F}^n f(x,t) \\ &= \rho_1(t)\int_{\mathbb{R}^2} e^{i(x\cdot\xi+t|\xi|)}(1+|\xi|^2)^{(\sigma-m)/2} a_n(\xi)\,\hat{f}(\xi)\,d\xi\end{aligned} \tag{143}$$

for $f \in \mathcal{S}(\mathbb{R}^2)$ with $a_n \in S^0(\mathbb{R}^2)$, $n \in \mathbb{N}_0^3$. Hence by Plancheral theorem

$$\begin{aligned}\|(1 - \Delta_x)^{(\sigma-m)/2}&\mathcal{F}^n f(\cdot,t)\|_{L^2(\mathbb{R}^2)} \\ &\leq |\rho_1(t)|\,\|(1+|\xi|^2)^{(\sigma-m)/2} a_n(\xi)\|_\infty \|f\|_{L^2(\mathbb{R}^2)}\end{aligned} \tag{144}$$

for each $t \in \mathbb{R}$, $n \in \mathbb{N}_0^3$. In view of (136) with $\alpha = 0$, we have

$$\|(1+|\xi|^2)^{(\sigma-m)/2} a_n(\xi)\|_\infty \leq \frac{C}{(1+|n|)^{3+1/4}} \tag{145}$$

for $\mathrm{Re}(\sigma) \leq 0$. Thus a further integration with respect to $t \in \mathbb{R}$ yields the L^2 estimate

$$\|(1 - \Delta_x)^{(\sigma-m)/2}\mathcal{F}^n f\|_{L^2(\mathbb{R}^2\times\mathbb{R})} \leq \frac{C}{(1+|n|)^{3+1/4}} \|f\|_{L^2(\mathbb{R}^2)} \tag{146}$$

valid for $\mathrm{Re}(\sigma) \leq 0$. Now by analytic interpolation between (142), and (146) yields

$$\|(1 - \Delta_x)^{(\sigma-m)/2}\mathcal{F}^n f\|_{L^p(\mathbb{R}^2\times\mathbb{R})} \leq \frac{C_\sigma}{(1+|n|)^{3+1/4}} \|f\|_{L^p(\mathbb{R}^2)} \tag{147}$$

for $\mathrm{Re}(\sigma) < 1/2(1/p - 1/2)$, $2 < p \leq 4$, which is equivalent to the inequality

$$\|\mathcal{F}^n f\|_{L^p(\mathbb{R}^2\times\mathbb{R})} \leq \frac{C_\sigma}{(1+|n|)^{3+1/4}} \|f\|_{L^p_{m-\sigma}(\mathbb{R}^2)} \tag{148}$$

by the commutativity of $(1-\Delta_x)^{(\sigma-m)/2}$ and $\mathcal{F}^n$. Since $\mathcal{F}f = \sum_{n\in\mathbb{N}_0^3} \mathcal{F}^n f$, (148) leads to the estimate

$$\|\mathcal{F}f\|_{L^p(\mathbb{R}^2\times\mathbb{R})} \leq C_\sigma \|f\|_{L^p_{m-\sigma}(\mathbb{R}^2)} \tag{149}$$

valid for $\mathrm{Re}(\sigma) < 1/2(1/p - 1/2)$, $2 < p \leq 4$. This completes the proof. □

Appendix

An Explicit Construction of a Homogeneous Partition of Unity

Here we give an explicit construction of a partition of unity $\{\chi_\nu\}_{\nu=0}^{N-1}$ on $\mathbb{R}^2 \setminus \{0\}$ for any $N \in \mathbb{N}$, satisfying the properties (22) and (23). We do this by constructing a partition of unity on $\mathbb{R}$ and transfer it to S^1 via the covering map $x \to (\cos x, \sin x)$, $x \in \mathbb{R}$.

For $N \in \mathbb{N}$, let $\{\xi_0, \ldots, \xi_{N-1}\}$ be a set of equally spaced points on S^1. In fact, we can take $\xi_\nu = \left(\cos(\frac{2\pi\,\nu}{N}), \sin(\frac{2\pi\,\nu}{N})\right)$, $\nu = 0, 1, \ldots, N-1$. Notice that the Euclidean distance between the consecutive points ξ_ν and $\xi_{\nu+1}$ is $|\xi_\nu - \xi_{\nu+1}| = 2\sin(\pi/N)$, $\nu = 0, \ldots, N-1$ where $\xi_N = \xi_0$.

Let Φ be a smooth, non negative even periodic function on $\mathbb{R}$ with period 2π, which is strictly increasing on $(-\pi, 0)$, with $\Phi(-\pi) = 0$ and $\Phi(0) = 1$. For instance, the function $\phi \in C_c^\infty(|x| \leq \pi)$ given by $\phi(x) = e^{-\frac{x^2}{\pi^2 - x^2}}$ for $|x| \leq \pi$, can be periodised to get one such. Then $x \to \Phi(Nx)$ is a periodic function on $\mathbb{R}$ with period $2\pi/N$.

Consider the family $\{\psi_\nu\}_{\nu\in\mathbb{Z}}$ of smooth compactly supported functions on $\mathbb{R}$ given by

$$\psi_\nu(\theta) = \frac{\Phi(N\theta - \nu\pi)}{\Phi(N\theta - \nu\pi) + \Phi(N\theta - (\nu+1)\pi)}, \quad \text{for } |\theta - \nu\pi/N| \leq \pi/N \tag{150}$$

and $\psi_\nu = 0$ for $|\theta - \nu\pi/N| \geq \pi/N$. The function ψ_ν is a translate of the function $\psi_0 \in C_c^\infty(|\theta| \leq \pi/N)$ by $\nu\pi/N$.

Note that $\psi_\nu \geq 0$, $\psi_\nu \in C_c^\infty\left(\left[\frac{(\nu-1)\pi}{N}, \frac{(\nu+1)\pi}{N}\right]\right)$ and satisfies $\psi_\nu(\nu\pi/N) = 1$ by choice of Φ. More over we have $\psi_\nu + \psi_{\nu+1} \equiv 1$, which follows from the 2π periodicity of Φ. At this point, we also note that $\psi_N = \psi_0$, $\psi_{-1} = \psi_{N-1}$ etc., again by the periodicity of Φ. Thus the family $\{\psi_\nu\}_{\nu\in\mathbb{Z}}$ defines a smooth partition of unity on $\mathbb{R}$, with supports of exactly two consecutive functions ψ_ν and $\psi_{\nu\pm1}$ intersect on a set of positive measure.

Through the covering map $\theta \to (\cos\theta, \sin\theta)$ of the unit circle, the family $\{\psi_\nu(\theta)\}_{\nu=0}^{N-1}$ gives a smooth partition of unity on the unit circle satisfying

$$|\partial_\theta^k \psi_\nu(\theta)| \leq CN^k,\ 0 \leq \nu \leq N-1,\ k \in \mathbb{N} \tag{151}$$

for some constant C independent of N. This is immediate from the fact that ψ_ν given by (150) is a function of $N\theta$. The functions $\psi_\nu, 0 \leq \nu \leq N-1$ defines a homogeneous partition of unity $\{\chi_\nu\}$ on $\mathbb{R}^2 \setminus \{0\}$ by setting

$$\chi_\nu(\xi) = \psi_\nu(\theta),\ \text{for } \xi = |\xi|(\cos\theta, \sin\theta), \tag{152}$$

which is non zero only for $\theta \in \left(\frac{(\nu-1)\pi}{N}, \frac{(\nu+1)\pi}{N}\right)$. Note that $\chi_\nu \in C^\infty(\mathbb{R}^2 \setminus \{0\})$. The estimates (151) for $\nu = 0$ give the following derivative estimates for χ_0:

$$|\partial_{\xi_1}^k \chi_0| \leq 1,\ |\partial_{\xi_2}^k \chi_0| \leq N^k \tag{153}$$

for all $k \in \mathbb{N}$, which is same as (23). This can be seen as follows. Taking powers of the tangential derivative ∂_θ in (152) on the unit circle and using (151), we see that for all $k \in \mathbb{N}$,

$$\left|(\xi^\perp \cdot \nabla)^k \chi_\nu(\xi)\right| \leq CN^k \ \text{ for } |\xi| = 1. \tag{154}$$

Note that for $\xi = (\cos\theta, \sin\theta)$

$$\xi^\perp \cdot \nabla\chi_\nu(\xi) = -\sin\theta\, \partial_{\xi_1}\chi_\nu(\xi) + \cos\theta\, \partial_{\xi_2}\chi_\nu(\xi). \tag{155}$$

Similarly, a differentiation in (152) in the radial direction gives $(\xi \cdot \nabla)\chi_\nu(\xi) = 0$. At the point $\xi = (\cos\theta, \sin\theta)$, this gives the relation $\cos\theta\, \partial_{\xi_1}\chi_\nu(\xi) = -\sin\theta\, \partial_{\xi_2}\chi_\nu(\xi)$. Using this relation in (155), we see that

$$\xi^\perp \cdot \nabla\chi_\nu = -\frac{1}{\sin\theta}\, \partial_{\xi_1}\chi_\nu = \frac{1}{\cos\theta}\, \partial_{\xi_2}\chi_\nu. \tag{156}$$

Since $\sin\theta = \xi_2/|\xi|$ and $\cos\theta = \xi_1/|\xi|$, we see that $\partial_{\xi_1}(1/\sin\theta) = \xi_1/\xi_2$, and $\partial_{\xi_2}(1/\cos\theta) = 2\xi_2/\xi_1$ on $|\xi| = 1$. A simple calculation using this, shows that the estimates (154) translate to the following inequalities

$$|\partial_{\xi_1}^k \chi_\nu(\xi)| \leq |N\sin\theta|^k \ \text{ and } |\partial_{\xi_2}^k \chi_\nu(\xi)| \leq |N\cos\theta|^k \tag{157}$$

for $\theta \in \left(\frac{(\nu-1)\pi}{N}, \frac{(\nu+1)\pi}{N}\right)$, the sector in $\mathbb{R}^2$ where χ_ν is supported. In particular, when $\nu = 0$ we have $|\sin\theta| \leq |\theta| \leq \pi/N$, and $|\cos\theta| \leq 1$. This gives the estimates (153).

Acknowledgments The authors wish to thank the Harish-Chandra Research institute, Dept. of Atomic Energy, Govt. of India, for providing excellent research facility. This research was supported in part by the 'INFOSYS scholarship for senior students'.

References

1. Asada, K., Fujiwara, D.: On some oscillatory integral transformations in $L^2(R^n)$. Jpn. J. Math. (N.S.) **4**(2), 299–361 (1978)
2. Bergh, J., Löfström, J.: Interpolation Spaces, an Introduction. Grundlehren der Mathematischen Wissenschaften, vol. 223. Springer, Berlin (1976)
3. Cordoba, A.: A note on Bochner–Riesz operators. Duke Math. J. **36**(3), 505–511 (1979)
4. Cordoba, A.: Geometric Fourier analysis. Ann. Inst. Fourier (Grenoble) **32**(3), 215–226 (1982)
5. Coriasco, S., Ruzhansky, M., Global L^p continuity of Fourier integral operators. Trans. Am. Math. Soc. **366**(5), 2575–2596 (2014)
6. Constantin, P., Saut, J.C.: Local smoothing properties of dispersive equations. J. Am. Math. Soc. **1**(2), 413–439 (1988)
7. Duistermaat, J.J.: Fourier Integral Operators. Progress in Mathematics, vol. 130. Birkhäuser, Basel (1996)
8. Duoandikoetxea, J.: Fourier Analysis. Graduate Studies in Mathematics, vol. 29. American Mathematical Society, Providence (2001)
9. Èskin, G.I.: Degenerate elliptic pseudodifferential equations of principal type (Russian). Mat. Sb. (N.S.) **82**(124), 585–628 (1970)
10. Fefferman, G.: A note on spherical summation multipliers. Israel J. Math. **15**, 44–52 (1973)
11. Folland, G.B.: Real analysis. Modern Techniques and Their Applications. Pure and Applied Mathematics Wiley, New York (1984)
12. Garcia-Cuerva, J., Rubio de Francia, J.L.: Weighted Norm Ineqalities and Related Topics. North-Holland Mathematics Studies, vol. 116. North-Holland, Amsterdam (1985)
13. Hörmander, L.: Fourier integral operators, I. Acta Math. **127**, 79–183 (1971)
14. Lebedev, N.N.: Special Functions and Their Applications. Dover Publications, New York (1972)
15. Lee, S., Vargas, A.: On the cone multiplier in $\mathbb{R}^3$. J. Funct. Anal. **263**, 925–940 (2012)
16. Peral, J.: L^p estimates for the wave equation. J. Funct. Anal. **36**, 114–145 (1980)
17. Miyachi, A.: On some estimates for the wave equation in L^p and H^p. J. Fac. Sci. Univ. Tokyo Sect. IA Math. **27**(2), 331–354 (1980)
18. Mockenhaupt, G., Seeger, A., Sogge, C.D.: Wave front sets, local smoothing and Bourgain's circular maximal theorem. Ann. Math. **136**(1), 207–218 (1992)
19. Mockenhaupt, G., Seeger, A., Sogge, C.D.: Local smoothing of Fourier integral operators and Carleson–Sjölin estimates. J. Amer. Math. Soc. **6**(1), 65–130 (1993)
20. Ruzhansky, M., Sugimoto, M.: Global L^2 boundedness theorems for a class of Fourier integral operators. Commun. PDE **31**(4–6), 547–569 (2006)
21. Seeger, A., Sogge, C.D., Stein, E.M.: Regularity properties of Fourier integral operators. Ann. Math. **134**(2), 231–251 (1991)
22. Sjölin, P.: Regularity properties of solutions to the Schrödinger equation. Duke Math J. **55**, 699–715 (1987)

23. Sogge, C.D.: Propagation of singularities and maximal functions in the plane. Invent. Math. **104**, 349–376 (1991)
24. Sogge, C.D.: Fourier Integrals in Classical Analysis. Cambridge Tracts in Mathematics, vol. 105. Cambridge University Press, Cambridge (1993)
25. Stein, E.M.: Singular Integrals and Differentiability Properties of Functions. Princeton Mathematical Series, vol. 30. Princeton University Press, Princeton (1970)
26. Stein, E.M.: Harmonic Analysis: Real-Variable Methods, Orthogonality, and Oscillatory Integrals, Princeton Mathematical Series, 43, Monographs in Harmonic Analysis, III. Princeton University Press, Princeton (1993)
27. Stein, E.M., Weiss, G.: Introduction to Fourier Analysis on Euclidean Spaces, Princeton Mathematical Series, vol. 32. Princeton University Press, Princeton (1971)
28. S. Thangavelu, Lectures on Hermite and Laguerre expansions, Mathematical Notes, vol. 42. Princeton University Press, Princeton (1993)
29. Vega, L., Schrödinger equations: pointwise convergence to the initial data. Proc. Am. Math. Soc. **102**(4), 874–878 (1988)
30. T. Wolff, Local smoothing type estimates on L^p for large p. Geom. Funct. Anal. **10**, 1237–1288 (2000)

On (λ, μ)-Classes on the Engel Group

Marianna Chatzakou

Abstract The purpose of this note is to compare the properties of the symbolic pseudo-differential calculus on the Heisenberg and on the Engel groups; nilpotent Lie groups of 2-step and 3-step, respectively. Here we provide a preliminary analysis of the structure and of the symbolic calculus with symbols parametrized by (λ, μ) on the Engel group, while for the case of the Heisenberg group we recall the analogous results on the λ-classes of symbols.

Keywords Engel group · Quantization · Classes of symbols

1 Introduction

In [4] the authors developed a global pseudo-differential calculus in the setting of a graded nilpotent Lie groups. Here we present the analogous preliminary results in the particular case of the Engel group $\mathcal{B}_4$.

We prove that the representation of $\mathcal{B}_4$ is associated with the Kohn–Nirenberg quantization on $\mathbb{R}^4$. This, together with the analogue of the Kohn–Nirenberg quantization on Lie groups (c.f [4–6]) gives rise to the development of the pseudo-differential calculus on $\mathcal{B}_4$ with scalar-valued symbols depending on the parameters (λ, μ)-the co-adjoint orbits.

In [6], M. Taylor describes a way one can develop a symbolic non-invariant calculus by defining a general quantization and the general symbols on any type-I Lie group, and explained his ideas in the setting of the Heisenberg group $\mathbb{H}_n$, with symbols defined by some asymptotic expansions. Particularising in the setting of a large class of nilpotent Lie groups; namely on the class of graded Lie groups, to the best of our knowledge, the development of a non-invariant calculus with

M. Chatzakou (✉)
Imperial College London, London, UK
e-mail: m.chatzakou16@imperial.ac.uk

© Springer Nature Switzerland AG 2020
V. Georgiev et al. (eds.), *Advances in Harmonic Analysis and Partial Differential Equations*, Trends in Mathematics, https://doi.org/10.1007/978-3-030-58215-9_2

scalar-valued symbols has been restricted to the case of the Heisenberg group (graded group of 2-step), see [1], or [3].

Besides the amount of work devoted to the case of the Heisenberg group, the same motivating aspects appear as well on any graded Lie group. In our consideration of $\mathcal{B}_4$ (graded group of 3-step) our approach differs from the one in [6] or in [1] in the sense that the symbols are operator valued. However, using the link between the Kohn–Nirenberg quantization and the representations on $\mathcal{B}_4$ they can be expressed on the Euclidean level. Concrete formulas for the difference operators in the setting of $\mathcal{B}_4$ are provided, laying down the necessary foundation for the characterisation of the symbol classes in our setting.

2 Preliminaries on the Engel Group $\mathcal{B}_4$ and Its Lie Algebra

We start by fixing the notation required for presenting our results. The map

$$
\begin{aligned}
&(x_1, x_2, x_3, x_4) \circ (y_1, y_2, y_3, y_4) \\
&\quad := (x_1 + y_1, x_2 + y_2, x_3 + y_3 - x_1 y_2, x_4 + y_4 + \frac{1}{2}x_1^2 y_2 - x_1 y_3)
\end{aligned}
\tag{1}
$$

endows $\mathbb{R}^4$ with a structure of a Lie group, and we shall refer to $\mathcal{B}_4 = (\mathbb{R}^4, \circ)$ as the *Engel group*. The Lie algebra of $\mathcal{B}_4$, say $\mathfrak{l}_4$, is, by the general theory, the vector space of (smooth) left invariant vector fields X on $\mathbb{R}^4$ characterised by the property

$$
(XI)(x) = \mathcal{J}_{\tau_x}(0) \cdot (XI)(0)\,, \forall x \in \mathcal{B}_4\,, \tag{2}
$$

where I is the identity map on $\mathbb{R}^4$ and $\mathcal{J}_{\tau_x}(0)$ denotes the Jacobian matrix at the origin of the map τ_x for $x \in \mathcal{B}_4$ where $\tau_x(y) := x \circ y$ is the left translation by x on $\mathcal{B}_4$. In particular we have

$$
\mathcal{J}_{\tau_x}(0) = \begin{pmatrix} 1 & 0 & 0 & 0 \\ 0 & 1 & 0 & 0 \\ 0 & -x_1 & 1 & 0 \\ 0 & \frac{x_1^2}{2} & -x_1 & 1 \end{pmatrix}, \tag{3}
$$

so that for example for $X_1 = \partial_{x_1}$, we can recognise that for every $x \in \mathcal{B}_4$,

$$
(X_1 I)(x) = \begin{pmatrix} 1 \\ 0 \\ 0 \\ 0 \end{pmatrix} = \begin{pmatrix} 1 & 0 & 0 & 0 \\ 0 & 1 & 0 & 0 \\ 0 & -x_1 & 1 & 0 \\ 0 & \frac{x_1^2}{2} & -x_1 & 1 \end{pmatrix} \cdot \begin{pmatrix} 1 \\ 0 \\ 0 \\ 0 \end{pmatrix} = \mathcal{J}_{\tau_x}(0) \cdot (X_1 I)(0)\,, \tag{4}
$$

while similarly for the vector fields $X_2 = \partial_{x_2} - x_1\partial_{x_3} + \frac{x_1^2}{2}\partial_{x_4}$, $X_3 = \partial_{x_3} - x_1\partial_{x_4}$ and $X_4 = \partial_{x_4}$. Simple calculations show that $[X_1, X_2] = X_3$, and $[X_1, X_3] = X_4$,[1] are the only non-zero relations, so that

$$\mathfrak{l}_4 = \text{span}\{X_1, X_2, [X_1, X_2], [X_1, X_3]\}, \tag{5}$$

and X_1, X_2 satisfy the so-called *Hörmander condition*. The lower center series of $\mathfrak{l}_4$ defined inductively by

$$\mathfrak{l}_{4(1)} := \mathfrak{l}_4, \quad \mathfrak{l}_{4(j)} = [\mathfrak{l}_4, \mathfrak{l}_{4(j-1)}],^2 \tag{6}$$

terminates at 0 after 3 steps, that is $\mathcal{B}_4$ is of 3-step. In addition, $\mathcal{B}_4$ is a homogeneous Lie group on $\mathbb{R}^4$ since the mapping

$$\delta_\lambda : \mathbb{R}^4 \to \mathbb{R}^4, \quad \delta_\lambda(x_1, x_2, x_3, x_4) = (\lambda x_1, \lambda x_2, \lambda^2 x_3, \lambda^3 x_4), \tag{7}$$

is an automorphism of $\mathcal{B}_4$ for every $\lambda > 0$, and so the natural gradation of its Lie algebra $\mathfrak{l}_4$ appears as

$$\mathfrak{l}_4 = V_1 \oplus V_2 \oplus V_3, \tag{8}$$

where $V_1 = \text{span}\{X_1, X_2\}$, $V_2 = \text{span}\{X_3\}$ and $V_3 = \text{span}\{X_4\}$, and is such that $[V_i, V_j] \subset V_{i+j}$, $i \neq j$.

Finally, let us note that, from the general theory of homogeneous Lie groups, the Lebesgue measure on $\mathbb{R}^4$ is invariant with respect to the left and right invariant translation on $\mathcal{B}_4$, that is the Lebesgue measure on $\mathbb{R}^4$ is the Haar measure for $\mathcal{B}_4$ and we can formulate as

$$\int_{\mathcal{B}_4} \cdots dx_1 dx_2 dx_3 dx_4 = \int_{\mathbb{R}^4} \cdots dx_1 dx_2 dx_3 dx_4. \tag{9}$$

3 Group Representation and Quantization of the Fourier Transform

The representations of the Engel group $\mathcal{B}_4$ are the infinite dimensional unitary (equivalence classes of) representations of $\mathcal{B}_4$. Parametrised by $\lambda \neq 0$ and $\mu \in \mathbb{R}$, following [2, p. 333], they act on $L^2(\mathbb{R}^n)$. We denote them by $\pi_{\lambda,\mu}$, and realise

[1] For smooth vectors X, Y in $\mathbb{R}^n$, we define the Lie-bracket $[X, Y] := YX - XY$.

[2] For V, W spaces of vector fields, we denote by $[V, W]$ the set $\{[v, w] : v \in V, w \in W\}$.

them as

$$\pi_{\lambda,\mu}(x_1, x_2, x_3, x_4)h(u) = exp\left(i\left(-\frac{\mu}{2\lambda}x_2 + \lambda x_4 - \lambda x_3 u + \frac{\lambda}{2}x_2 u^2\right)\right) h(u + x_1), \tag{10}$$

for $h \in L^2(\mathbb{R})$, $u \in \mathbb{R}$. The group Fourier transform of a function $f \in L^1(\mathcal{B}_4)$ is by definition the linear endomorphism on $L^2(\mathbb{R})$

$$\mathcal{F}_{\mathcal{B}_4}(f)(\pi_{\lambda,\mu}) \equiv \hat{f}(\pi_{\lambda,\mu}) \equiv \pi_{\lambda,\mu}(f) := \int_{\mathcal{B}_4} f(x)\pi_{\lambda,\mu}(x)^* \, dx \,. \tag{11}$$

Rigorous computations show that $\hat{f}(\pi_{\lambda,\mu})h(u)$ can be written as

$$\begin{aligned}
&\int_{\mathbb{R}^4}\Bigg[f(x_1, x_2, x_3, x_4) \exp\left(i\left(\frac{\mu}{2\lambda}x_2 - \lambda x_4 + \lambda x_3(u - x_1) - \frac{\lambda}{2}x_2(u - x_1)^2\right)\right) \\
&\quad \times h(u - x_1)\Bigg] dx_1\, dx_2\, dx_3\, dx_4 \\
&=(2\pi)^{-2}\int_{\mathbb{R}^4}\int_{\mathbb{R}^4}\Bigg[\mathcal{F}_{\mathbb{R}^4}(f)(\xi, \eta, \tau, \omega) \cdot e^{ix_1\xi} \cdot e^{ix_2\eta} \cdot e^{ix_3\tau} \cdot e^{ix_4\omega} \\
&\quad \times \exp\left(i\left(\frac{\mu}{2\lambda}x_2 - \lambda x_4 + \lambda x_3(u - x_1) - \frac{\lambda}{2}x_2(u - x_1)^2\right)\right) \\
&\quad \times h(u - x_1)\Bigg] dx_1\, dx_2\, dx_3\, dx_4 d\xi\, d\eta\, d\tau\, d\omega \\
&= - (2\pi)\int_{\mathbb{R}}\int_{\mathbb{R}}\Bigg[e^{ix_1\xi} \\
&\quad \times \mathcal{F}(f)_{\mathbb{R}^4}(\xi, \frac{\lambda}{2}(u - x_1)^2 - \frac{\mu}{2\lambda}, \lambda(x_1 - u), \lambda)h(u - x_1)\Bigg] dx_1\, d\xi \\
&=(2\pi)\int_{\mathbb{R}}\int_{\mathbb{R}}\Bigg[e^{i(u-v)\xi}\mathcal{F}(f)_{\mathbb{R}^4}(\xi, \frac{\lambda}{2}v^2 - \frac{\mu}{2\lambda}, -\lambda v, \lambda)h(v)\Bigg] dv\, d\xi \,,
\end{aligned} \tag{12}$$

for $h \in L^2(\mathbb{R})$ and $u \in \mathbb{R}$, that is

$$\mathcal{F}_{\mathcal{B}_4}(f)(\pi_{\lambda,\mu}) = Op[a_{f,\lambda,\mu}(\cdot, \cdot)] \,, \tag{13}$$

where

$$a_{f,\lambda,\mu}(v, \xi) = (2\pi)^2\mathcal{F}_{\mathbb{R}^4}(f)(\xi, \frac{\lambda}{2}v^2 - \frac{\mu}{2\lambda}, -\lambda v, \lambda) \,. \tag{14}$$

Here the Fourier transform $\mathcal{F}_{\mathbb{R}^4}$ is defined via:

$$\mathcal{F}_{\mathbb{R}^4} f(\xi) = (2\pi)^{-2} \int_{\mathbb{R}^4} f(x) e^{-ix\xi}\, dx \quad (\xi \in \mathbb{R}^4,\ f \in L^1(\mathbb{R}^4)), \tag{15}$$

and Op denotes the Kohn–Nirenberg quantization, that is for a smooth symbol a on $\mathbb{R} \times \mathbb{R}$ the operator

$$Op(a) f(u) = (2\pi)^{-1} \int_{\mathbb{R}} \int_{\mathbb{R}} e^{i(u-v)\xi} a(v, \xi) f(v)\, dv\, d\xi\,, \tag{16}$$

for $f \in \mathcal{S}(\mathbb{R})$ and $u \in \mathbb{R}$.

We note that for the case of the Heisenberg group $\mathbb{H}_n$ the group Fourier transform has been computed in [3] as being the operator

$$\mathcal{F}_{\mathbb{H}_n}(f)(\pi_\lambda) = (2\pi)^{\frac{n}{2}} Op^W[\mathcal{F}_{\mathbb{R}^{2n+1}}(f)(\sqrt{|\lambda|}\cdot, \sqrt{\lambda}\cdot, \lambda)]\,, \tag{17}$$

where Op^W denotes the Weyl-quantization, i.e.,

$$Op^W(a) f(u) = (2\pi)^{-n} \int_{\mathbb{R}^n} \int_{\mathbb{R}^n} e^{i(u-v)\xi} a\left(\xi, \frac{u+v}{2}\right) f(v)\, dv\, d\xi\,, \tag{18}$$

for $f \in \mathcal{S}(\mathbb{R}^n)$ and $u \in \mathbb{R}^n$, where π_λ denotes the Schrödinger representations of the Heisenberg group $\mathbb{H}_n$,

Going back to our case, of one keeps the same notation $\pi_{\lambda,\mu}$ for the infinitesimal representation, we compute that:

$$\begin{aligned}
\pi_{\lambda,\mu}(X_1) &= \partial_u = Op(i\xi)\,, \\
\pi_{\lambda,\mu}(X_2) &= \frac{i}{2}\left(\lambda u^2 - \frac{\mu}{\lambda}\right) = Op\left(\frac{i\lambda u^2}{2} - \frac{i\mu}{2\lambda}\right), \\
\pi_{\lambda,\mu}(X_3) &= -i\lambda u = Op(-i\lambda u)\,, \\
\pi_{\lambda,\mu}(X_4) &= i\lambda = Op(i\lambda)\,,
\end{aligned} \tag{19}$$

thus

$$\begin{aligned}
\pi_{\lambda,\mu}(\mathcal{L}) = \pi_{\lambda,\mu}(X_1)^2 + \pi_{\lambda,\mu}(X_2)^2 &= \frac{d^2}{du^2} - \frac{1}{4}\left(\lambda u^2 - \frac{\mu}{\lambda}\right)^2 \\
&= -Op\left(\xi^2 + \frac{1}{4}\left(\lambda u^2 - \frac{\mu}{\lambda}\right)^2\right).
\end{aligned} \tag{20}$$

With our choice of notation, the Plancherel measure of the Engel group $\mathcal{B}_4$ is $(2^{-3}\pi^{-4})d\lambda\, d\mu$, in the sense that following expression for the Plancherel formula

$$\begin{aligned}&\int_{\mathcal{B}_4} |f(x_1, x_2, x_3, x_4)|^2\, dx_1\, dx_2\, dx_3\, dx_4 \\ &\qquad\qquad = 2^{-3}\pi^{-4}\int_{\lambda\neq 0}\int_{\mu\in\mathbb{R}} \|\pi_{\lambda,\mu}(f)\|^2_{\text{HS}}\, d\mu\, d\lambda\,,\end{aligned} \tag{21}$$

holds for any $f \in \mathcal{S}(\mathbb{R})$, where $\|\cdot\|_{\text{HS}}$ denotes the Hilbert-Schmidt norm of an operator on L^2, that is $\|A\|_{\text{HS}} := Tr(A^*A)$. The last allows for an extension of the group Fourier transform to $L^2(\mathcal{B}_4)$, and in particular formula (21) holds true for any $f \in L^2(\mathcal{B}_4)$.

Indeed, by using (13) the operator $\pi_{\lambda,\mu}(f)$ has integral kernel

$$\mathcal{K}_{f,\lambda,\mu}(u, v) = 2\pi \int_{\mathbb{R}} e^{i(u-v)\xi}\mathcal{F}_{\mathbb{R}^4}(f)(\xi, \frac{\lambda}{2}v^2 - \frac{\mu}{2\lambda}, -\lambda v, \lambda)\, d\xi\,, \tag{22}$$

or equivalently

$$\mathcal{K}_{f,\lambda,\mu}(u, v) = (2\pi)^{\frac{3}{2}}\mathcal{F}_{\mathbb{R}^3}(f)(v - u, \frac{\lambda}{2}v^2 - \frac{\mu}{2\lambda}, -\lambda v, \lambda)\,, \tag{23}$$

where the Fourier transform is taken with respect to the second, the third and the fourth variable of f. Integrating the $L^2(\mathbb{R}\times\mathbb{R})$-norm of $\mathcal{K}_{f,\lambda,\mu}$ (or the Hilbert-Schmidt norm of $\pi_{\lambda,\mu}(f)$) against $d\lambda$, $d\mu$ we obtain

$$\begin{aligned}&\int_{\mathbb{R}\setminus\{0\}}\int_{\mathbb{R}}\int_{\mathbb{R}}\int_{\mathbb{R}} |\mathcal{K}_{f,\lambda,\mu}(u, v)|^2\, du\, dv\, d\mu\, d\lambda \\ &= (2\pi)^3\int_{\mathbb{R}\setminus\{0\}}\int_{\mathbb{R}}\int_{\mathbb{R}}\int_{\mathbb{R}} |\mathcal{F}_{\mathbb{R}^3}(f)(u - v, \frac{\lambda}{2}v^2 - \frac{\mu}{2\lambda}, -2\lambda, \lambda)|^2\, du\, dv\, d\lambda\, d\mu \\ &= (2\pi)^3\int_{\mathbb{R}\setminus\{0\}}\int_{\mathbb{R}}\int_{\mathbb{R}}\int_{\mathbb{R}} |\mathcal{F}_{\mathbb{R}^3}(f)(x_1, w_2, w_3, w_4)|^2\,\frac{1}{2}\, dw_2\, dw_3\, dw_4\, dx_1\,,\end{aligned} \tag{24}$$

where the constant $\frac{1}{2}$ comes from the calculation of the determinant of the Jacobian matrix of the linear transformation $F(u, v, \lambda, \mu) = (w_1 = u - v, w_2 = \frac{\lambda}{2}v^2 - \frac{\mu}{2\lambda}, w_3 = -\lambda v, w_4 = \lambda)$. Finally, the Plancherel formula on $\mathbb{R}^3$ in the variable

(w_2, w_3, w_4) with dual variable (x_2, x_3, x_4) gives

$$\int_{\mathbb{R}\setminus\{0\}} \int_{\mathbb{R}^3} |\mathcal{K}_{f,\lambda,\mu}(u, v)|^2 \, dv \, du \, d\mu \, d\lambda$$
$$= 2^2\pi^3 \int_{\mathbb{R}^4} |f(x_1, x_2, x_3, x_4)|^2 \, dx_1 \, dx_2 \, dx_3 \, dx_4 \,, \tag{25}$$

and the last implies (21).

4 Difference Operators

Difference operators on the setting of a compact Lie group introduced in [5] as acting on Fourier coefficients, while on graded Lie groups in [4]. In the setting of the Engel group $\mathcal{B}_4$ this yields the definition of the difference operators Δ_{x_i} as:

$$\Delta_{x_i}\hat{\kappa}(\pi_{\lambda,\mu}) := \pi_{\lambda,\mu}(x_i\kappa)\,, \quad i = 1, \cdots, 4\,, \tag{26}$$

for suitable distributions κ on $\mathcal{B}_4$.

To find the explicit expressions of the difference operators Δ_{x_i} we make use of the following property: For X and $\tilde{X}$ being a left and a right invariant vector field, respectively, in the Lie algebra $\mathfrak{l}_4$, and for a distribution κ on $\mathcal{B}_4$ we have

$$\pi_{\lambda,\mu}(X\kappa) = \pi_{\lambda,\mu}(X)\pi_{\lambda,\mu}(\kappa)\,, \quad \pi_{\lambda,\mu}(\tilde{X}\kappa) = \pi_{\lambda,\mu}(\kappa)\pi_{\lambda,\mu}(X)\,. \tag{27}$$

Notice that the right invariant vector fields that generate $\mathfrak{l}_4$ can be calculated as:

$$\tilde{X}_1 = \partial_{x_1} - x_2\partial_{x_3} - x_3\partial_{x_4}\,, \tilde{X}_2 = \partial_{x_2}\,, \tilde{X}_3 = \partial_{x_3}\,, \tilde{X}_4 = \partial_{x_4}\,. \tag{28}$$

Proposition 4.1 *For suitable distribution κ on $\mathcal{B}_4$ we have:*

$$\Delta_{x_1}\hat{\kappa}(\pi_{\lambda,\mu}) = \frac{i}{\lambda}(\pi_{\lambda,\mu}(X_3)\pi_{\lambda,\mu}(\kappa) - \pi_{\lambda,\mu}(\kappa)\pi_{\lambda,\mu}(X_3))\,, \tag{29}$$

where $\pi_{\lambda,\mu}(X_3) = -i\lambda u$, and

$$\Delta_{x_2}\hat{\kappa}(\pi_{\lambda,\mu}) = \frac{2\lambda}{i}\partial_\mu\pi_{\lambda,\mu}(\kappa)\,. \tag{30}$$

Proof Since $\pi_{\lambda,\mu}(X_4) = i\lambda$, and $\tilde{X}_3 - X_3 = X_4 x_1$, we have

$$\begin{aligned}\pi_{\lambda,\mu}(x_1\kappa) &= \frac{1}{i\lambda}\pi_{\lambda,\mu}(X_4 x_1 \kappa) = \frac{1}{i\lambda}((\tilde{X}_3 - X_3)\kappa) \\ &= \frac{i}{\lambda}(\pi_{\lambda,\mu}(X_3)\pi_{\lambda,\mu}(\kappa) - \pi_{\lambda,\mu}(\kappa)\pi_{\lambda,\mu}(X_3)) .\end{aligned} \tag{31}$$

Now, for the difference operator corresponding to x_2, we differentiate the group Fourier transform of κ as in (12) at h with respect to μ and get

$$\begin{aligned}&\partial_\mu\{\pi_{\lambda,\mu}(\kappa)h(u)\} \\ &\quad = \partial_\mu\Big\{\int_{\mathbb{R}^4}\kappa(x)\exp\left(i\left(\frac{\mu}{2\lambda}x_2 - \lambda x_4\right)\right) \\ &\quad \times \exp\left(i\left(\lambda x_3(u-x_1) - \frac{\lambda}{2}x_2(u-x_1)^2\right)\right)h(u-x_1)dx\Big\} \\ &= \int_{\mathbb{R}^4}\kappa(x)\exp\left(i\left(\frac{\mu}{2\lambda}x_2 - \lambda x_4\right)\right) \\ &\quad \times \exp\left(i\left(\lambda x_3(u-x_1) - \frac{\lambda}{2}x_2(u-x_1)^2\right)\right)h(u-x_1)\left(\frac{i}{2\lambda}x_2\right)dx ,\end{aligned} \tag{32}$$

or in terms of difference operators,

$$\partial_\mu \pi_{\lambda,\mu}(\kappa) = \pi_{\lambda,\mu}\left(\frac{i}{2\lambda}x_2\kappa\right) = \frac{i}{2\lambda}\Delta_{x_2}\pi_{\lambda,\mu}(\kappa) . \tag{33}$$

□

Proposition 4.2 *For a suitable distribution κ we have:*

$$\begin{aligned}&\Delta_{x_3}\hat{\kappa}(\pi_{\lambda,\mu}) \\ &\quad = \frac{i}{\lambda}(\Delta_{x_2}\pi_{\lambda,\mu}(\kappa)\pi_{\lambda,\mu}(X_3) + \pi_{\lambda,\mu}(\kappa)\pi_{\lambda,\mu}(X_1) - \pi_{\lambda,\mu}(X_1)\pi_{\lambda,\mu}(\kappa)) ,\end{aligned} \tag{34}$$

where $\Delta_{x_2|\pi_{\lambda,\mu}}$ is given in Proposition 4.1 and $\pi_{\lambda,\mu}(X_1) = \partial_u$, $\pi_{\lambda,\mu}(X_3) = -i\lambda u$.

Proof Since $X_1 - \tilde{X}_1 - x_2 X_3 = \partial_{x_4} x_3$ we have

$$\begin{aligned}\pi_{\lambda,\mu}(x_3\kappa) &= \frac{1}{i\lambda}(X_4 x_3 \kappa) = \frac{1}{i\lambda}((X_1 - \tilde{X}_1 - x_2\tilde{X}_3)\kappa)\\ &= \frac{1}{i\lambda}(\pi_{\lambda,\mu}(X_1)\pi_{\lambda,\mu}(\kappa) - \pi_{\lambda,\mu}(\kappa)\pi_{\lambda,\mu}(X_1) - \Delta_{x_2}\pi_{\lambda,\mu}(\kappa)\pi_{\lambda,\mu}(X_3)\\ &= \frac{i}{\lambda}(\Delta_{x_2}\pi_{\lambda,\mu}(\kappa)\pi_{\lambda,\mu}(X_3) + \pi_{\lambda,\mu}(\kappa)\pi_{\lambda,\mu}(X_1) - \pi_{\lambda,\mu}(X_1)\pi_{\lambda,\mu}(\kappa)),\end{aligned} \tag{35}$$

completing the proof. □

Proposition 4.3 *For a suitable distribution κ on $\mathcal{B}_4$ we have:*

$$\begin{aligned}(\Delta_{x_4}\pi_{\lambda,\mu}(\kappa))h(u) &= i\partial_\lambda\{\pi_{\lambda,\mu}(\kappa)h(u)\} - \left(\frac{\mu}{2\lambda^2} + \frac{u^2}{2}\right)\left\{\Delta_{x_2}\pi_{\lambda,\mu}(\kappa)h(u)\right\}\\ &\quad + u\left\{\Delta_{x_3}\pi_{\lambda,\mu}(\kappa)h(u)\right\} - \left\{\Delta_{x_3}\Delta_{x_1}\pi_{\lambda,\mu}(\kappa)h(u)\right\}\\ &\quad + u\left\{\Delta_{x_2}\Delta_{x_1}\pi_{\lambda,\mu}(\kappa)h(u)\right\} - \frac{1}{2}\left\{\Delta_{x_2}\Delta_{x_1}^2\pi_{\lambda,\mu}(\kappa)h(u)\right\},\end{aligned} \tag{36}$$

where the difference operators $\Delta_{x_i|\pi_{\lambda,\mu}}$, $i = 1, 2, 3$, are given in Propositions 4.1 and 4.2, respectively.

Proof Differentiating the group Fourier transform of κ as in (12) at h with respect to λ yields

$$\begin{aligned}\partial_\lambda\{\pi_{\lambda,\mu}(\kappa)h(u)\} &= \partial_\lambda\Big\{\int_{\mathbb{R}^4}\kappa(x)\exp\left(i\left(\frac{\mu}{2\lambda}x_2 - \lambda x_4\right)\right)\\ &\quad\times\exp\left(i\left(\lambda x_3(u-x_1) - \frac{\lambda}{2}x_2(u-x_1)^2\right)\right)h(u-x_1)\,dx\Big\}\\ &= \int_{\mathbb{R}^4}\kappa(x)\exp\left(i\left(\frac{\mu}{2\lambda}x_2 - \lambda x_4 + \lambda x_3(u-x_1) - \frac{\lambda}{2}x_2(u-x_1)^2\right)\right)\\ &\quad\times h(u-x_1)\Big\{i\left(-\frac{\mu}{2\lambda^2}x_2 - x_4 + x_3(u-x_1) - \frac{x_2}{2}(u-x_1)^2\right)\Big\}dx\,.\end{aligned} \tag{37}$$

Rewriting the above formula in terms of difference operators we obtain

$$\begin{aligned}
&\partial_\lambda\{\pi_{\lambda,\mu}(\kappa)h(u)\}\\
&=i\Big[-\left(\frac{\mu}{2\lambda^2}+\frac{u^2}{2}\right)\left\{\Delta_{x_2}\pi_{\lambda,\mu}(\kappa)h(u)\right\}-\left\{\Delta_{x_4}\pi_{\lambda,\mu}(\kappa)h(u)\right\}\\
&\quad+u\left\{\Delta_{x_3}\pi_{\lambda,\mu}(\kappa)h(u)\right\}-\left\{\Delta_{x_3}\Delta_{x_1}\pi_{\lambda,\mu}(\kappa)h(u)\right\}\\
&\quad+u\left\{\Delta_{x_2}\Delta_{x_1}\pi_{\lambda,\mu}(\kappa)h(u)\right\}-\frac{1}{2}\left\{\Delta_{x_2}\Delta_{x_1}^2\pi_{\lambda,\mu}(\kappa)h(u)\right\}\Big],
\end{aligned} \tag{38}$$

completing the proof. □

For example we have

$$\begin{aligned}
&\Delta_{x_1}\pi_{\lambda,\mu}(X_1)=-I,\\
&\Delta_{x_1}\pi_{\lambda,\mu}(X_2)=\Delta_{x_1}\pi_{\lambda,\mu}(X_3)=\Delta_{x_1}\pi_{\lambda,\mu}(X_4)=0
\end{aligned} \tag{39a}$$

$$\begin{aligned}
&\Delta_{x_2}\pi_{\lambda,\mu}(X_1)=\Delta_{x_2}\pi_{\lambda,\mu}(X_3)=\Delta_{x_2}\pi_{\lambda,\mu}(X_4)=0,\\
&\Delta_{x_2}\pi_{\lambda,\mu}(X_2)=-\lambda I
\end{aligned} \tag{39b}$$

$$\begin{aligned}
\Delta_{x_3}\pi_{\lambda,\mu}(X_1)&=\Delta_{x_3}\pi_{\lambda,\mu}(X_4)=0,\\
\Delta_{x_3}\pi_{\lambda,\mu}(X_2)&=-\lambda u+u,\\
\Delta_{x_3}\pi_{\lambda,\mu}(X_3)&=-I,
\end{aligned} \tag{39c}$$

$$\begin{aligned}
\Delta_{x_4}\pi_{\lambda,\mu}(X_1)&=\Delta_{x_4}\pi_{\lambda,\mu}(X_4)=0,\\
\Delta_{x_4}\pi_{\lambda,\mu}(X_2)&=\frac{u^2}{2}(1-\lambda)+\frac{\mu}{2\lambda},\\
\Delta_{x_4}\pi_{\lambda,\mu}(X_4)&=-I,
\end{aligned} \tag{39d}$$

where the difference operators $\Delta_{x_i}\pi_{\lambda,\mu}(X_j)$ can be understood as the group Fourier transform of the distribution $x_i X_j \delta_0$.

5 Quantization and Symbol Classes

In this note, we may slightly change the notation of the symbol introduced in [4]. We keep the notation

$$x=(x_1,x_2,x_3,x_4)\in\mathcal{B}_4\,, \tag{40}$$

to denote the coordinates of an element in the Engel group $\mathcal{B}_4$, and we may denote by

$$\sigma(x, \lambda, \mu) := \sigma(x, \pi_{\lambda,\mu}), \quad (x, \lambda, \mu) \in \mathcal{B}_4 \times \mathbb{R} \setminus \{0\} \times \mathbb{R}, \tag{41}$$

the symbol σ parametrised by (x, λ, μ). In addition, if the multi-index $\alpha \in \mathbb{N}_0^4$ is written as

$$\alpha = (\alpha_1, \alpha_2, \alpha_3, \alpha_4), \quad \alpha_i \in \mathbb{N}_0, \tag{42}$$

then the homogeneous degree of α is given by:

$$[\alpha] = \alpha_1 + \alpha_2 + 2\alpha_3 + 3\alpha_4. \tag{43}$$

For each α we may write:

$$x^\alpha = x_1^{\alpha_1} x_2^{\alpha_2} x_3^{\alpha_3} x_4^{\alpha_4}, \tag{44}$$

so that the corresponding difference operator can be defined as:

$$\Delta'^{\alpha} = \Delta_{x_1}^{\alpha_1} \Delta_{x_2}^{\alpha_2} \Delta_{x_3}^{\alpha_3} \Delta_{x_4}^{\alpha_4}. \tag{45}$$

Finally for the vector field X we write X^α to denote the following composition of vector fields:

$$X_1^{\alpha_1} X_2^{\alpha_2} X_3^{\alpha_3} X_4^{\alpha_4}. \tag{46}$$

Following [4] we define the symbol classes $S^m_{\rho,\delta}(\mathcal{B}_4)$, where $0 \leq \delta < \rho \leq 1$ and $m \in \mathbb{R}$, as the set of symbols σ for which the following quantities are finite:

$$\|\sigma\|_{S^m_{\rho,\delta},a,b,c} := \sup_{\lambda \in \mathbb{R}\setminus\{0\}, \mu \in \mathbb{R}, x \in \mathcal{B}_4} \|\sigma(x, \lambda, \mu)\|_{S^m_{\rho,\delta},a,b,c}, \quad a, b, c \in \mathbb{N}_0, \tag{47}$$

where $\|\sigma(x, \lambda, \mu)\|_{S^m_{\rho,\delta},a,b,c}$ is defined as

$$\sup_{\substack{[a] \leq a \\ [\beta] \leq b, |\gamma| \leq c}} \|\pi_{\lambda,\mu}(I - \mathcal{L})^{\frac{\rho[\alpha] - m - \delta[\beta] + \gamma}{2}} X^\beta \Delta'^{\alpha} \sigma(x, \lambda, \mu) \pi_{\lambda,\mu}(I - \mathcal{L})^{-\frac{\gamma}{2}}\|_{op}. \tag{48}$$

There is a natural quantization on any type-I Lie group introduced by Taylor [6] that can be served as the analogue of the Kohn–Nirenberg quantization on $\mathbb{R}^n$. In particular, the *quantization*, i.e., the mapping $\sigma \mapsto Op(\sigma)$ produces operators associated with a symbol σ (for example in the class of symbols $S^m_{\rho,\delta}(\mathcal{B}_4)$) on $\mathcal{S}(\mathcal{B}_4)$ given by:

$$Op(\sigma)\phi(x) = 2^{-3}\pi^{-4} \int_{\lambda\neq 0}\int_{\mu\in\mathbb{R}} Tr\left(\pi_{\lambda,\mu}(x)\sigma(x,\lambda,\mu)\pi_{\lambda,\mu}(\phi)\right) d\mu\, d\lambda\,. \tag{49}$$

Here we have used our notation for the description of the dual, as well as for the symbol and the Plancherel measure, see (21).

Let us note that by (13), we see that for the symbol σ quantized as:

$$\sigma(x,\lambda,\mu) = Op(a_{\kappa_x,\lambda,\mu})\,, \tag{50}$$

then its symbol that is given by

$$a_{\kappa_x,\lambda,\mu}(v,\xi) = (2\pi)^2\mathcal{F}_{\mathbb{R}^4}(\kappa_x)(\xi, \frac{\lambda}{2}v^2 - \frac{\mu}{2\lambda}, -\lambda v, \lambda)\,, \tag{51}$$

shall be called the (λ,μ)-symbol, where $\{\kappa_x(y)\}$ is the kernel of the symbol $\{\sigma(x,\lambda,\mu)\}$, i.e.,

$$\sigma(x,\lambda,\mu) = \pi_{\lambda,\mu}(\kappa_x)\,. \tag{52}$$

The above, together with the property of the Fourier transform

$$\hat{\phi}(\pi_{\lambda,\mu})\pi_{\lambda,\mu}(x) = \mathcal{F}_{\mathcal{B}_4}(\phi(x\cdot))(\pi_{\lambda,\mu})\,, \tag{53}$$

and the properties of the trace yield the following alternative formula for the quantization given in (49):

$$Op(\sigma)(\phi)(x) = 2^{-3}\pi^{-4} \int_{\lambda\neq 0}\int_{\mu\in\mathbb{R}} Tr\left(Op(a_{\kappa_x,\lambda,\mu})Op(a_{\phi(x\cdot),\lambda,\mu})\right) d\mu\, d\lambda\,. \tag{54}$$

The last formula shows that the quantization formula (49) can be expressed in terms of composition of quantization of symbols in the Euclidean space.

Similarly, for the case of the Heisenberg group $\mathbb{H}_n$, (17) implies that the operator $Op(\sigma)$ on $\mathcal{S}(\mathbb{H}_n)$ involves "Euclidean objects", and in particular:

$$\begin{aligned} &Op(\sigma)\phi(x) \\ &= c_n \int_{\lambda\neq 0} Tr\left(Op^W(a_{x,\lambda})Op^W[\mathcal{F}_{\mathbb{R}^{2n+1}}(\phi(x\cdot))(\sqrt{|\lambda|}\cdot, \sqrt{\lambda}\cdot, \lambda)]\right) |\lambda|^n\, d\lambda\,, \end{aligned} \tag{55}$$

where the symbol $a_{x,\lambda}$ (also called the λ-symbol) given by:

$$a_{x,\lambda}(\xi, u) = \mathcal{F}_{\mathbb{R}^{2n+1}}(\kappa_x^{'})(\sqrt{|\lambda|}\xi, \sqrt{\lambda}u, \lambda), \tag{56}$$

where $\{\kappa_x^{'}(y)\}$ is the kernel of the symbol σ, is such that

$$\sigma(x, \lambda) := \sigma(x, \pi_\lambda) = Op^W(a_{x,\lambda}). \tag{57}$$

For our notation, especially for the Plancherel measure $c_n|\lambda|^n$ on $\mathbb{H}_n$, see [4, Chapter 6].

In contrast with the case of the Engel group $\mathcal{B}_4$, in the setting of the Heisenberg group $\mathbb{H}_n$, one can renormalise $a_{x,\lambda}$ as

$$a_{x,\lambda}(\xi, u) := \tilde{a}_{x,\lambda}(\sqrt{|\lambda|}\xi, \sqrt{\lambda}u), \tag{58}$$

and therefore, one can characterise the symbol classes $S^m_{\rho,\delta}(\mathbb{H}_n)$ by the property that these λ-symbols belong to some Shubin spaces, called λ-type version of the usual Shubin classes, leading to sufficient criteria for ellipticity and hypoellipticity of operators on $\mathbb{H}_n$ in terms of the invertibility properties of their λ-symbols, see [2, Chapter 6].

Acknowledgments We wish to thank Professor Michael Ruzhansky for the useful discussions and suggestions that helped to improve the present work.

References

1. Bahouri, H., Fermanian-Kammerer, C., Gallagher, I.: Phase-space analysis and pseudo-differential calculus on the Heisenberg group. Astérisque **342** (2012). See also revised version of March 2013 of arXiv:0904.4746
2. Dixmier, J.: Sur les représentations unitaires des groupes de Lie nilpotents. III. Can. J. Math **10**, 321–348 (1957)
3. Fischer, V., Ruzhansky, M.: A pseudo-differential calculus on the Heisenberg group. C. R. Math. Acad. Sci. Paris **352(3)**, 197–204 (2014)
4. Fischer, V., Ruzhansky, M.: Quantization on Nilpotent Lie Groups, vol. 314. Progress in Mathematics. Birkhäuser/Springer (2016). Open access book
5. Ruzhansky, M., Turunen, V.: Pseudo-Differential Operators and Symmetries: Background Analysis and Advanced Topics. Pseudo-Differential Operators: Theory and Applications, vol. 2. Birkäuser, Basel (2010)
6. Taylor, M.E.: Noncommutative microlocal analysis. I. Memoirs of the American Mathematical Society, vol. 52. American Mathematical Society, Providence (1984)

Gelfand Triples for the Kohn–Nirenberg Quantization on Homogeneous Lie Groups

Jonas Brinker and Jens Wirth

Abstract We study the action of the group Fourier transform and of the Kohn–Nirenberg quantization (Fischer and Ruzhansky, Quantization on Nilpotent Lie Groups. Birkäuser, Boston, 2016) on certain Gelfand triples for homogeneous Lie groups G. Even for the Heisenberg group $G = \mathbb{H}$ there seems to be no simple intrinsic characterization for the Fourier image of the Schwartz space of rapidly decreasing smooth functions $\mathcal{S}(G)$, see (Geller, J Funct Anal 36(2), 205–254, 1980; Astengo et al., Stud Math 214, 201–222, 2013). But we may derive a simple characterization of the Fourier image for a certain subspace $\mathcal{S}_*(G)$ of $\mathcal{S}(G)$. We restrict our considerations to the case, where G admits irreducible unitary representations, that are square integrable modulo the center $Z(G)$ of G, and where $\dim Z(G) = 1$. This enables us to use an especially applicable characterization of these irreducible unitary representations that are square integrable modulo $Z(G)$ (Moore and Wolf, Trans Am Math Soc 185, 445–445, 1973; Măntoiu and Ruzhansky, J Geom Anal 29(2), 2823–2861, 2018; Gröchenig and Rottensteiner, J Funct Anal 275(12), 3338–3379, 2018). Also, Pedersen's machinery (Pedersen, Invent. Math. 118, 1–36, 1994) combines very well with this setting (Măntoiu and Ruzhansky, J Geom Anal 29(2), 2823–2861, 2018). Starting with $\mathcal{S}_*(G)$, we are able to construct distributions and Gelfand triples around $L^2(G, \mu)$ and its Fourier image $L^2(\widehat{G}, \widehat{\mu})$, such that the Fourier transform becomes a Gelfand triple isomorphism. In this context we show for the Fourier side, that multiplication of distributions with a large class of vector valued smooth functions is possible and well behaved. Furthermore, we rewrite the Kohn–Nirenberg quantization as an isomorphism for our new Gelfand triples and prove a formula for the Kohn–Nirenberg symbol, which is known from the compact group case (Ruzhansky and Turunen, Pseudo-Differential Operators and Symmetries. Birkäuser, Boston, 2010).

Keywords Homogeneus Lie groups; Group Fourier transform; Kohn-Nirenberg quantization; Distribution spaces

---

J. Brinker (✉) · J. Wirth
Institut für Analysis, Dynamik und Modellierung, Universität Stuttgart, Stuttgart, Germany
e-mail: jonas.brinker@mathematik.uni-stuttgart.de; jens.wirth@mathematik.uni-stuttgart.de

© Springer Nature Switzerland AG 2020
V. Georgiev et al. (eds.), *Advances in Harmonic Analysis and Partial Differential Equations*, Trends in Mathematics, https://doi.org/10.1007/978-3-030-58215-9_3

1 Introduction

There is an imbalance between the Schwartz space of rapidly decreasing smooth functions $\mathcal{S}(G)$ and its Fourier image $\mathcal{S}(\widehat{G})$ for simply connected nilpotent Lie groups G. For the space $\mathcal{S}(\widehat{G})$ is a space of operator valued functions on the irreducible unitary representations of G, that is not easily characterized without relying on the Fourier transform itself, see [1, 2]. Of course we can see $\mathcal{S}(G)$ as a projective limit of suitable Hilbert spaces, which results in a corresponding representation of $\mathcal{S}(\widehat{G})$ as a projective limit. But this approach is rather cumbersome, if we want to identify multiplication operators on $\mathcal{S}(\widehat{G})$. Though, if G is also a homogeneous group with one-dimensional center, then the dual $\widehat{G}$ can be parametrized (up to a null set of the Plancherel measure) by $\mathbb{R} \setminus \{0\} = \mathbb{R}^{\times}$. The setting of homogeneous Lie groups with square integrable (modulo the kernel) representations seems to be convenient in general, see e.g. [3, 4]. Now the main idea is to define a subspace $\mathcal{S}_*(G)$ of $\mathcal{S}(G)$, such that its Fourier image can be identified with a tensor product of rapidly decreasing functions on $\mathbb{R}^{\times}$ and a well behaved space of operators isomorphic to $\mathcal{L}(\mathcal{S}'(\mathbb{R}^n), \mathcal{S}(\mathbb{R}^n))$. Here we choose $\mathcal{S}_*(G)$, such that $\mathcal{S}_*(G)$ is still densely embedded into $L^2(G, \mu)$. Hence, we may construct a Gelfand triple $\mathcal{G}_*(G)$ from $\mathcal{S}_*(G)$ and $L^2(G, \mu)$ and show that the Fourier transform acts as a Gelfand triple isomorphism. By using the theory of vector valued distributions of L. Schwartz and related results in [5], we may define multiplication operators on the Fourier side. We will employ the concept of polynomial manifolds, which were used in [6], for the corresponding spaces $\mathcal{S}(\mathbb{R}^{\times})$ and $\mathcal{O}_{\mathrm{M}}(\mathbb{R}^{\times})$ of smooth functions on $\mathbb{R}^{\times}$ with growth conditions.

Using the Fourier transform on $\mathcal{S}_*(G)$, we will also examine the Kohn–Nirenberg quantization, defined in [7]. We incorporate our new function spaces into Gelfand triples of symbols and operators, on which the Kohn–Nirenberg quantization acts as a Gelfand triple isomorphism. Finally, we will prove a formula for the Kohn–Nirenberg symbol of operators $A \in \mathcal{L}(\mathcal{O}_{\mathrm{M}}(G))$, that is motivated by the corresponding formula $a(x, \xi) = \xi^* \cdot A(\xi)$ for symbols of operators $A \in \mathcal{L}(\mathcal{D}(H))$ for compact Lie groups H from [8].

The paper is structured as follows: Sects. 1.1 and 1.2 are dedicated to recalling common facts about harmonic analysis and functional analysis. At the end of Sect. 1.2 we cite theorems about the continuation of bilinear maps to completions of topological tensor products, which will be of importance for the multiplication of vector valued functions and distributions. In Sect. 1.3 we define Gelfand triples with additional real structure and discuss direct sums, tensor products and kernel maps in this context.

Section 2 is dedicated to the definition of new Gelfand triples for the Fourier transform, by using polynomial manifolds and Pedersen's quantization procedure. We start by recalling basic concepts from the theory of vector valued functions and by defining polynomial manifolds. Then, in Sect. 2.1, we pay special attention to the polynomial manifold $\mathbb{R}^\times$, the space of Schwartz functions $\mathcal{S}(\mathbb{R}^\times)$ and the dual space of $\mathcal{S}(\mathbb{R}^\times)\,\hat{\otimes}\,E$. In the succeeding subsection we recall the characterization of irreducible representations by the orbit method and Pedersen's machinery and apply both to homogenous Lie groups G with irreducible representations, that are square integrable modulo the center $Z(G)$. In Sect. 2.3, using the methods developed in the previous two subsections, we define an adjusted group Fourier transform $\mathcal{F}_\pi$. Finally, in Sect. 2.4, we define the test function space $\mathcal{S}_*(G)$, define the corresponding Gelfand triple $\mathcal{G}_*(G)$ and characterize the Gelfand triple $\mathcal{G}(\mathbb{R}^\times;\pi)$, that is isomorphic to $\mathcal{G}_*(G)$ via the adjusted Fourier transform. We finish this section by discussing multiplication operators on $\mathcal{S}(\mathbb{R}^\times;\pi)$, the Fourier side of $\mathcal{S}_*(G)$.

In the last section we show, in what way the Kohn–Nirenberg quantization extends to a Gelfand triple isomorphism from the space of operator valued test functions $\mathcal{S}(G)\,\hat{\otimes}\,\mathcal{S}(\mathbb{R}^\times;\pi)$. Finally in Sect. 3.2 we prove the formula for the Kohn–Nirenberg symbol of an operator $A \in \mathcal{L}(\mathcal{O}_{\mathrm{M}}(G))$.

Now we start by reminding ourselves of some standard notations and concepts.

1.1 Generalities

By G we will always denote a simply connected nilpotent lie group. Since G is diffeomorphic to its Lie algebra $\mathfrak{g}$ via the exponential map, we will model G to be $\mathfrak{g}$ as set. I.e. $G = \mathfrak{g}$ is a Lie algebra and a group with multiplication $(x, y) \mapsto xy$ given by the Baker–Campbell–Hausdorff formula. We will use the symbol G or $\mathfrak{g}$ depending on which property we want to emphasise. The center of $G = \mathfrak{g}$ will be denoted by $Z(G) = \mathfrak{z}$. We will denote the space of left invariant differential operators on G by $\mathfrak{u}(\mathfrak{g}_{\mathrm{L}})$. We choose a fixed Haar measure μ on the group $\mathfrak{g} = G$. This measure μ is both a Haar measure with respect to the multiplication and addition. Note, that for each $P \in \mathfrak{u}(\mathfrak{g}_{\mathrm{L}})$, there is a unique left invariant differential operator P^{t}, such that

$$\int_G (P\varphi)\,\psi\,\mathrm{d}\mu = \int_G \varphi\,P^{\mathrm{t}}\psi\,\mathrm{d}\mu, \quad \text{for all } \varphi, \psi \in \mathcal{D}(G), \tag{1}$$

where $\mathcal{D}(M)$ denotes the space of smooth compactly supported functions on a manifold M.

Let E, F be locally convex spaces (always assumed to be Hausdorff) over $\mathbb{C}$. In general, we will denote the strong dual space of E, by E'. The dual pairing between E and E', will be denoted by $\langle e', e\rangle := e'(e)$ for $e' \in E'$, $e \in E$. Similarly we will equip the space of continuous operators from E to F with the topology of uniform convergence on bounded sets of E and denote it by $\mathcal{L}(E, F)$, resp. $\mathcal{L}(E)$ for $E = F$. For any subspace $A \subset E$ we denote its annihilator by A°. If E and F are Hilbert spaces, the Hilbert-Schmidt operators from E to F will be denoted by $\mathcal{HS}(E, F)$ (again $\mathcal{HS}(E) := \mathcal{HS}(E, E)$). As usual A^* is the adjoint of $A \in \mathcal{L}(E)$. The trace of some nuclear operator $A \in \mathcal{L}(E)$, will be denoted by $\mathrm{Tr}[A]$. Furthermore $\mathcal{HS}(E, F)$ is equipped with the inner product $(A, B) \mapsto \mathrm{Tr}[AB^*]$.

Often we need to integrate vector valued functions. For this purpose we will use the concept of weak integrals.

Definition 1.1 Suppose (X, ν) is a measure space and E is a locally convex vector space. We will call a function $f\colon X \to E$ is integrable, iff there is some $e \in E$, such that for each $e' \in E'$ we have $e' \circ f \in L^1(X, \nu)$ and

$$e'(e) = \int_X e' \circ f \, \mathrm{d}\nu. \tag{2}$$

The element $\int_X f \, \mathrm{d}\mu := e$ is called integral over f. Usually we will just say, that $\int_X f \, \mathrm{d}\mu$ converges in E.

From this definition automatically follows, that

$$A \int_X f \, \mathrm{d}\nu = \int_X A \circ f \, \mathrm{d}\nu. \tag{3}$$

for any continuous linear or antilinear operator $A\colon E \to F$ into another locally convex space F. Here, the integrability of f implies the integrability of $A \circ f$.

We denote by $\mathrm{Irr}(G)$ the set of strongly continuous, unitary and irreducible representations of the group G. The dual of G is the quotient of $\mathrm{Irr}(G)$ under the equivalence relation of unitary equivalence and is denoted by $\widehat{G}$. The Plancherel measure to μ, will be $\widehat{\mu}$. The representation space of some $\pi \in \mathrm{Irr}(G)$ is denoted by H_π and the corresponding space of smooth vectors will be denoted by H_π^∞. It is equipped with a Fréchet topology defined by the seminorms

$$v \mapsto \|\pi(P)v\|_{\mathcal{H}_\pi}, \quad \text{for } P \in \mathfrak{u}(\mathfrak{g}_\mathrm{L}), \quad \text{where } \pi(P)v := P_x \pi(x)v\big|_{x=0}. \tag{4}$$

Finally we will write $H_\pi^{-\infty}$ for the strong dual space of H_π^∞, i.e. the dual space of H_π^∞ equipped with the topology of uniform convergence on bounded sets of H_π^∞. The group Fourier transform is defined by

$$\mathcal{F}_G\varphi(\pi) := \int_G \varphi(x)\pi(x)^* \, \mathrm{d}\mu(x), \quad \text{for } \varphi \in \mathcal{S}(G),\ \pi \in \mathrm{Irr}(G), \tag{5}$$

and its inverse reads

$$\varphi(x) := \int_{\widehat{G}} \mathrm{Tr}[\pi(x)\mathcal{F}_G\varphi(\pi)] \, \mathrm{d}\widehat{\mu}([\pi]), \quad \text{for } \varphi \in \mathcal{S}(G),\ x \in G. \tag{6}$$

Notice, that for each $\pi \in \mathrm{Irr}(G)$ and each $\varphi \in \mathcal{S}(G)$ the operator $\mathcal{F}_G\varphi(\pi)$ is nuclear on H_π [9, Theorem 4.2.1]. Let $\mathcal{S}(\widehat{G}) = \mathcal{F}_G\mathcal{S}(G)$ with the final topology induced by $\mathcal{F}_G$. Since $\mathcal{F}_G$ is injective, this is the unique topology that makes $\mathcal{F}_G\colon \mathcal{S}(G) \to$

$\mathcal{S}(\widehat{G})$ an isomorphism. The space $L^2(\widehat{G}, \widehat{\mu})$ is defined to be the completion of $\mathcal{S}(\widehat{G})$ with respect to the inner product

$$(\widehat{\varphi}, \widehat{\psi})_{L^2(\widehat{G},\widehat{\mu})} := \int_{\widehat{G}} \mathrm{Tr}[\widehat{\varphi}(\pi)\, \widehat{\psi}(\pi)^*]\, \mathrm{d}\widehat{\mu}([\pi]), \quad \text{for } \widehat{\varphi}, \widehat{\psi} \in \mathcal{S}(\widehat{G}). \tag{7}$$

The Fourier transform extends to a unitary operator between $L^2(G, \mu)$ and $L^2(\widehat{G}, \widehat{\mu})$. See e.g. [7] for more details.

1.2 Tensor Products

By $E \otimes F$ we denote the algebraic tensor product between E and F. Their complete injective tensor product will be denoted by $E \hat{\otimes}_\varepsilon F$ and their complete projective tensor product by $E \hat{\otimes}_\pi F$. If E_j and F_j, $j = 1, 2$, are locally convex spaces and $A_j \in \mathcal{L}(E_j, F_j)$, then

$$A_1 \otimes A_2 \colon E_1 \hat{\otimes}_\varepsilon E_2 \to\colon F_1 \hat{\otimes}_\varepsilon F_2 \tag{8}$$

denotes the tensor product map of A_1 and A_2. The linear map $A_1 \otimes A_2$ is continuous and is even an isomorphism, if A_1 and A_2 are isomorphisms [10, Proposition 43.7]. Notice, $A_1 \otimes A_2$ can also be defined, if A_1 and A_2 are continuous anti-linear operators. Later, we will need the following Lemma.

Lemma 1.1 *Let E, F and G be locally convex spaces, then*

$$\mathcal{L}(E, G) \to \mathcal{L}(E \hat{\otimes}_\varepsilon F, G \hat{\otimes}_\varepsilon F) \colon A \mapsto A \otimes 1 \tag{9}$$

is continuous.

Proof The topology on $\mathcal{L}(E \hat{\otimes}_\varepsilon F, G \hat{\otimes}_\varepsilon F)$ is induced by seminorms of the form

$$A \mapsto \sup_{z \in B} p(Az) \tag{10}$$

where B is a bounded set in $E \hat{\otimes}_\varepsilon F$ and p is a continuous seminorm on $G \hat{\otimes}_\varepsilon F$. For p it is sufficient to take any semi norm of the form

$$p(z) := \sup_{\phi \in V} q((1 \otimes \phi)(z)) \tag{11}$$

where V is an equicontinuous subset of F' and q a continuous seminorm on G [10, Definition 43.1 and Proposition 36.1]. Notice that the set $B \subset E \hat{\otimes}_\varepsilon F$ is bounded, iff for all equicontinuous sets $W \subset F'$ and all continuous seminorms r on E

$$\sup_{\phi\in W}\sup_{z\in B} r(1\otimes\phi(z)) < \infty. \tag{12}$$

In general, a subset of E is bounded, iff all continuous seminorms r are bounded on the set. Hence the set $B_V := \cup_{\phi\in V}(1\otimes\phi)(B)$ is a bounded subset of E. We arrive at

$$\begin{aligned}\sup_{z\in B} p((A\otimes 1)z) &= \sup_{z\in B}\sup_{\phi\in V} q((A\otimes\phi)(z))\\ &= \sup_{\phi\in V}\sup_{e\in(1\otimes\phi)B} q(Ae) = \sup_{e\in B_V} q(Ae),\end{aligned} \tag{13}$$

where the right hand side defines a continuous seminorm on $\mathcal{L}(E, G)$. □

If either E or F are nuclear, both tensor product topologies result in the same locally convex space and we may just write $E\,\hat{\otimes}\,F$ for either of the complete tensor products of E and F [10, Theorem 50.1]. If both E and F are nuclear Fréchet spaces, then so is $E\,\hat{\otimes}\,F$ [10, Proposition 50.1] and [11, Chapter III corollary to 6.3]. One reason for the usage of nuclear spaces is the following abstract kernel theorem [10, Propositions 50.5–50.7].

Theorem 1.1 *If F is a complete locally convex space and E is a nuclear Fréchet space or dual to a nuclear Fréchet space, then $F\,\hat{\otimes}\,E' \cong \mathcal{L}(E, F)$ via the extension of the canonical map*

$$\sum_j f_j\otimes e'_j \mapsto \big[e\mapsto \sum_j f_j\, e'_j(e)\big]. \tag{14}$$

Suppose additionally both E and F are Fréchet spaces, then $F'\,\hat{\otimes}\,E' \simeq (F\,\hat{\otimes}\,E)'$, via the extension of the map

$$\sum_k f'_k\otimes e'_k \mapsto \big[\sum_j f_j\otimes e_j \mapsto \sum_{j,k} f'_k(f_j)\, e'_k(e_j)\big]. \tag{15}$$

The multiplication between spaces of smooth functions and spaces of distributions is rarely a continuous bilinear map. But often it is hypocontinuous. Here, a bilinear map $u\colon E\times F\to G$ is defined to be hypocontinuous between the locally convex spaces E, F and G, if for all bounded sets $B_E\subset E$ and $B_F\subset F$, the two sets of linear maps

$$\{u(e,\cdot)\mid e\in B_E\}\quad\text{and}\quad\{u(\cdot,f)\mid f\in B_F\} \tag{16}$$

are equicontinuous.

Linear maps on tensor factors can easily be combined to construct a linear map on the complete tensor product. The situation for bilinear maps is not as simple. However, in the context of nuclear spaces, we may use the following theorem, which is an amalgamation of a proposition of C. Bargetz and N. Ortner and a corollary of L. Schwartz.

Theorem 1.2 *Let $\mathcal{H}$, $\mathcal{K}$ and $\mathcal{L}$ be complete nuclear locally convex spaces with nuclear strong duals and let E, F and G be complete locally convex spaces. Suppose that*

$$u\colon \mathcal{H} \times \mathcal{K} \to \mathcal{L} \quad \text{and} \quad b\colon E \times F \to G \tag{17}$$

are two hypocontinuous bilinear maps. Suppose furthermore, that either one of the three properties

- *$\mathcal{H}$ and E are Fréchet spaces*
- *$\mathcal{H}$ and E are strong duals of Fréchet spaces*
- *the bilinear map b is continuous*

are fulfilled. Then there is a unique hypocontinuous bilinear map

$$\overset{b}{\underset{u}{}}\colon (\mathcal{H} \hat{\otimes} E) \times (\mathcal{K} \hat{\otimes} F) \to \mathcal{L} \hat{\otimes} G, \tag{18}$$

that fulfils the consistency property

$$\overset{b}{\underset{u}{}}(S \otimes e, T \otimes f) = u(S, T) \otimes b(e, f). \tag{19}$$

Proof For the cases, where $\mathcal{H}$ and E are both Fréchet or both duals to Fréchet spaces, this statement can be found in [5, Proposition 1]. For the case, where b is continuous, we find the statement in [12, Corollair and Remarques on page 38]. However, in the sources the notation $\mathcal{H}(E) := \mathcal{H} \varepsilon E$ for the ε-product of Schwartz is used. The nuclearity and completeness of $\mathcal{H}$ and the completeness of E make sure, that $\mathcal{H} \varepsilon E = \mathcal{H} \hat{\otimes} E$ by [13, Satz 10.17 and Satz 11.18], which fits our notation. □

Examples of spaces fulfilling the conditions for $\mathcal{H}$, $\mathcal{K}$ and $\mathcal{L}$, are $\mathcal{S}(\mathbb{R}^n)$ $\mathcal{S}'(\mathbb{R}^n)$, $\mathcal{O}_{\mathrm{M}}(\mathbb{R}^n)$ [14, Chapitre II Théorème 16], $\mathcal{L}(\mathcal{S}(\mathbb{R}^n)) \simeq \mathcal{L}(\mathcal{S}'(\mathbb{R}^n))$ and $\mathcal{O}_{\mathrm{M}}(\mathbb{R}^n) \hat{\otimes} \mathcal{L}(\mathcal{S}(\mathbb{R}^m))$ [15].

The Hilbert space tensor product of E with F will be denoted by $E \hat{\otimes}_2 F$. By a slight abuse of notation, we will also denote

$$A_1 \otimes A_2\colon E_1 \hat{\otimes}_2 F_2 \to F_1 \hat{\otimes}_2 E_2 \tag{20}$$

to be the tensor product map of continuous linear maps A_j between the Hilbert spaces E_j and F_j. If A_1 and A_2 are unitary, then so is $A_1 \otimes A_2$.

1.3 Gelfand Triples

Gelfand triples are a convenient setting for both distributions and the Fourier transform. We start by defining the class of Gelfand triples we are going to use.

Definition 1.2 A Gelfand triple (with a real structure) is a tuple of spaces $\mathcal{G} = (E, H, E')$, and a structure map $C\colon H \to H$ fulfilling the following properties:

(a) E is a nuclear Fréchet space, with strong dual E'
(b) H is a Hilbert space, with dense and continuous embedding $E \hookrightarrow H$
(c) C is antiunitary, $C^2 = 1$ and $C|_E\colon E \to E$ is a homeomorphism.

The map C will be called the real structure of $\mathcal{G}$.

Notice, that by the definition of Gelfand triples we automatically have a continuous dense embedding $H' \hookrightarrow E'$ dual to the embedding $E \hookrightarrow H$. Classically Gelfand triples are defined without the structure map C [16]. Here E and H are antilinearly embedded into E' via the Fréchet-Riesz isomorphism $\mathcal{R}\colon H \overset{\simeq}{\to} H'$. But this approach would be unwieldy, because we will use Gelfand triples in concert with tensor products. Since there is no canonical unitary map between H and H', we are going to use C to fix one. The corresponding induced embedding $\mathcal{I}\colon H \hookrightarrow E'$ is defined via

$$\mathcal{I}\colon H \xrightarrow{\mathcal{R}\circ C} H' \hookrightarrow E'. \tag{21}$$

The structure map C has a natural continuation to a homeomorphism $C\colon E' \to E'$ by $(Ce')(e) := \overline{e'(Ce)}$ for $e \in E$, $e' \in E'$. Hence C induces compatible real structures on E, H and E'.

In the sequel we will always use the term Gelfand triple for the concept of a Gelfand triple with real structure equipped with embeddings as described above.

Definition 1.3 Let $\mathcal{G}_j = (E_j, H_j, E_j')$, $j = 1, 2$, be Gelfand triples. For a map $T\colon E' \to E'$, we write

$$T\colon \mathcal{G}_1 \to \mathcal{G}_2, \tag{22}$$

if $T(E_1) \subset E_2$ and $T(H_1) \subset H_2$ with respect to the above described embeddings. We will call T a Gelfand triple isomorphism, if $T|_{E_1}\colon E_1 \to E_2$, $T\colon E_1' \to E_2'$ are homeomorphisms and $T|_{H_1}\colon H_1 \to H_2$ is unitary.

The above definition implies, that writing $T\colon \mathcal{G}_1 \to \mathcal{G}_2$ is equivalent to saying, that the diagram

$$\begin{array}{ccccc} E_1 & \hookrightarrow & H_1 & \hookrightarrow & E_1' \\ {\scriptstyle T}\downarrow & & {\scriptstyle T}\downarrow & & {\scriptstyle T}\downarrow \\ E_2 & \hookrightarrow & H_2 & \hookrightarrow & E_2' \end{array}$$

is commutative.

Now we will describe, how we may construct new Gelfand triples.

Definition 1.4 Let $\mathcal{G}_j = (E_j, H_j, E_j')$ be Gelfand triples with structure maps C_j for $j = 1, 2, 3, 4$. Using the identifications $E_1' \oplus E_2' \simeq (E_1 \oplus E_2)'$ and $E_1' \hat{\otimes} E_2' \simeq (E_1 \oplus E_2)'$ resp. $\mathcal{L}(E_2, E_1') \simeq \mathcal{L}(E_2', E_1)'$ via Theorem 1.1 we may define the following Gelfand triples.

The sum resp. tensor product of $\mathcal{G}_1$ and $\mathcal{G}_2$ is defined by

$$\mathcal{G}_1 \oplus \mathcal{G}_2 := \begin{pmatrix} E_1 \oplus E_2 \\ H_1 \oplus H_2 \\ E_1' \oplus E_2' \end{pmatrix} \quad \text{resp.} \quad \mathcal{G}_1 \otimes \mathcal{G}_2 := \begin{pmatrix} E_1 \hat{\otimes} E_2 \\ H_1 \hat{\otimes}_2 H_2 \\ E_1' \hat{\otimes} E_2' \end{pmatrix} \tag{23}$$

with structure maps $C_1 \oplus C_2$ resp. $C_1 \otimes C_2$. Here $C_1 \otimes C_2$ is the continuation of

$$H_1 \otimes H_2 \to H_1 \otimes H_2 \colon \sum_{j,k} h_{1,j} \otimes h_{2,k} \mapsto \sum_{j,k} (C_1 h_{1,j}) \otimes (C_2 h_{2,k}). \tag{24}$$

The operator Gelfand triple from $\mathcal{G}_2$ to $\mathcal{G}_1$ is defined as

$$\mathcal{L}(\mathcal{G}_2, \mathcal{G}_1) := \begin{pmatrix} \mathcal{L}(E_2', E_2) \\ \mathcal{HS}(H_2, H_1') \\ \mathcal{L}(E_2, E_1') \end{pmatrix} \tag{25}$$

with structure map $T \mapsto C_1 T C_2$.

Let us now discuss, why $\mathcal{G}_1 \otimes \mathcal{G}_2$ and $\mathcal{L}(\mathcal{G}_2, \mathcal{G}_1)$ are indeed Gelfand triples. Notice, that by using the unitary isomorphism $\mathcal{I}_2 \colon H_2 \xrightarrow{\simeq} H_2'$ determined by C_2, we get an isomorphism resp. a unitary isomorphism

$$\mathcal{N}(H_2, H_1) \simeq H_1 \hat{\otimes}_\pi H_2 \quad \text{resp.} \quad \mathcal{HS}(H_2, H_1) \simeq H_1 \hat{\otimes}_2 H_2, \tag{26}$$

by extending the linear map

$$\begin{aligned} &\mathcal{K}^{-1} \colon H_1 \otimes H_2 \to \mathcal{N}(H_2, H_1), \\ &\text{where} \quad \mathcal{K}^{-1}(h_1 \otimes h_2)(\tilde{h}_2) := h_1 \cdot (\mathcal{I}_2 h_2,)(\tilde{h}_2) \end{aligned} \tag{27}$$

and $\mathcal{N}(H_2, H_1)$ is the space of nuclear operator from H_2 to H_1. These isomorphisms, together with Theorem 1.1, can be used to construct the following chain of continuous maps with dense ranges

$$\begin{array}{ccccc} E_1 \hat{\otimes} E_2 & \hookrightarrow & H_1 \hat{\otimes}_\pi H_2 & & H_1 \hat{\otimes}_2 H_2 \\ \mathcal{K}\big\uparrow\simeq & & \mathcal{K}^{-1}\big\downarrow\simeq & & \mathcal{K}\big\uparrow\simeq \\ \mathcal{L}(E_2', E_1) & & \mathcal{N}(H_2, H_1) & \hookrightarrow & \mathcal{HS}(H_2, H_1) \end{array}$$

We get a commutative diagram by completing the rows of this diagram. This both shows, that $\mathcal{L}(\mathcal{G}_2, \mathcal{G}_1)$ and $\mathcal{G}_1 \otimes \mathcal{G}_2$ are Gelfand triples, and proves the following Lemma.

Lemma 1.2 *Suppose $\mathcal{G}_1$ and $\mathcal{G}_2$ are the Gelfand triples from above, then the canonical linear map*

$$\mathcal{K}^{-1}\colon E_1 \otimes E_2 \to \mathcal{L}(E_2', E_1), \quad \text{where} \quad \mathcal{K}^{-1}(e_1 \otimes e_2)(e_2') := e_1 \cdot e_2'(e_2), \tag{28}$$

induces a Gelfand triple isomorphism $\mathcal{K}\colon \mathcal{L}(\mathcal{G}_2, \mathcal{G}_1) \to \mathcal{G}_1 \otimes \mathcal{G}_2$.

Since we choose G simply connected and nilpotent, the space H_π^∞ is a nuclear Fréchet space [9, Corollary 4.1.2] for any $\pi \in \mathrm{Irr}(G)$. With respect to π we may define the Gelfand triples

$$\mathcal{G}(\pi) := \begin{pmatrix} H_\pi^\infty \\ H_\pi \\ H_\pi^{-\infty} \end{pmatrix} \quad \text{and} \quad \mathcal{G}_{\mathrm{op}}(\pi) := \mathcal{L}(\mathcal{G}(\pi), \mathcal{G}(\pi)) = \begin{pmatrix} \mathcal{L}(H_\pi^{-\infty}, H_\pi^\infty) \\ \mathcal{HS}(H_\pi) \\ \mathcal{L}(H_\pi^\infty, H_\pi^{-\infty}) \end{pmatrix}, \tag{29}$$

if we associate a real structure C_π to $\mathcal{G}(\pi)$. Of course, this real structure is not unique. Instead, we will define $\mathrm{Irr}^{\mathbb{R}}(G)$ to be pairs consisting of $\pi \in \mathrm{Irr}(G)$ and an associated real structure C_π on $\mathcal{G}(\pi)$. Usually we will just write $\pi \in \mathrm{Irr}^{\mathbb{R}}(G)$ and mean, that we took a choice of C_π for π.

2 Polynomial Manifolds and Gelfand Triples for the Fourier Transform

We will need polynomial manifolds for the Pedersen quantization, a generalization of the Weyl quantization, and for the generalizations of the spaces $\mathcal{S}(\mathbb{R}^n; E)$ and $\mathcal{O}_{\mathrm{M}}(\mathbb{R}^n; E)$ for a complete locally convex space E, see [14, 17]. For it will be convenient to have one notion of Schwartz functions and slowly increasing functions, that can be applied to simply connected nilpotent Lie groups, Lie algebras, coadjoint orbits, and that is also compatible with their relation to each other. Furthermore this will lead to a notion of Schwartz functions and slowly increasing function on $\mathbb{R}^\times = \mathbb{R} \setminus \{0\}$, that we will rely on heavily.

Before we define polynomial manifolds, let us fix some basic notation and recall a few definitions from the theory of vector valued smooth functions.

As usual we will just write $\mathcal{S}(\mathbb{R}^n) := \mathcal{S}(\mathbb{R}^n; E)$ and $\mathcal{O}_{\mathrm{M}}(\mathbb{R}^n) := \mathcal{O}_{\mathrm{M}}(\mathbb{R}^n; E)$ for scalar valued case $E = \mathbb{C}$.

A function $f : \mathbb{R}^n \to E$ is differentiable, if for each $j = 1, \dots, n$ and each $x \in \mathbb{R}^n$ the limit

$$\partial_j f(x) := \lim_{t \to 0} \frac{1}{t}(f(x + te_j) - f(x)) \tag{30}$$

exists in E, where $(e_j)_j$ is the standard basis in $\mathbb{R}^n$, and each partial derivative, $\partial_j f$, is a continuous function. f is called smooth, if partial derivatives of arbitrary order are again differentiable. I.e. for all $\alpha \in \mathbb{N}_0^n$ the functions $\partial^\alpha f = \partial_1^{\alpha_1} \cdots \partial_n^{\alpha_n} f$ exist and are continuous.

Denote by $\mathcal{P}(\mathbb{R}^n)$ the vector space of polynomial functions from $\mathbb{R}^n$ to $\mathbb{C}$ and by $\mathrm{Diff}_{\mathcal{P}}(\mathbb{R}^n)$ the set of differential operators with polynomial coefficients on $\mathbb{R}^n$. The space of E-valued Schwartz functions, $\mathcal{S}(\mathbb{R}^n; E)$, is the space of smooth functions $\varphi : \mathbb{R}^n \to E$, such that

$$\sup_{x \in \mathbb{R}^n} p(P\varphi(x)) < \infty \tag{31}$$

for each continuous seminorm p on E and each $P \in \mathrm{Diff}_{\mathcal{P}}(\mathbb{R}^n)$. The above expression also defines a set of seminorms which define the topology on $\mathcal{S}(\mathbb{R}^n; E)$.

The scalar valued slowly increasing functions, $\mathcal{O}_{\mathrm{M}}(\mathbb{R}^n)$, is the space of smooth functions $f : \mathbb{R}^n \to C$, such that for each $P \in \mathrm{Diff}_{\mathcal{P}}(\mathbb{R}^n)$ there is a real valued $q \in \mathcal{P}(\mathbb{R}^n)$, such that $|Pf| \leq q$.

The space of E-valued slowly increasing functions, $\mathcal{O}_{\mathrm{M}}(\mathbb{R}^n; E)$, is the space of smooth functions $f : \mathbb{R}^n \to E$, such that

$$[\varphi \mapsto f \cdot \varphi] \in \mathcal{L}(\mathcal{S}(\mathbb{R}^n), \mathcal{S}(\mathbb{R}^n; E)), \tag{32}$$

equipped with the subspace topology in $\mathcal{L}(\mathcal{S}(\mathbb{R}^n), \mathcal{S}(\mathbb{R}^n; E))$. For $E = \mathbb{C}$ the two given definitions are equivalent [10, Theorem 25.5].

Furthermore the above defined spaces of E-valued functions have a very useful characterization by tensor products. To be precise, we have $\mathcal{S}(\mathbb{R}^n; E) = \mathcal{S}(\mathbb{R}^n) \hat{\otimes} E$ and $\mathcal{O}_{\mathrm{M}}(\mathbb{R}^n; E) = \mathcal{O}_{\mathrm{M}}(\mathbb{R}^n) \hat{\otimes} E$ as topological vector spaces.

The definition given below is a slight generalization of the polynomial manifolds used by Pedersen in [6].

Definition 2.1 Suppose now M is an n-dimensional smooth manifold with finitely many connected components. An atlas $\mathcal{A}$ of M will be called a polynomial atlas, iff each two charts $(\phi, U), (\psi, V) \in \mathcal{A}$ fulfil

(i) U, V are connected components of M and $\phi(U) = \psi(V) = \mathbb{R}^n$,
(ii) and if $U = V$, then $\phi \circ \psi^{-1}$ is a polynomial function on $\mathbb{R}^n$.

Two polynomial atlases $\mathcal{A}, \mathcal{A}'$ are said to be equivalent, iff $\mathcal{A} \cup \mathcal{A}'$ is a polynomial atlas. A polynomial structure is an equivalence class of polynomial atlases.

Together with a polynomial structure M will be called a polynomial manifold. A chart of a polynomial structure of M will be called a polynomial chart on M.

The following list provides some basic examples of polynomial manifolds:

- a finite dimensional vector space, with respect to the linear charts
- a finite dimensional affine spaces, with respect to the affine linear charts
- a simply connected nilpotent Lie group, with respect to the exponential map
- a coadjoint orbit to a simply connected nilpotent Lie group [6]

By using polynomial charts, we may generalize polynomials and definitions that depend on the set of polynomials.

Definition 2.2 Suppose M, N are polynomial manifolds and E a complete locally convex space. For $\mathcal{X}(\bullet) \in \{\mathcal{P}(\bullet), \mathcal{S}(\bullet; E), \mathcal{O}_{\mathrm{M}}(\bullet; E)\}$ we define

$$\mathcal{X}(M) := \{f : M \to E \mid f \circ \phi^{-1} \in \mathcal{X}(\mathbb{R}^n) \text{ for all polynomial charts } \phi\} \tag{33}$$

We equip $\mathcal{S}(M; E)$ resp. $\mathcal{O}_{\mathrm{M}}(M; E)$ with the projective topology from the maps $f \mapsto f \circ \phi^{-1}$ into $\mathcal{S}(\mathbb{R}^n; E)$ resp. $\mathcal{O}_{\mathrm{M}}(\mathbb{R}^n; E)$. As usual we set $\mathcal{S}(M) := \mathcal{S}(M; \mathbb{C})$ and $\mathcal{O}_{\mathrm{M}}(M) := \mathcal{O}_{\mathrm{M}}(M; \mathbb{C})$.

The set of polynomial differential operators on M is defined to be

$$\mathrm{Diff}_{\mathcal{P}}(M) := \left\{ P \in \mathcal{L}(\mathcal{D}(M)) : \begin{matrix} \left[\varphi \mapsto P(\varphi \circ \phi) \circ \phi^{-1}\right] \in \mathrm{Diff}_{\mathcal{P}}(\mathbb{R}^n) \\ \text{for all polynomial charts } \phi \end{matrix} \right\}. \tag{34}$$

A function $f : M \to N$ will be called polynomial resp. slowly increasing, iff $\psi \circ f \circ \phi^{-1}$ is a polynomial resp. slowly increasing (ϕ, U) on M and (ψ, V) on N with $U \subset f^{-1}(V)$. The function f will be called polynomial resp. tempered diffeomorphism, iff f is bijective and both f and f^{-1} are polynomial resp. slowly increasing.

As for smooth manifolds, we may construct new polynomial manifolds by disjoint unions $M \dot{\cup} N$ of polynomial manifolds N, M with the same dimension and products $M \times N$ of arbitrary polynomial manifolds N, M. The corresponding polynomial structure on $M \dot{\cup} N$ is induced by the polynomial charts on M and N. On $M \times N$ we choose the canonical polynomial structure defined by combining charts ϕ on M and ψ on N to polynomial charts (ϕ, ψ) on $M \times N$. Directly from our definition follows, that similar to the Euclidean case

$$\mathcal{S}(M \dot{\cup} N) = \mathcal{S}(M) \oplus \mathcal{S}(N) \quad \text{and} \quad \mathcal{S}(M \times N; E) = \mathcal{S}(M) \hat{\otimes} \mathcal{S}(N) \hat{\otimes} E, \tag{35}$$

where E is a complete locally convex space. The identities also hold, if we exchange $\mathcal{S}$ with $\mathcal{O}_{\mathrm{M}}$. Similar identities are true for $\mathcal{P}(M)$ and $\mathrm{Diff}_{\mathcal{P}}(M)$.

We will call a Radon measure ν on $\mathbb{R}^n$ tempered, iff it is equivalent to the Lebesgue measure $\mathrm{d}x$ and the Radon–Nikodym derivatives $\frac{\mathrm{d}x}{\mathrm{d}\nu}$ and $\frac{\mathrm{d}\nu}{\mathrm{d}x}$ are slowly increasing almost everywhere. A Radon measure on a polynomial manifold $\mathbb{R}^n$ will be called tempered, if each pushforward by a polynomial chart is tempered.

Definition 2.3 Suppose M is a polynomial manifold and ν a tempered measure on M. Then $\mathcal{G}(M, \nu)$ is defined to be the Gelfand triple

$$\mathcal{S}(M) \hookrightarrow L^2(M, \nu) \hookrightarrow \mathcal{S}'(M), \tag{36}$$

equipped with the real structure defined by the usual complex conjugation $\varphi \mapsto \overline{\varphi}$.

If $f\colon M_1 \to M$ is a tempered diffeomorphism, then for each $\phi \in \mathcal{S}'(M)$ the pull back $\wp_f \phi(\varphi) := \phi(\varphi \circ f^{-1})$ is well defined and induces a Gelfand triple isomorphism

$$\mathcal{G}(M, \nu) \to \mathcal{G}(M_1, \nu \circ f). \tag{37}$$

Indeed, we defined tempered measures and polynomial manifolds in such a way, that we have a very simple Gelfand-Triple isomorphism

$$\mathcal{G}(M, \nu) \simeq \bigoplus_{j=1}^{k} \mathcal{G}(\mathbb{R}^n, \mathrm{d}x), \tag{38}$$

given by pullbacks and multiplications with slowly increasing functions, provided that M is a n-dimensional polynomial manifold with k connected components.

2.1 *The Polynomial Manifold* $\mathbb{R}^\times$

For us the two most important examples of polynomial manifolds are the half lines $\mathbb{R}^+$ and $\mathbb{R}^-$. Here the polynomial structure is induced by the chart $\sigma : \lambda \mapsto |\lambda| - 1/|\lambda|$. On $\mathbb{R}^+$ the inverse reads $\sigma^{-1}(y) = (y + \sqrt{y^2 + 4})/2$.

Lemma 2.1 *If we extend each function in* $\mathcal{S}(\mathbb{R}^\pm)$ *by zero to the whole real line, then*

$$\mathcal{S}(\mathbb{R}^\pm) = \{\varphi \in \mathcal{S}(\mathbb{R}) \mid \varphi \equiv 0 \textit{ on } \mathbb{R}^\mp\} \tag{39}$$

and $\mathcal{S}(\mathbb{R}^\pm)$ *carries the subspace topology with respect to* $\mathcal{S}(\mathbb{R})$.

Proof We will prove the statement for the $\mathbb{R}^+$ case, for $\mathbb{R}^-$ the proof is analogous. Since σ is a polynomial diffeomorphism from $\mathbb{R}^+$ to $\mathbb{R}$, the map

$$\varphi \mapsto \varphi \circ \sigma \tag{40}$$

is a linear homeomorphism between $\mathcal{D}(\mathbb{R})$ and $\mathcal{D}(\mathbb{R}^+)$ resp. between $\mathcal{S}(\mathbb{R})$ and $\mathcal{S}(\mathbb{R}^+)$. Hence $\mathcal{D}(\mathbb{R}^+)$ is dense in $\mathcal{S}(\mathbb{R}^+)$. Let us define

$$\mathcal{S}_+(\mathbb{R}) := \{\varphi \restriction_{\mathbb{R}^+} \mid \varphi \in \mathcal{S}(\mathbb{R}) \text{ with } \varphi \equiv 0 \text{ on } \mathbb{R}^-\}, \tag{41}$$

equipped with the subspace topology with respect to $\mathcal{S}(\mathbb{R})$. Let $f\colon \mathbb{R} \to [0,1]$ be smooth, such that $\operatorname{supp} f \subset \mathbb{R}^+$ and $f \equiv 1$ on $[1,\infty)$. For each $\varphi \in \mathcal{S}_+(\mathbb{R})$ and $\alpha \in \mathbb{N}_0$ we have $\partial^\alpha\varphi(x) = o(x^N)$ for $x \to 0$ of arbitrary high order $N \in \mathbb{N}$. Hence each $\alpha, \beta \in \mathbb{N}_0$ there is some $C_1, C_2 > 0$ and $N > \alpha$ with

$$\begin{aligned}\sup_{x\in\mathbb{R}^+} &|x^\beta \partial_x^\alpha (f(nx)\,\varphi(x) - \varphi(x))| \\ &\leq C_1 \sum_{0<\gamma\leq\alpha} n^\gamma \sup_{y\in\mathbb{R}} |\partial^\gamma f(y)| \sup_{0<x\leq 1/n} x^\beta |\partial^{\alpha-\gamma}\varphi(x)| \\ &\leq C_2 \sum_{0<\gamma\leq\alpha} n^{\gamma-\beta-N} \xrightarrow{n\to\infty} 0.\end{aligned} \tag{42}$$

By employing the usual cut-off functions, we realize that $\mathcal{D}(\mathbb{R}^+)$ is dense in $\mathcal{S}_+(\mathbb{R})$, too. Thus it is enough to show that the topologies of $\mathcal{S}(\mathbb{R}^+)$ and $\mathcal{S}_+(\mathbb{R})$ coincide on $\mathcal{D}(\mathbb{R}^+)$. The $\mathcal{S}(\mathbb{R}^+)$-topology is induced by seminorms of the form

$$\mathcal{D}(\mathbb{R}^+) \to \mathbb{R}\colon \varphi \mapsto \sup_{x>0} |A^k B^j \varphi(x)|, \quad k, j \in \mathbb{N}_0, \tag{43}$$

where $A\varphi := (\partial(\varphi \circ \sigma^{-1})) \circ \sigma$ and $B\varphi := (\boldsymbol{m}(\varphi \circ \sigma^{-1})) \circ \sigma = \sigma \cdot \varphi$. First of all, we have

$$A\varphi(x) = \frac{x^2}{x^2+1}\partial_x\varphi(x) =: \eta(x)\cdot\partial_x\varphi(x), \qquad \varphi \in \mathcal{D}(\mathbb{R}_+), x \in \mathbb{R}_+. \tag{44}$$

The rational function η and all of its derivatives are bounded. Hence A can be extended to an operator in $\mathcal{L}(\mathcal{S}_+(\mathbb{R}))$.

We show, that B has an extension in $\mathcal{L}(\mathcal{S}_+(\mathbb{R}))$. For this purpose it is enough to prove that $\frac{1}{\boldsymbol{m}} \in \mathcal{L}(\mathcal{S}_+(\mathbb{R}))$, where $\frac{1}{\boldsymbol{m}}\varphi(x) = \varphi(x)/x$. First of all, for each $\varphi \in \mathcal{D}(\mathbb{R}^+)$ and each $x > 1$

$$\begin{aligned}|x^k\partial_x^n(\frac{1}{x}\varphi(x))| &\leq \sum_{j=0}^n \frac{n!}{(n-j)!} x^{k-j-1} |\varphi^{(n-j)}(x)| \\ &\leq \sum_{j=0}^n \frac{n!}{(n-j)!} \sup_y |y^k \varphi^{(n-j)}(y)|,\end{aligned} \tag{45}$$

for arbitrary $k, n \in \mathbb{N}_0$. Now we only need to bound the left-hand side for $0 < x < 1$. For $k > n$ almost the same inequality as above can be used. We assume now $n \geq k$. For $0 < x < 1$ and each $m \in \mathbb{N}$

$$|\varphi(x)/x^m| = |\frac{1}{x^m}\int_0^x \frac{(x-t)^{m-1}}{(m-1)!}\varphi^{(m)}(t)\,\mathrm{d}t| \leq \frac{1}{m!}\sup_y |\varphi^{(m)}(y)|. \tag{46}$$

Hence

$$
\begin{aligned}
|x^k \partial_x^n (\frac{1}{x}\varphi(x))| &\le \sum_{j=0}^{n} \frac{n!}{(n-j)!} x^{k-j-1} |\varphi^{(n-j)}(x)| \\
&\le \sum_{j=0}^{n} \frac{n!}{(n-j)!} x^{-n-1} |\varphi^{(n-j)}(x)| \\
&\le \sum_{j=0}^{n} \frac{1}{(n-j)!(n+1)} \sup_y |\varphi^{(2n+1-j)}(y)|,
\end{aligned}
\tag{47}
$$

for all $0 < x < 1$, $n \le k$ and $\varphi \in \mathcal{D}(\mathbb{R}^+)$. In conclusion $\frac{1}{\boldsymbol{m}} \in \mathcal{L}(\mathcal{S}_+(\mathbb{R}))$ and thus also $B \in \mathcal{L}(\mathcal{S}_+(\mathbb{R}))$. Due to the continuity of A and B we arrive at

$$
\mathcal{S}_+(\mathbb{R}) \hookrightarrow \mathcal{S}(\mathbb{R}^+), \tag{48}
$$

i.e. the $\mathcal{S}_+(\mathbb{R})$-topology is finer than the $\mathcal{S}(\mathbb{R}^+)$-topology.

For the reverse embedding we will transport our situation to the whole real line by

$$
\varphi \mapsto \varphi \circ \sigma^{-1}, \tag{49}
$$

which is an isomorphism $\mathcal{D}(\mathbb{R}^+) \to \mathcal{D}(\mathbb{R})$ and $\mathcal{S}(\mathbb{R}^+) \to \mathcal{S}(\mathbb{R})$. We denote the image of $\mathcal{S}_+(\mathbb{R})$ by this map by $\mathcal{S}_\oplus(\mathbb{R})$ and equip it with the transported $\mathcal{S}_+(\mathbb{R})$-topology. Then $\mathcal{S}_\oplus(\mathbb{R})$ is a space of smooth functions on $\mathbb{R}$ with

$$
\mathcal{D}(\mathbb{R}) \hookrightarrow \mathcal{S}_\oplus(\mathbb{R}) \hookrightarrow \mathcal{S}(\mathbb{R}), \tag{50}
$$

where both embeddings are dense. The topology in $\mathcal{S}_\oplus(\mathbb{R})$ is induced by seminorms of the form

$$
\mathcal{S}_\oplus(\mathbb{R}) \to \mathbb{R} \colon \varphi \mapsto \sup_{y \in \mathbb{R}} |C^k E^j \varphi(y)|, \quad k, j \in \mathbb{N}_0, \tag{51}
$$

where $C\varphi := (\partial(\varphi \circ \sigma)) \circ \sigma^{-1}$ and $E\varphi := (\boldsymbol{m}(\varphi \circ \sigma)) \circ \sigma^{-1} = \sigma^{-1} \cdot \varphi$. The operator C can be rewritten as

$$
C\varphi(y) = \left(1 + \frac{2}{(y + \sqrt{y^2+4})^2}\right)\varphi'(y) =: \psi(y) \cdot \varphi'(y), \quad \varphi \in \mathcal{S}_\oplus(\mathbb{R}), y \in \mathbb{R}. \tag{52}
$$

Because $\sigma^{-1}, \psi \in O_M(\mathbb{R})$, both C and E have extensions in $\mathcal{L}(\mathcal{S}(\mathbb{R}))$. Thus $\mathcal{S}_\oplus(\mathbb{R}) = \mathcal{S}(\mathbb{R})$ and finally $\mathcal{S}_+(\mathbb{R}) = \mathcal{S}(\mathbb{R}^+)$. □

The most important property of $\mathcal{S}(\mathbb{R}^\pm)$ (next to being a closed subspace of $\mathcal{S}(\mathbb{R})$) is stated in the following corollary.

Corollary 2.1 *The map $x \mapsto |x|^v$ is in $O_M(\mathbb{R}^\pm)$ for each $v \in \mathbb{R}$.*

Proof The continuity $\frac{1}{\boldsymbol{m}}$ was already shown in the proof to the last lemma with inequalities (47). Of course $\boldsymbol{m}\varphi(x) := x\varphi(x)$ defines a continuous operator on $\mathcal{S}(\mathbb{R}^\pm)$, as well. The derivatives of $x \mapsto |x|^v$ can be bounded by terms of the form $x \mapsto x^k$ for $k \in \mathbb{Z}$, which concludes the proof. □

We now find a characterisation for the functions in $O_M(\mathbb{R}^\pm \times M; E)$. This space will be of importance later on, when we examine the Fourier image of $\mathcal{S}(G)$ in further detail and when we want to discuss the integral formula for the Kohn–Nirenberg quantization.

Corollary 2.2 *A smooth function $f : \mathbb{R}^\pm \times M \to E$ is in $O_M(\mathbb{R}_\pm \times M; E)$, iff for each $k \in \mathbb{N}_0$, each $P \in \mathrm{Diff}_{\mathcal{P}}(M)$ and each continuous seminorm p on E, there exists an $l \in \mathbb{N}$ and an $q \in \mathcal{P}(M)$, such that $p(\partial_\lambda^k P_x f(\lambda, x)) \le (1 + |\lambda|^l + |\lambda|^{-l})q(x)$.*

Proof We know, that $O_M(\mathbb{R}^+ \times M; E)$ is the space of all smooth functions f on $\mathbb{R}^+$, such that

$$[\varphi \mapsto f \cdot \varphi] \in \mathcal{L}(\mathcal{S}(\mathbb{R}^\pm \times M), \mathcal{S}(\mathbb{R}^\pm \times M; E)). \tag{53}$$

We prove the statement for $\mathbb{R}^+$, then the other statement follows at once, since $\mathbb{R}^-$ is isomorphic to $\mathbb{R}^+$ by $x \mapsto -x$. Also, it is enough to consider $M = \mathbb{R}^n$, as the more general case follows by just using polynomial coordinate charts.

Suppose $f \in O_M(\mathbb{R}^+ \times \mathbb{R}^n; E)$. Because f induces a continuous multiplication operator and because $\mathcal{S}(\mathbb{R}^+)$ is a subspace of $\mathcal{S}(\mathbb{R})$, for each $k \in \mathbb{N}_0$, $\alpha \in \mathbb{N}_0^n$ and each continuous seminorm p on E, there is some $m \in \mathbb{N}$ and $C > 0$ with

$$\begin{aligned} &\sup_{\lambda \in \mathbb{R}^+, x \in M} p(\partial_\lambda^k \partial_x^\alpha (f(\lambda, x)\varphi(\lambda, x))) \\ &\qquad \le C \max_{|\beta|, l \le m} \sup_{\lambda \in \mathbb{R}^+, x \in M} (1 + |\lambda|^m)(1 + |x|^2)^m |\partial_\lambda^l \partial_x^\beta \varphi(\lambda, x)|, \end{aligned} \tag{54}$$

for all $\varphi \in \mathcal{S}(\mathbb{R}^+ \times \mathbb{R}^n)$. We choose some $\varphi \in \mathcal{S}(\mathbb{R}^+ \times \mathbb{R}^n)$, such that $\varphi \equiv 1$ on some neighbourhood around $(\lambda, x) = (1, 0)$, and define $\varphi_{a,y}(x) := \varphi(xa^{-1}, x - y)$

for $a > 0$, $y \in \mathbb{R}^n$. Then

$$\begin{aligned}
p(\partial^{(k,\alpha)} f(a, y)) \\
&= p(\partial_\lambda^k \partial_x^\alpha (f(\lambda, x)\varphi_{a,y}(\lambda, x)))\big|_{(\lambda,x)=(a,y)} \\
&\leq C \max_{|\beta|, l \leq m} \sup_{\lambda \in \mathbb{R}^+, x \in \mathbb{R}^n} (1 + |\lambda|^m)(1 + |x|^2)^m |\partial_\lambda^l \partial_x^\beta \varphi_{a,y}(\lambda, x)| \\
&= C \max_{|\beta|, l \leq m} \sup_{\lambda \in \mathbb{R}^+, x \in \mathbb{R}^n} a^{-l}(1 + |a\lambda|^m)(1 + |x + y|^2)^m |\partial_\lambda^l \partial_x^\beta \varphi(\lambda, x)| \\
&\leq C'(1 + a^m + a^{-m})(1 + |y|^2)^m,
\end{aligned} \tag{55}$$

where k, α, m and C are as above. Of course this implies, that for each $k \in \mathbb{N}_0$, $P \in \mathrm{Diff}_{\mathcal{P}}(\mathbb{R}^n)$ and each continuous seminorm p on E, there exists an $l \in \mathbb{N}$ and a $q \in \mathcal{P}(\mathbb{R}^n)$, such that

$$p(\partial_\lambda^k P_x f(\lambda, x)) \leq (1 + |\lambda|^l + |\lambda|^{-l})q(x). \tag{56}$$

Now for the converse implication. Let $f\colon \mathbb{R}^+ \times \mathbb{R}^n \to \mathbb{C}$ be any smooth function, such that for p, k and P we find m and q for the inequality (56). Then for arbitrary $\varphi \in \mathcal{S}(\mathbb{R}^+ \times M)$,

$$\begin{aligned}
\sup_{\lambda \in \mathbb{R}^+, x \in \mathbb{R}^n} (1 + |\lambda|^k)(1 + |x|^2)^k p(\partial^\alpha (f\varphi)(\lambda, x)) \\
&\leq C \sup_{\lambda \in \mathbb{R}^+, x \in \mathbb{R}^n} (1 + \lambda^k)(1 + |x|^2)^k \sum_{\beta \leq \alpha} |\partial^{\alpha-\beta} f(\lambda, x)\, \partial^\beta \varphi(\lambda, x)| \\
&\leq C' \sup_{x \in \mathbb{R}_+} (1 + |x|^2)^{k+m}(1 + \lambda^{k+m} + \lambda^{k-m}) \sum_{\beta \leq \alpha} |\partial^\beta \varphi^{(j)}(x)|.
\end{aligned} \tag{57}$$

Since $\frac{1}{m}$ is a continuous operator on $\mathcal{S}(\mathbb{R}^+)$, the last line defines a continuous seminorm on $\mathcal{S}(\mathbb{R}^+ \times \mathbb{R}^n)$. Thus the operator $\varphi \mapsto f \cdot \varphi$ is continuous. □

From the polynomial structures on $\mathbb{R}^+$ and $\mathbb{R}^-$, we construct the polynomial manifold $\mathbb{R}^\times = \mathbb{R}^+ \dot{\cup} \mathbb{R}^-$. Its Schwartz space $\mathcal{S}(\mathbb{R}^\times) = \mathcal{S}(\mathbb{R}^+) \oplus \mathcal{S}(\mathbb{R}^-)$ can be seen as the closed subspace of $\mathcal{S}(\mathbb{R})$ of functions f, which vanish of arbitrary order in 0, i.e. $\partial^k f(0) = 0$ for all $k \in \mathbb{N}_0$. The dual space and the Fourier image of $\mathcal{S}(\mathbb{R}^\times)$ will play an important role in the coming discussion. The first statement requires no further proof.

Lemma 2.2 *The image of $\mathcal{S}(\mathbb{R}^\times)$ under the Fourier transform on $\mathbb{R}$, $\mathcal{F}_\mathbb{R}$, is $\mathcal{S}_*(\mathbb{R})$, which is defined to be the subspace of Schwartz functions f with vanishing moments of arbitrary order, i.e.*

$$\int_\mathbb{R} f(x)\, p(x)\, \mathrm{d}x = 0, \quad \textit{for all} \quad p \in \mathcal{P}(\mathbb{R}). \tag{58}$$

The next Lemma is less obvious. It is an extension of the well know fact, that $\mathcal{S}'_*(\mathbb{R})$, as a vector space, can be identified with the quotient $\mathcal{S}'(\mathbb{R})/\mathcal{P}(\mathbb{R})$ e.g. [18, Proposition 1.1.3].

Lemma 2.3 *Let E be a nuclear Fréchet space and $\mathcal{E}'_0(\mathbb{R})$ the space of distributions on $\mathbb{R}$ with support in $\{0\}$. Then*

$$(\mathcal{S}(\mathbb{R}^\times)\,\hat{\otimes}\, E)' \simeq (\mathcal{S}'(\mathbb{R})\,\hat{\otimes}\, E')/(\mathcal{E}'_0(\mathbb{R}) \otimes E'), \tag{59}$$

especially $\mathcal{E}'_0(\mathbb{R}) \otimes E'$ is a closed subspace of $\mathcal{S}'(\mathbb{R})\,\hat{\otimes}\, E'$.

Proof First we will prove, that $Z := \mathcal{E}'_0(\mathbb{R}) \otimes E'$ is a closed subspace of $X' \simeq \mathcal{S}'(\mathbb{R})\,\hat{\otimes}\, E'$, where $X := \mathcal{S}(\mathbb{R})\,\hat{\otimes}\, E$. The family $(\partial^k\delta_0)_{k\in\mathbb{N}_0}$ is a basis for $\mathcal{E}'_0(\mathbb{R})$ where δ_0 is the delta distribution. We use Lemma 1.1 on the sequence P_N of projections onto the subspaces spanned by $\{\delta_0, \dots, \partial^N\delta_0\}$ and conclude, that Z is sequentially dense in its closure $\overline{Z}$. Furthermore we realize, that for any $\phi \in \overline{Z}$ there is a sequence $(e'_k) \subset E'$, such that

$$\phi = \lim_{N\to\infty} \phi_N := \lim_{N\to\infty} \sum_{k=0}^{N} (\partial^k\delta_0) \otimes e'_k. \tag{60}$$

Because X is a Fréchet space and $Z \subset X'$, we can apply the Banach-Steinhaus Theorem. Hence there exists a continuous seminorm q on E and $M \in \mathbb{N}$, such that

$$|\phi_N(f)| \le \max_{k\le M} \sup_{x\in\mathbb{R}} \langle x\rangle^M q(\partial_x^k f(x)) \tag{61}$$

for all functions $f \in X = \mathcal{S}(\mathbb{R})\,\hat{\otimes}\, E$ and all $N \in \mathbb{N}$.

Now suppose there is one $l > M$, such that $e'_l \neq 0$. Let us define the sequence of Schwartz functions $f_m(x) := \mathrm{e}^{\mathrm{i}mx}\psi(x)e/m^{l-1}$, where ψ is a Schwartz function equal to one near zero and $e \in E$ with $e'_l(e) = 1$. We arrive at

$$|\phi_l(f_m)| = \left| \sum_{k=0}^{l} \frac{(\mathrm{i}m)^k}{m^{(l-1)}} e'_k(e) \right| \xrightarrow{m\to\infty} \infty. \tag{62}$$

But also

$$\sup_{m\in\mathbb{N}}\max_{k\leq M}\sup_{x\in\mathbb{R}}\langle x\rangle^M q(\partial_x^k f_m(x)) < \infty, \tag{63}$$

which is a contradiction. Hence $\phi \in Z$, i.e. ϕ is in the finite span of the $\partial^k\delta$ and e'_k.

Now let $Y := Z^\circ$ be the polar of Z. Because X is reflexive, we may identify $Y \subset X$. Since Z is a closed subspace, we also have $Y^\circ = Z^{\circ\circ} = Z$. Since $\partial^k\delta_0 \otimes e' \in Z$ for all $k \in \mathbb{N}_0$, $e' \in E'$ and

$$(\partial^k\delta_0 \otimes e')(\varphi) = e'(\partial^k\varphi(0)), \quad \text{for} \quad \varphi \in X = \mathcal{S}(\mathbb{R}; E), \tag{64}$$

it is quite obvious, that $Y = \mathcal{S}(\mathbb{R}^\times)\,\hat{\otimes}\,E$.

Since E is a nuclear Fréchet space, X is a nuclear Fréchet space. That also means, that X is an (FS) space. I.e. it is the projective limit

$$X_1 \leftarrow X_2 \leftarrow \cdots \leftarrow X \tag{65}$$

of a sequence of Banach spaces $(X_k)_k$ with compact maps $X_k \leftarrow X_{k+1}$ [11, Chapter 3, Corollary 3 to Theorem 7.3]. Notice that the maps $X_k \leftarrow X_{k+1}$ are weakly compact, too. Now we may conclude the proof, by using Theorem 13 of [19]. The theorem states, that in our situation—Y is closed and X is an (FS) space—we have $Y' \simeq X'/Y^\circ$. □

By using the Euclidean Fourier transform in combination with the last lemma, we get the following corollary.

Corollary 2.3 *Let E be a nuclear Fréchet space, then*

$$(\mathcal{S}_*(\mathbb{R})\,\hat{\otimes}\,E)' \simeq (\mathcal{S}'(\mathbb{R})\,\hat{\otimes}\,E')/(\mathcal{P}(\mathbb{R}) \otimes E') \tag{66}$$

and $\mathcal{P}(\mathbb{R}) \otimes E'$ is closed in $\mathcal{S}'(\mathbb{R})\,\hat{\otimes}\,E'$.

Furthermore, this characterization for the dual spaces of $\mathcal{S}(\mathbb{R}^\times)\,\hat{\otimes}\,E$ and $\mathcal{S}_*(\mathbb{R})\,\hat{\otimes}\,E$ by quotient spaces, enables us to find subspaces of $\mathcal{S}'(\mathbb{R})\,\hat{\otimes}\,E'$ which are embedded into these dual spaces. Suppose F is a Banach space, such that there is a continuous embedding $E \hookrightarrow F$ with dense range. Then we may see, that the Lebesgue-Bochner spaces $L^p(\mathbb{R}; F')$ are embedded into $\mathcal{S}'_*(\mathbb{R})\,\hat{\otimes}\,E'$ and into $\mathcal{S}'(\mathbb{R}^\times)\,\hat{\otimes}\,E'$ for $p \in (1, \infty)$. Here we define the distribution corresponding to $f \in L^p(\mathbb{R}; F')$ by

$$T_f(\varphi) := \int_{\mathbb{R}} \langle f(x), \varphi(x)\rangle\,\mathrm{d}x, \quad \varphi \in \mathcal{S}(\mathbb{R}; E), \tag{67}$$

where $\langle\cdot,\cdot\rangle$ denotes the dual pairing on $F' \times F$. Notice, that T_f is indeed an injective map into $\mathcal{S}'(\mathbb{R}; E')$, since $f = 0$ almost everywhere, iff $T_f(\varphi \otimes e) = 0$ for all $\varphi \in \mathcal{S}(\mathbb{R})$ and all $e \in E$.

Though, we can make a much more general claim. For this purpose, we define the following subspaces of $\mathcal{S}'(\mathbb{R}) \hat{\otimes} E'$.

$$\begin{aligned}\dot{\mathcal{B}}'(\mathbb{R}; E') &:= \{\phi \in \mathcal{S}'(\mathbb{R}) \hat{\otimes} E' \mid \forall_{\varphi \in \mathcal{S}(\mathbb{R}) \hat{\otimes} E}\ \phi(\varphi(\cdot - x)) \xrightarrow{|x| \to \infty} 0\} \\ \widetilde{\mathcal{B}}'(\mathbb{R}; E') &:= \{\phi \in \mathcal{S}'(\mathbb{R}) \hat{\otimes} E' \mid \forall_{\varphi \in \mathcal{S}(\mathbb{R}) \hat{\otimes} E}\ \phi(\varphi(\lambda^{-1} \cdot)) \xrightarrow{\lambda \to 0} 0\}\end{aligned} \tag{68}$$

Lemma 2.4 *Let F be a Banach space as described above. The Lebesgue-Bochner space $L^p(\mathbb{R}; F')$ is a subspace of $\dot{\mathcal{B}}'(\mathbb{R}; E')$ for $p \in [1, \infty)$ and a subspace of $\widetilde{\mathcal{B}}'(\mathbb{R}; E')$ for $p \in [1, \infty]$ with respect to the embedding $f \mapsto T_f$.*

Proof Let $f \in L^p(\mathbb{R}; F')$ and let $\varphi \in \mathcal{S}(\mathbb{R}) \hat{\otimes} E$ then also

$$[x \mapsto (1 + x^2)\, \varphi(x)] \in L^q(\mathrm{R}; F) \tag{69}$$

for each $1 = 1/p + 1/q$. Suppose first $p \in [1, \infty)$, then for some $C > 0$ independent of $x \in \mathbb{R}$

$$|T_f(\varphi(\cdot - x))| \leq \int_{\mathbb{R}} |\langle f(y), \varphi(y - x)\rangle| \,\mathrm{d}y \leq C \left(\int_{\mathbb{R}} \frac{\|f(y)\|_{F'}^p}{(1 + (x - y)^2)^p} \,\mathrm{d}y \right)^{\frac{1}{p}}. \tag{70}$$

Now let $\varepsilon > 0$ be arbitrary and let $R > 0$ be big enough, such that

$$\int\limits_{|y| \geq R} \|f(y)\|_{F'}^p \,\mathrm{d}y \leq \varepsilon, \tag{71}$$

With this inequality, we get

$$\left(\int_{\mathbb{R}} \frac{\|f(y)\|_{F'}^p}{(1 + (x - y)^2)^p} \,\mathrm{d}y \right)^{\frac{1}{p}} \leq \left(\varepsilon + \int\limits_{|y| \leq R} \frac{\|f(y)\|_{F'}^p}{(1 + (x - y)^2)^p} \,\mathrm{d}y \right)^{\frac{1}{p}} \xrightarrow{x \to \pm\infty} \varepsilon^{\frac{1}{p}}. \tag{72}$$

Hence $T_f \in \dot{\mathcal{B}}'(\mathbb{R}; E')$, because $\varepsilon > 0$ can be arbitrarily small. With the same calculation as before, we get

$$|T_f(\varphi(\cdot/\lambda))| \leq C \left(\varepsilon + \int\limits_{|y| \leq R} \frac{\|f(y)\|_{F'}^p}{(1 + (y/\lambda)^2)^p} \,\mathrm{d}y \right)^{\frac{1}{p}} \xrightarrow{\lambda \to 0} C\varepsilon^{\frac{1}{p}}. \tag{73}$$

Thus $T_f \in \widetilde{\mathcal{B}}'(\mathbb{R}; E')$. Now suppose $p = \infty$. Here we have

$$|T_f(\varphi(\cdot/\lambda))| \leq \lambda \operatorname*{ess\,sup}_{x\in\mathbb{R}} \|f(x)\|_{F'} \int_{\mathbb{R}} \|\varphi(y)\| \,\mathrm{d}y \xrightarrow{\lambda\to 0} 0. \tag{74}$$

Hence also $T_f \in \widetilde{\mathcal{B}}'(\mathbb{R}; E')$ for this case. □

Note, that the distributions in $\dot{\mathcal{B}}'(\mathbb{R}; E')$ can have any form in a bounded region, whereas distributions in $\widetilde{\mathcal{B}}'(\mathbb{R}; E')$ can have any form away from zero, as long as they are tempered.

Proposition 2.1 *The quotient maps*

$$\begin{aligned} &\mathcal{S}'(\mathbb{R}) \hat{\otimes} E' \to \mathcal{S}'(\mathbb{R}^\times) \hat{\otimes} E', \\ &\mathcal{S}'(\mathbb{R}) \hat{\otimes} E' \to \mathcal{S}'_*(\mathbb{R}) \hat{\otimes} E', \end{aligned} \tag{75}$$

restrict to embeddings

$$\begin{aligned} &\widetilde{\mathcal{B}}'(\mathbb{R}; E') \hookrightarrow \mathcal{S}'(\mathbb{R}^\times) \hat{\otimes} E', \\ &\dot{\mathcal{B}}'(\mathbb{R}; E') \hookrightarrow \mathcal{S}'_*(\mathbb{R}) \hat{\otimes} E'. \end{aligned} \tag{76}$$

Proof A short calculation yields

$$\widetilde{\mathcal{B}}'(\mathbb{R}; E') \cap \mathcal{E}'_0(\mathbb{R}) \otimes E' = \{0\} = \dot{\mathcal{B}}'(\mathbb{R}; E') \cap \mathcal{P}(\mathbb{R}) \otimes E'. \tag{77}$$

Together with the above lemma and corollary, this already concludes the proof. □

2.2 Flat Orbits of Homogeneous Lie Groups

Let Ad be the adjoint action of G on $\mathfrak{g}$. Denote by $\mathrm{Ca}_x\, \xi := \xi \circ \mathrm{Ad}_{x^{-1}}$ the coadjoint action of $x \in G$ on linear functionals $\xi \in \mathfrak{g}'$. A subalgebra $\mathfrak{m} \subset \mathfrak{g}$ is called polarizing to $\ell \in \mathfrak{g}'$, iff $\ell([\mathfrak{m}, \mathfrak{m}]) = \{0\}$ and $\mathfrak{m}$ is a maximal algebra fulfilling this condition. For any $\xi \in \mathfrak{g}'$ we can find at least one polarizing algebra. There is a bijection between the coadjoint Orbits and the irreducible unitary representations of G. It can be described by $[\pi] \leftrightarrow \Omega = \mathrm{Ca}_G\, \xi$, where π is unitarily equivalent to the induced representation of $\chi(m) = \mathrm{e}^{2\pi \mathrm{i} \xi(m)}$ for $m \in \mathfrak{m} \subset G$ for some maximal subordinate algebra $\mathfrak{m}$ of ℓ [9, Theorems 2.2.1–2.2.4]. This correspondence only depends on the orbit Ω and not on the choice of element ξ spanning Ω or the choice of polarizing algebra $\mathfrak{m}$. We will write $\pi \sim \xi$ or $\pi \sim \Omega$, if the equivalence class of π corresponds to the orbit $\Omega = \mathrm{Ca}_G\, \xi$. We equip $\widehat{G}$ with the initial topology with respect to the bijection $[\pi] \mapsto \Omega$ for $\pi \sim \Omega$ from $\widehat{G}$ to $\mathfrak{g}'/G$. For any ξ the orbit $\Omega = \mathrm{Ca}_G\,\xi$ is an even dimensional polynomial manifold [6, page 521] and [9, Lemma 1.3.2].

A Jordan–Hölder basis of $\mathfrak{g}$, is a basis $(e_j)_j$, such that the linear hull $\mathfrak{g}_k = \operatorname{span}\{e_1, \dots e_k\}$, is an ideal in $\mathfrak{g}$ for each $k \leq \dim G$. Let q_k be the quotient map $\mathfrak{g}' \to \mathfrak{g}'/\mathfrak{g}_k^\circ$. The set of jump indices J is the set of $j > 1$, such that

$$\dim q_j(\Omega) - \dim q_{j-1}(\Omega) = 1 \tag{78}$$

Let us denote $\mathfrak{g}_J := \operatorname{span}\{e_j \mid j \in J\}$. From Corollary 3.1.5 of [9] follows, that a polynomial chart of Ω is given by

$$\sigma_\Omega \colon \Omega \to \mathfrak{g}'_J \colon \xi \mapsto \xi \restriction_{\mathfrak{g}_J} . \tag{79}$$

This equivalence between orbits and the corresponding subspaces $\mathfrak{g}_J$, leads to the definition of the orbital Fourier transform as the integral

$$\mathcal{F}_\Omega \varphi(x) := \int_\Omega \mathrm{e}^{-2\pi \mathrm{i}\xi(x)} \varphi(\xi)\, \mathrm{d}\theta_\Omega(\xi), \quad x \in \mathfrak{g}_J,\ \varphi \in \mathcal{S}(\Omega), \tag{80}$$

where $\theta_\Omega \circ \sigma_\Omega^{-1}$ is a Haar measure on $\mathfrak{g}'_J$. The Pedersen quantization [20] is the equivalent of the Weyl quantization for general simply connected nilpotent Lie groups. It is defined by the integral

$$\mathrm{op}_\pi(\varphi) := \int_{\mathfrak{g}_J} \pi(x) \int_\Omega \mathrm{e}^{-2\pi \mathrm{i}\xi(x)} \varphi(\xi)\, \mathrm{d}\theta_\Omega(\xi)\, \mathrm{d}\nu_\Omega(x), \tag{81}$$

for some representation $\pi \sim \Omega$ and a fitting Haar measure ν_Ω on $\mathfrak{g}_J$. We can easily see, that the outermost integral converges in $\mathcal{L}(H_\pi)$. The following theorem fixes the choice of ν_Ω.

Theorem 2.1 *For each θ_Ω as above, there is a unique ν_Ω, such that the Pedersen quantization to $\pi \sim \Omega$ extends to a Gelfand triple isomorphism*

$$\mathrm{op}_\pi \colon \mathcal{G}(\Omega, \theta_\Omega) \to \mathcal{G}_{\mathrm{op}}(\pi). \tag{82}$$

Proof This is essentially stated in [20, Theorem 4.1.4]. Here Pedersen proves, that

$$\mathcal{S}(\Omega) \to \mathcal{B}(H_\pi)_\infty \colon a \mapsto \mathrm{op}_\pi(a) \tag{83}$$

is a homeomorphism, where $\mathcal{B}(H_\pi)_\infty$ is the space of smooth operators with respect to π. The spaces of smooth operators is defined to be $\mathcal{B}(H_\pi)_\infty = H_\Pi^\infty$, where Π is the unitary representation of $G \times G$ on $\mathcal{HS}(H_\pi)$ defined by $\Pi(x, y)T = \pi(x) \circ T \circ \pi(y)^{-1}$. Furthermore Pedersen shows that

$$\int_\Omega a\,\overline{b}\,\mathrm{d}\theta_\Omega = \mathrm{Tr}[\mathrm{op}_\pi(a)\ \mathrm{op}_\pi(b)^*], \quad \text{for } a, b \in \mathcal{S}(\Omega) \tag{84}$$

for a suitable choice of ν_Ω.

In order to fit this result in our scheme we will make sure, that $\mathcal{L}(H_\pi^{-\infty}, H_\pi^\infty) = \mathcal{B}(H_\pi)_\infty$ as topological vector spaces. It is easy to see, that this identity holds in the sense that each $T \mapsto T \circ \mathcal{I}$ is a bijection from the left-hand side to the right-hand side, where $\mathcal{I}\colon H_\pi \hookrightarrow H_\pi^{-\infty}$ is the embedding defined by the real structure on $\mathcal{G}(\pi)$.

It is left to check that the topologies on both sides coincide. For $P \in \mathfrak{u}(\mathfrak{g}_\mathrm{L})$ denote by $\pi(P)_* \in \mathcal{L}(H_\pi^{-\infty})$ the unique operator fulfilling $\mathcal{I}\pi(P)v = \pi(P)_*\mathcal{I}v$ for $v \in H_\pi^\infty$. By Cartier [21] there is $P \in \mathfrak{u}(\mathfrak{g}_\mathrm{L})$, such that $\pi(P)$ is invertible on H_π^∞, $\pi(P)^{-1}$ can be extended to a nuclear operator on H_π and $H_\pi^{-\infty}$ is the compact inductive limit of the Hilbert spaces $H_\pi^{-k} := \pi(P)_*^k \mathcal{I} H_\pi$, equipped with the norm $\|w\|_{-k} := \|\mathcal{I}^{-1}\pi(P)^{-k}w\|_{H_\pi}$. But this means each bounded $B \subset H_\pi^{-\infty}$ is in fact a bounded set in some H_π^{-k}, i.e. $B = \pi(P)_*^k \mathcal{I}\tilde{B}$ for $\tilde{B}$ bounded in H_π. The topology in $\mathcal{L}(H_\pi^{-\infty}, H_\pi^\infty)$ is defined by the seminorms

$$T \mapsto \sup_{\phi \in B} \|\pi(P)^j T\phi\|, \quad \text{for} \quad B \subset H_\pi^{-\infty} \text{ bounded and } j \in \mathbb{N}_0. \tag{85}$$

The above implies, that actually it is enough to consider the seminorms

$$T \mapsto \|\pi(P)^j\, T\, \mathcal{I}\, \pi(P)^k\|, \quad \text{for } j, k \in \mathbb{N}_0. \tag{86}$$

Now the above seminorms are also continuous on $\mathcal{B}(H_\pi)_\infty$, so by the open mapping theorem for Fréchet spaces, $\mathcal{L}(H_\pi^{-\infty}, H_\pi^\infty) = \mathcal{B}(H_\pi)_\infty$ as topological vector spaces and

$$\mathcal{S}(\Omega) \to \mathcal{L}(H_\pi^{-\infty}, H_\pi^\infty)_\infty\colon a \mapsto \mathrm{op}_\pi(a) \tag{87}$$

is a homeomorphism.

Finally by the dense and continuous embeddings

$$\mathcal{S}(\Omega) \hookrightarrow L^2(\Omega, \theta_\Omega) \quad \text{and} \quad \mathcal{L}(H_\pi^{-\infty}, H_\pi^\infty) \hookrightarrow \mathcal{HS}(H_\pi) \tag{88}$$

we may extend op_π to a unitary operator between $L^2(\Omega, \theta_\Omega)$ and $\mathcal{HS}(H_\pi)$, and as such, even to a Gelfand triple isomorphism

$$\mathrm{op}_\pi : \mathcal{G}(\Omega, \theta_\Omega) \to \mathcal{G}_{\mathrm{op}}(\pi). \tag{89}$$

Note, that Pedersen uses the convention $\xi \leftrightarrow \chi(\cdot) = e^{i\xi(\cdot)}$ for bijection between functionals and characters. Though adjusting the formulas just results in additional constants, that may be hidden away inside the measures ν_Ω and θ_Ω. □

Though in our case, we can simplify this process by a lot, since we are only interested in representations derived from generic resp. flat orbits.

An orbit is called *generic*, if for each k the dimension of $q_k(\Omega)$ is maximal compared to all other orbits. Let us denote the set of equivalence classes derived from generic orbits by $\widehat{G}_{\text{gen}} \subset \widehat{G}$. Note, that the Plancherel measure $\widehat{\mu}$ is concentrated on $\widehat{G}_{\text{gen}}$.

A representation $\pi \in \text{Irr}(G)$ is *square integrable modulo the center*, if $x \mapsto |(\pi(x)v, w)_{H_\pi}|$ is square integrable on $\mathfrak{g}/\mathfrak{z}$ with respect to the Haar measure for all $v, w \in H_\pi$. Let us denote the set of irreducible representations, that are square integrable modulo the center, by $\text{SI/Z}(G) \subset \text{Irr}(G)$ and pairs of such representations together with some matching real structure by $\text{SI/Z}_{\mathbb{R}}(G)$. Suppose $\pi \sim \Omega = \text{Ca}_G\,\xi$, then $\pi \in \text{SI/Z}(G)$, if and only if $\Omega = \xi + \mathfrak{z}^\circ$ [22]. Furthermore, if $\text{SI/Z}(G) \neq \emptyset$, then the orbits to representations in $\text{SI/Z}(G)$ are exactly those having the maximal possible dimension [9, Corollary 4.5.6]. Also, the jump indices for $\pi \in \text{SI/Z}(G)$ are given by $J = \{k+1, k+2, \dots, \dim G\}$, where $k = \dim \mathfrak{z}$, and the equivalence class $[\pi] \in \widehat{G}$ is uniquely determined by the central character $\pi \restriction_{\mathfrak{z}} = e^{2\pi i \xi(\cdot)}\,\text{id}_{H_\pi}$, where $\xi \in \omega^\circ \simeq \mathfrak{z}'$. For this fact see [9, Corollaries 4.5.3 and 4.5.4].

For all $\pi \in \text{SI/Z}(G)$ the Pedersen quantization is simpler, for we can just take *one* Haar measure θ on $\mathfrak{z}^\circ$ and translate it to a measure θ_Ω on $\Omega \sim \pi$ for each $\pi \in \text{SI/Z}(G)$. The subspace $\omega := \mathfrak{g}_J$ complements $\mathfrak{z}$ in $\mathfrak{g}$ and is the same for each representation in $\text{SI/Z}(G)$. We get a Gelfand triple isomorphisms

$$\mathcal{G}(\mathfrak{z}^\circ, \theta) \to \mathcal{G}(\Omega, \theta_\Omega) \colon \phi \mapsto \phi \circ P_{\mathfrak{z}^\circ}, \tag{90}$$

where $P_{\mathfrak{z}^\circ}$ is the projection onto $\mathfrak{z}^\circ$ along ω°. Using this isomorphism, we adjust the Pedersen quantization.

Definition 2.4 We will use the Pedersen quantization $\mathfrak{op}_\pi$ on $\mathcal{G}(\mathfrak{z}^\circ, \theta)$ with respect to $\pi \in \text{SI/Z}(G)$, defined by

$$\mathfrak{op}_\pi : \mathcal{G}(\mathfrak{z}^\circ, \theta) \to \mathcal{G}_{\text{op}}(\pi),\ \phi \mapsto \text{op}_\pi(\phi \circ P_{\mathfrak{z}^\circ}). \tag{91}$$

This version of Pedersen quantization takes on the form

$$\mathfrak{op}_\pi(\varphi) = \int_\omega \pi(x) \int_{\mathfrak{z}^\circ} e^{-2\pi i \xi(x)} \varphi(\xi)\, d\theta(\xi)\, d\nu(x), \tag{92}$$

where $\nu = \nu_\Omega$ depends on θ. Of course $\mathfrak{op}_\pi$ is a Gelfand triple isomorphism, as well.

Now we will discuss the concept of generic orbits and square integrable (modulo the center) representation in context with homogeneous groups. The Lie group $G = \mathfrak{g}$ is called a homogeneous Lie group, if it is equipped with a group of dilations

$$(0, \infty) \to \mathrm{Hom}(G) \colon \lambda \mapsto \delta_\lambda, \tag{93}$$

where $\delta_\lambda x = e^{\log(\lambda) A} x$ is also a Lie algebra isomorphism and A is a diagonalizable map with positive eigenvalues. The number $Q := \mathrm{Tr}[A]$ is the *homogeneous dimension* of G.

We may always decompose $\mathfrak{g}$ into eigenspaces $\mathcal{E}_\kappa$ of A to Eigenvalues $\kappa > 0$, i.e.

$$\mathfrak{g} = \bigoplus_{\kappa > 0} \mathcal{E}_\kappa, \quad \text{where} \quad [\mathcal{E}_\kappa, \mathcal{E}_{\kappa'}] \subset \mathcal{E}_{\kappa+\kappa'}. \tag{94}$$

Notice that the center $\mathfrak{z}$ of $\mathfrak{g}$ is always an eigenspace to both δ_λ and A, since

$$[\delta_\lambda z, x] = \delta_\lambda [z, \delta_{\lambda^{-1}} x] = 0 \quad \text{for all} \quad \lambda > 0, z \in \mathfrak{z} \text{ and } x \in \mathfrak{g}. \tag{95}$$

For every $\mu > 0$ the space $\bigoplus_{\kappa \geq \mu} \mathcal{E}_\kappa$ is an ideal in $\mathfrak{g}$. We may always choose a Jordan–Hölder basis $(e_j)_j$ through these ideals [9, Theorem 1.1.13].

If $\dim \mathfrak{z} = 1$, then the center fulfils $\mathfrak{z} = \mathcal{E}_\mu$ for $\mu = \max\{\kappa > 0 \mid \mathcal{E}_\kappa \neq \{0\}\}$. Hence, one vector of our chosen Jordan–Hölder basis of eigenvectors will always lie in the center $\mathfrak{z}$. We also have the unique decomposition

$$\mathfrak{g} = \mathfrak{z} \oplus \omega, \quad \omega \text{ is } A\text{-invariant}. \tag{96}$$

Now for $\lambda < 0$ denote

$$\delta_\lambda x := -\delta_{|\lambda|} x \text{ for } x \in \mathfrak{z}, \quad \text{and} \quad \delta_\lambda x := \delta_{|\lambda|} x \text{ for } x \in \omega. \tag{97}$$

Furthermore, let also $\delta_\lambda \xi := \xi \circ \delta_\lambda$ for $\lambda \in \mathbb{R}^\times$ and $\xi \in \mathfrak{g}'$.

The question arises whether generic orbits are mapped to generic orbits by δ_λ. The dilation δ_λ on $\mathfrak{g}'/\mathfrak{g}_k^\circ$ is a well defined vector space isomorphism by $\delta_\lambda \circ q_j := q_j \circ \delta_\lambda$, since $\mathfrak{g}_k$ and thus also $\mathfrak{g}_k^\circ$ are δ_λ-invariant. Furthermore

$$\dim q_j(\delta_\lambda \Omega) = \dim \delta_\lambda \circ q_j(\Omega) = \dim q_j(\Omega). \tag{98}$$

Thus $\delta_\lambda \Omega$ is generic for each $\lambda \in \mathbb{R}^\times$.

Now take any $\pi \in \mathrm{Irr}_\mathbb{R}(G)$ with real structure C_π and define $\overline{\pi} := C_\pi \pi C_\pi \in \mathrm{Irr}_\mathbb{R}(G)$ equipped with the same real structure. The representation $\overline{\pi}$ is equivalent to the dual representation of π. Now denote

$$\pi_\lambda(x) := \pi(\delta_\lambda x) \text{ for } \lambda > 0, \quad \text{and} \quad \pi_\lambda(x) := \overline{\pi}_{|\lambda|}(x) := \overline{\pi}(\delta_{|\lambda|} g) \text{ for } \lambda < 0. \tag{99}$$

All the representations π_λ are irreducible unitary representations acting on $\mathcal{H}_\pi$ resp. acting smoothly on $\mathcal{H}_\pi^\infty$. With these definitions and the discussion above, we get the equivalence of the three statements

- $\pi \in \mathrm{SI/Z}(G)$, if and only if $\pi_\lambda \in \mathrm{SI/Z}(G)$,
- $[\pi] \in \widehat{G}_{\mathrm{gen}}$, if and only if $[\pi_\lambda] \in \widehat{G}_{\mathrm{gen}}$,
- $\pi \sim \xi$, if and only if $\pi_\lambda \sim \delta_\lambda \xi$.

Suppose that $\mathrm{SI/Z}(G) \neq \emptyset$ and $\dim \mathfrak{z} = 1$. Furthermore, suppose we chose a Jordan–Hölder basis of eigenvectors to A. Let $\pi \in \mathrm{SI/Z}(G)$. As every equivalence class of representations in $\widehat{G}_{\mathrm{gen}}$ only depends on its central character, we get a bijection between $\mathbb{R}^\times$ and $\widehat{G}_{\mathrm{gen}}$. Hence, $\pi \in \mathrm{SI/Z}(G)$, if and only if $[\pi] \in \widehat{G}_{\mathrm{gen}}$.

We can even go one step further. The dilations δ_λ help us to understand $\widehat{G}$ as measure space. For this purpose we need the Pfaffian $Pf(\xi)$ to a coadjoint orbit $\Omega = \mathrm{Ca}_G\, \xi$, which is defined by $Pf(\xi)^2 = \det B_\xi$ up to a sign. Here $B_\xi := (\xi([e_j, e_i]))_{j,i\in J}$ where the $(e_j)_{j\in J}$ span ω.

We define $\kappa > 0$ and $B \in \mathcal{L}(\omega)$, by $\delta_\lambda \eta := \mathrm{sgn}(\lambda)|\lambda|^\kappa \eta$, for $\eta \in \omega^\circ$ and $A|_\omega = B$.

Proposition 2.2 *Suppose G is a homogeneous group, $\pi \in \mathrm{SI/Z}_{\mathbb{R}}(G)$ and $\dim \mathfrak{z} = 1$, then*

$$(\mathbb{R}^\times, \kappa|\lambda|^{Q-1}|Pf(\ell)|\, \mathrm{d}\lambda) \to (\widehat{G}_{\mathrm{gen}}, \widehat{\mu}) \colon \lambda \mapsto [\pi_\lambda], \tag{100}$$

where $\pi \sim \ell \in \omega^\circ$, is a homeomorphism resp. a strict isomorphism between the Borel measure spaces. Furthermore, if Ω is a fixed generic orbit, then $\lambda \mapsto \delta_\lambda \Omega$ defines a bijection between $\mathbb{R}^\times$ and the generic orbits.

Proof Let U be the Zariski open set of functionals $\xi \in \mathfrak{g}'$, such that $\mathrm{Ca}_G\, \xi$ is a generic orbit with respect to our basis. For $\xi \in U$ we have $\delta_\lambda \xi \in U$ for each $\lambda \in \mathbb{R}^\times$ by Eq. (98). Each orbit meets $U \cap \omega^\circ$ in exactly one point [9, Theorem 3.1.9 and Theorem 4.5.5]. Furthermore, for any $\xi \in \omega^\times := \omega^\circ \setminus \{0\}$, we have that

$$\mathbb{R}^\times \to \omega^\times \colon \lambda \mapsto \delta_\lambda \xi \tag{101}$$

is a homeomoprhism. Thus also $U \cap \omega^\circ = \omega^\times = \{\delta_\lambda \ell \mid \lambda \in \mathbb{R}^\times\}$. But $\omega^\times$ also induces all maximal flat orbits, so they coincide with the generic orbits. Since the correspondence $\mathfrak{g}'/G \simeq \widehat{G}$ is a homeomorphism, we also have $U/G \simeq \widehat{G}_{\mathrm{gen}}$ with respect to the subspace topologies. Let $q \colon U \to U/G$ be the quotient map. Now $q|_{\omega^\times}$ is a continuous bijection. We show, that it is also open. By [9, Theorem 3.1.9], there is a well define map $\psi \colon \omega^\times \times \mathfrak{z}^\circ \to U$, such that

$$\psi(u, v) = w \quad \Leftrightarrow \quad w \in \mathrm{Ca}_G\, u \text{ and } P_{\mathfrak{z}^\circ} w = v, \tag{102}$$

where $P_{\mathfrak{z}^\circ}$ is the projection onto $\mathfrak{z}^\circ$ along ω°. The map ψ is a rational, non singular bijection with rational non singular inverse. Hence ψ is a homeomorphism. If

$V \subset \omega^\times$ is open in $\omega^\times$, then $\mathrm{Ca}_G\, V$ is open in U, since

$$\psi(V \times \mathfrak{z}^\circ) = \mathrm{Ca}_G\, V. \tag{103}$$

Now, since q is open and $q(\mathrm{Ca}_G\, V) = q(V)$, the restriction $q|_{\omega^\times}$ is an open map and thus a homeomorphism. If we now denote

$$\sigma : \mathbb{R}^\times \to \widehat{G}_{\mathrm{gen}} : \lambda \mapsto [\delta_\lambda \pi], \tag{104}$$

then σ is a homeomorphism by the discussion above. Let $\varphi : \widehat{G} \to [0, \infty)$ be Borel measurable. Then by Theorem 4.3.10 and the subsequent discussion in [9]

$$\int_{\widehat{G}} \varphi([\pi])\, \mathrm{d}\widehat{\mu}([\pi]) = \int_{U \cap \omega^\circ} \varphi([\pi_\xi]) |Pf(\xi)|\, \mathrm{d}\widetilde{\mu}(\xi), \tag{105}$$

where $\widetilde{\mu}$ is the measure on $U \cap \omega^\circ$, such that $\{t\ell \mid t \in [0, 1]\}$ has measure equal to one and $\pi_\xi \sim \mathrm{Ca}_G\, \xi$. Also, since our chosen Jordan–Hölder basis is an eigenbasis to A resp. δ_λ, we have

$$\begin{aligned} |Pf(\delta_\lambda \ell)| &= |\det(\delta_\lambda \ell([e_j, e_i]))_{j,i}|^{\frac{1}{2}} \\ &= |\det(|\lambda|^{n_i + n_j} \ell([e_j, e_i]))_{j,i}|^{\frac{1}{2}} = |\lambda|^{\mathrm{Tr}\, B} |Pf(\ell)|, \end{aligned} \tag{106}$$

where $|\lambda|^{n_j}$ is the eigenvalue of e_j to δ_λ for $j \in J$. Both σ and σ^{-1} are measurable and we have $\mathrm{d}(\widetilde{\mu} \circ \sigma)(\lambda) = \kappa |\lambda|^{\kappa - 1}\, \mathrm{d}\lambda$. Hence

$$\begin{aligned} \int_{U \cap \omega^\circ} \varphi([\pi_\xi]) |Pf(\xi)|\, \mathrm{d}\widetilde{\mu}(\xi) &= \int_{\mathbb{R}^\times} \varphi([\pi_\lambda]) |Pf(\delta_\lambda \ell)|\, \mathrm{d}(\widetilde{\mu} \circ \sigma)(\lambda) \\ &= \int_{\mathbb{R}^\times} \varphi([\pi_\lambda]) \kappa |\lambda|^{-1 + \mathrm{Tr}\, A} |Pf(\ell)|\, \mathrm{d}\lambda \end{aligned} \tag{107}$$

and σ is a strict isomorphism of measure spaces. □

We will denote the Euclidean Fourier transform on $\mathfrak{g}$ by

$$\mathcal{F}_{\mathfrak{g}} \varphi(\xi) = \int_{\mathfrak{g}} \mathrm{e}^{2\pi \mathrm{i} \xi(x)} \varphi(x)\, \mathrm{d}\mu(x), \quad \varphi \in \mathcal{S}(\mathfrak{g}),\ \xi \in \mathfrak{g}'. \tag{108}$$

Of course, there is exactly one Haar measure μ' on $\mathfrak{g}'$, such that the Fourier transform is a Gelfand triple isomorphism $\mathcal{G}(\mathfrak{g}, \mu) \to \mathcal{G}(\mathfrak{g}', \mu')$. Suppose $\ell \in \omega^\times$. The map

$$\wp_\ell f(\lambda, \xi) := f(\delta_\lambda(\ell + \xi)) \quad \text{for} \quad \xi \in \mathfrak{z}^\circ, \lambda \in \mathbb{R}^\times \text{ and } f : \mathfrak{g}' \to \mathbb{C}, \tag{109}$$

together with the Euclidean Fourier transform and the Pedersen quantization will enable us to describe the group Fourier transform on G (see also [3] for a similar statement).

2.3 The Group Fourier Transform on Homogeneous Groups

Let us from now on always denote by G a homogeneous Lie group with $\dim \mathfrak{z} = 1$ and $\mathrm{SI}/\mathrm{Z}(G) \neq \emptyset$. Trivially, the group Fourier transform is an isomorphism between $\mathcal{S}(G)$ and $\mathcal{S}(\widehat{G})$. Also, the group Fourier transform is a unitary map from $L^2(G, \mu)$ to $L^2(\widehat{G}, \widehat{\mu})$. Of course, we may define a Gelfand triple

$$\mathcal{G}(\widehat{G}, \widehat{\mu}) := (\mathcal{S}(\widehat{G}), L^2(\widehat{G}, \widehat{\mu}), \mathcal{S}'(\widehat{G})), \tag{110}$$

such that $\mathcal{F}_G$ becomes a Gelfand triple isomorphism. Now we will use the isomorphism from Proposition 2.2 in order to find a new representation of the group Fourier transform on $L^2(G)$. This will be the basis for the definition of our new Gelfand triples and a Gelfand triple isomorphism in the form of an equivalent Fourier transform.

Proposition 2.3 *Suppose $\varphi \in \mathcal{S}(G)$ and $\pi \in \mathrm{SI}/\mathrm{Z}_{\mathbb{R}}(G)$ with $\pi \sim \ell \in \omega^\times$, then*

$$\mathcal{F}_G\varphi(\pi_\lambda) = \begin{cases} \mathrm{op}_\pi\big(\wp_\ell \mathcal{F}_{\mathfrak{g}}\varphi(\lambda, \cdot)\big), & \lambda > 0, \\ \mathrm{op}_{\overline{\pi}}\big(\wp_\ell \mathcal{F}_{\mathfrak{g}}\varphi(\lambda, \cdot)\big), & \lambda < 0. \end{cases} \tag{111}$$

Proof First of all, for any $\varphi \in \mathcal{S}(G)$, we have

$$\mathcal{F}_G\varphi(\pi_\lambda) = \int_G \lambda^{-\operatorname{Tr} A}\varphi(\delta_\lambda^{-1}x)\pi(x)^*\, \mathrm{d}\mu(x) = \lambda^{-\operatorname{Tr} A}\mathcal{F}_G(\varphi \circ \delta_\lambda^{-1})(\pi), \tag{112}$$

for $\lambda > 0$. Also

$$\mathcal{F}_{\mathfrak{g}}(\varphi \circ \delta_\lambda^{-1}) = \lambda^{\operatorname{Tr} A}(\mathcal{F}_{\mathfrak{g}}\varphi) \circ \delta_\lambda, \tag{113}$$

for $\lambda > 0$. Notice, that for $x \in \mathfrak{g}$ and $z \in \mathfrak{z}$, we have $x \cdot z = x + z$ and thus

$$\mathrm{e}^{2\pi \mathrm{i}\ell(z)}\pi(x) = \pi(z)\pi(x) = \pi(z \cdot x) = \pi(z + x). \tag{114}$$

Let $\mu_{\mathfrak{z}}$ resp. ν be Haar measures on $\mathfrak{z}$ resp ω, such that $\mu = \mu_{\mathfrak{z}} \otimes \nu$, then by the above calculation

$$\begin{aligned} \mathcal{F}_G\varphi(\pi) &= \int_\omega \pi(x) \int_{\mathfrak{z}} \mathrm{e}^{-2\pi \mathrm{i}\ell(z)} \varphi(z-x)\, \mathrm{d}\mu_{\mathfrak{z}}(z)\, \mathrm{d}\nu(x) \\ &= \int_\omega \pi(x) \int_{\mathfrak{z}^\circ} \mathrm{e}^{-2\pi \mathrm{i}\xi(X)} \mathcal{F}_{\mathfrak{g}}\varphi(\xi)\, \mathrm{d}\theta(\xi)\, \mathrm{d}\nu(x). \end{aligned} \tag{115}$$

Here θ is the measure associated to ν as described in Definition 2.4. This formula indeed holds pointwise. Hence

$$\begin{aligned} \mathcal{F}_G\varphi(\pi_\lambda) &= \lambda^{-\operatorname{Tr} A} \mathfrak{op}_\pi(\mathcal{F}_{\mathfrak{g}}(\varphi \circ \delta_\lambda^{-1})) = \mathfrak{op}_\pi((\mathcal{F}_{\mathfrak{g}}\varphi) \circ \delta_\lambda) \\ &= \mathfrak{op}_\pi\big(\wp_\ell \mathcal{F}_{\mathfrak{g}}\varphi(\lambda, \cdot)\big) \end{aligned} \tag{116}$$

for all $\lambda > 0$. For $\lambda < 0$ we get

$$\mathcal{F}_G\varphi(\pi_\lambda) = \mathcal{F}_G(\overline{\pi}_{-\lambda}) = \mathfrak{op}_{\overline{\pi}}\big(\wp_{-\ell}\mathcal{F}_{\mathfrak{g}}\varphi(-\lambda, \cdot)\big), \tag{117}$$

since $\overline{\pi} \sim -\ell$. Now we can conclude the proof, by using $\delta_{-\lambda}(-\ell + \xi) = \delta_\lambda(\ell + \xi)$, for any $\xi \in \mathfrak{z}^\circ$. □

The above proposition (c.f. [3, Theorem 3.3]) shows that the group Fourier transform splits into operators which are easy to handle in the L^2-setting, if $\dim \mathfrak{z} = 1$. If we use the isomorphism $(\widehat{G}, \widehat{\mu}) \simeq (\mathbb{R}^\times, \kappa|\lambda|^{Q-1}|Pf(\ell)|\,\mathrm{d}\lambda)$ then we can see $\mathcal{F}_G$ as the composition of unitary operators

$$\begin{aligned} \mathcal{F}_{\mathfrak{g}}&\colon L^2(G, \mu) \to L^2(\mathfrak{g}', \mu'), \\ \wp_\ell&\colon L^2(\mathfrak{g}', \mu') \to L^2(\mathbb{R}^\times \times \mathfrak{z}^\circ; \kappa|\lambda|^{Q-1}|Pf(\ell)|\,\mathrm{d}\lambda\,\mathrm{d}\theta(\xi)), \\ \mathfrak{Op}_\pi &: L^2\big(\mathbb{R}^\times, \kappa|\lambda|^{Q-1}|Pf(\ell)|\,\mathrm{d}\lambda; L^2(\mathfrak{z}^\circ, \theta)\big) \\ &\qquad \to L^2\big(\mathbb{R}^\times, \kappa|\lambda|^{Q-1}|Pf(\ell)|\,\mathrm{d}\lambda; \mathcal{HS}(H_\pi)\big), \end{aligned} \tag{118}$$

where $\mathfrak{Op}_\pi = P_+ \otimes \mathfrak{op}_\pi + P_- \otimes \mathfrak{op}_{\overline{\pi}}$, for the projection $P_\pm$ of $L^2(\mathbb{R}^\times)$ onto $L^2(\mathbb{R}^\pm)$. It is very convenient, that here the operator component emerges as a tensor product factor, which in turn enables us to understand multiplication operators more easily. This motivates us to define the following alternative spaces of test functions.

2.4 The Fourier Transform on $\mathcal{S}_*(G)$

In order to know which function space is a good choice, we will first take a look at the pull back $\wp_\ell$. Here our earlier discussion of polynomial manifolds comes into play again. Remember that $\mathbb{R}^\times$ is equipped with a polynomial structure defined by $\mathbb{R}^\times = \mathbb{R}_+ \dot{\cup} \mathbb{R}_-$, i.e. defined by the polynomial structures on $\mathbb{R}^\pm$. Similarly, we define $\mathfrak{g}_\ell^+$, $\mathfrak{g}_\ell^-$ and $\mathfrak{g}^\times$ by

$$\mathfrak{g}_\ell^\pm := \{t\ell + \eta \mid t \in \mathbb{R}_\pm, \eta \in \mathfrak{z}^\circ\}, \quad \text{for } \ell \in \omega^\times = \omega^\circ \setminus \{0\} \tag{119}$$

and $\mathfrak{g}^\times = \mathfrak{g}_\ell^+ \dot{\cup} \mathfrak{g}_\ell^-$ and equip $\mathfrak{g}_\ell^\pm$ with the polynomial structure analogously to the one on $\mathbb{R}^\pm$, i.e. the polynomial structure induced by the map

$$\mathfrak{g}_\ell^\pm \to \mathbb{R} \times \mathfrak{z}^\circ : (t\ell + \eta) \mapsto (t - 1/t, \eta). \tag{120}$$

Then δ_λ induces a tempered diffeomorphism, as written in the following Lemma. The polynomial structure on $\mathfrak{g}^\times$ is just the one induced by its connected components. Notice that we just have $\mathfrak{g}^\times = \mathfrak{g}' \setminus \mathfrak{z}^\circ$ as a set.

Lemma 2.5 *Let $\ell \in \omega^\times$. The Map $w_\ell : \mathbb{R}^\pm \times \mathfrak{z}^\circ \to \mathfrak{g}_\ell^\pm : (\lambda, \xi) \mapsto \delta_\lambda(\ell + \xi)$ is a tempered diffeomorphism.*

Proof We prove that $\mathbb{R}^+ \times \mathfrak{z}^\circ \simeq \mathfrak{g}_\ell^+$ via w_ℓ. The proof to the second statement is analogous. Suppose $(\xi^j)_j$ is the dual basis to our Jordan–Hölder basis $(e_j)_{j=0}^{2n}$ of eigenvectors. Here $(\xi_j)_{j=1}^{2n}$ is the basis of $\mathfrak{z}^\circ$. Let κ_j be the positive number, such that $\delta_\lambda \xi^j = \lambda^{\kappa_j} \xi^j$ for $\lambda > 0$.

We use the charts σ resp. σ_1, defined by

$$(\lambda, \sum_{j=1}^{2n} c_j \xi^j) \text{ resp. } (\lambda\ell + \sum_{j=1}^{2n} c_j \xi^j) \mapsto (\lambda - 1/\lambda, c_1, \dots, c_{2n}). \tag{121}$$

Then

$$\begin{aligned} &\sigma_1 \circ w_\ell \circ \sigma^{-1}(t, c_1, \dots, c_{2n}) \\ &= \left(\frac{(t + \sqrt{t^2 + 4})^{\kappa_0}}{2^{\kappa_0}} - \frac{2^{\kappa_0}}{(t + \sqrt{t^2 + 4})^{\kappa_0}}, \right. \\ &\qquad \left. \frac{(t + \sqrt{t^2 + 4})^{\kappa_1}}{2^{\kappa_1}} c_1, \dots, \frac{(t + \sqrt{t^2 + 4})^{\kappa_{2n}}}{2^{\kappa_{2n}}} c_{2n} \right), \end{aligned} \tag{122}$$

which is a slowly increasing function. Similarly

$$\sigma \circ w_\ell^{-1} \circ \sigma_1^{-1}(t, c_1, \ldots, c_{2n})$$
$$= \left(\frac{(t+\sqrt{t^2+4})^{\frac{1}{\kappa_0}}}{2^{\frac{1}{\kappa_0}}} - \frac{2^{\frac{1}{\kappa_0}}}{(t+\sqrt{t^2+4})^{\frac{1}{\kappa_0}}}, \right. \tag{123}$$
$$\left. \frac{(t+\sqrt{t^2+4})^{\frac{-\kappa_1}{\kappa_0}}}{2^{\frac{-\kappa_1}{\kappa_0}}} c_1, \ldots, \frac{(t+\sqrt{t^2+4})^{\frac{-\kappa_{2n}}{\kappa_0}}}{2^{\frac{-\kappa_{2n}}{\kappa_0}}} c_{2n} \right)$$

is slowly increasing. □

By Lemma 2.1, we can see $\mathcal{S}(\mathfrak{g}_\ell^{\pm})$ as the space

$$\mathcal{S}(\mathfrak{g}_\ell^{\pm}) = \{\varphi \in \mathcal{S}(\mathfrak{g}') \mid \varphi \equiv 0 \text{ on } \mathfrak{g}_\ell^{\mp}\}, \tag{124}$$

equipped with the subspace topology in $\mathcal{S}(\mathfrak{g}')$.

The tempered diffeomorphism from the last lemma induces a Gelfand triple isomorphism.

Lemma 2.6 *The pullback $\wp_\ell f := f \circ w_\ell$ defines a Gelfand triple isomorphism*

$$\wp_\ell \colon \mathcal{G}(\mathfrak{g}_\ell^{\pm}, \mu') \to \mathcal{G}(\mathbb{R}^{\pm}, \kappa |Pf(\ell)|\, |\lambda|^{Q-1}\, \mathrm{d}\lambda) \otimes \mathcal{G}(\mathfrak{z}^\circ, \theta), \tag{125}$$

where κ, $|Pf(\ell)|$ are the constants introduced in Proposition 2.2 and the preceding remarks and Q is the homogenous dimension of G.

Proof We take an arbitrary $f \in C_c(\mathfrak{g}_\ell^{\pm})$. Define $\omega^{\pm} := \mathbb{R}^{\pm} \cdot \ell$, then

$$\begin{aligned}
&\int_{\mathbb{R}^{\pm}} \int_{\mathfrak{z}^\circ} f(\delta_\lambda(\ell + \xi))\kappa |Pf(\ell)|\, |\lambda|^{Q-1}\, \mathrm{d}\lambda\, \mathrm{d}\theta(\xi) \\
&\qquad = \int_{\mathbb{R}^{\pm}} \int_{\mathfrak{z}^\circ} f((\delta_\lambda \ell) + \xi)\kappa |Pf(\ell)|\, |\lambda|^{\kappa_0 - 1}\, d\lambda\, \mathrm{d}\theta(\xi) \\
&\qquad = \int_{\omega^{\pm}} \int_{\mathfrak{z}^\circ} f(\eta + \xi))\, |Pf(\ell)|\, d\mu_{\omega^\circ}(\eta)\, \mathrm{d}\theta(\xi) \\
&\qquad = \int_{\mathfrak{g}_\ell^{\pm}} f(\xi)\, \mathrm{d}\mu'(\xi).
\end{aligned} \tag{126}$$

For the last two lines we used that the measure μ_{ω° on ω° is defined by the Lebesgue measure and ℓ and that θ is defined by $\mu' = |Pf(\ell)|\, \mu_{\omega^\circ} \otimes \theta$. The rest follows with the fact, that $\wp_\ell f = f \circ w_\ell$, where w_ℓ is the tempered diffeomorphism from Lemma 2.5. □

We also proved that the restriction of the Haar measure μ' to $\mathfrak{g}_\ell^\pm$ is actually a tempered measure with respect to our chosen polynomial structure.

Now we are ready to define Gelfand triples, with respect to which we get a convenient theory for the group Fourier transform.

Definition 2.5 We define the following reduced Schwartz space

$$\mathcal{S}_*(G) := \{\varphi \in \mathcal{S}(G) \mid [(\lambda, x) \mapsto \varphi(\lambda z + x)] \in \mathcal{S}_*(\mathbb{R}) \hat{\otimes} \mathcal{S}(\omega)\} \tag{127}$$

for any choice $z \in \mathfrak{z} \setminus \{0\}$, equipped with the subspace topology in $\mathcal{S}(G)$, and the corresponding Gelfand triple

$$\mathcal{G}_*(G, \mu) := (\mathcal{S}_*(G), L^2(G, \mu), \mathcal{S}'_*(G)), \tag{128}$$

equipped with the real structure given by the pointwise complex conjugation. Furthermore, we define the Gelfand triple

$$\mathcal{G}(\mathbb{R}^\times; \pi) := \begin{pmatrix} \mathcal{S}(\mathbb{R}^\times; \pi) \\ L^2(\mathbb{R}^\times; \pi) \\ \mathcal{S}'(\mathbb{R}^\times; \pi) \end{pmatrix} := \mathcal{G}(\mathbb{R}^\times, \kappa\, |Pf(\ell)|\, |\lambda|^{Q-1}\, \mathrm{d}\lambda) \otimes \mathcal{G}_{\mathrm{op}}(\pi). \tag{129}$$

for each $\mathrm{SI}/\mathrm{Z}_{\mathbb{R}}(G) \ni \pi \sim \ell \in \omega^\circ$.

That $\mathcal{G}_*(G)$ is indeed a Gelfand triple can be seen by using Proposition 2.1. We use any linear isomorphism $\mathbb{R} \simeq \mathfrak{z}$ to define $\dot{\mathcal{B}}'_*(\mathfrak{z}; \mathcal{S}'(\omega))$, then we see, since $L^2(G, \mu) \subset \dot{\mathcal{B}}'_*(\mathfrak{z}; \mathcal{S}'(\omega))$ by Lemma 2.4, that the space $L^2(G, \mu)$ is embedded into $\mathcal{S}'_*(G) = \mathcal{S}'_*(\mathfrak{z}; \mathcal{S}'(\omega))$. This embedding is continuous, since the embedding $L^2(G, \mu) \hookrightarrow \mathcal{S}'(G)$ is continuous. Of course, the canonical map of $\mathcal{S}_*(G)$ into $L^2(G, \mu)$ is a continuous embedding as well. Now the Hahn–Banach theorem implies that both embeddings are also dense, for they are dual to each other.

To be more precise, if $\mathcal{S}_*(G)^\perp$ is the polar of $\mathcal{S}_*(G)$ in $L^2(G, \mu)$, then it is also the kernel of the dual map $L^2(G, \mu) \to \mathcal{S}'_*(G)$. But this map has a trivial kernel by Lemma 2.4. Hence $\mathcal{S}_*(G)^\perp = \{0\}$ and $\mathcal{S}_*(G)$ is dense in $L^2(G, \mu)$. Now denote by Y the image of $L^2(G, \mu)$ in $\mathcal{S}'_*(G)$. Since $\mathcal{S}_*(G)$ is reflexive, Y° can be identified with the kernel of the embedding $\mathcal{S}_*(G) \hookrightarrow L^2(G, \mu)$, which is trivial. Hence $Y \subset \mathcal{S}'_*(G)$ is dense as well.

Notice that $\mathcal{G}_*(G, \mu)$ does not depend either on the choice of some $\pi \in \mathrm{SI}/\mathrm{Z}_{\mathbb{R}}(G)$ or some $z \in \mathfrak{z}$. The Gelfand triple $\mathcal{G}(\mathbb{R}^\times; \pi)$ does depend on $\pi \in \mathrm{SI}/\mathrm{Z}_{\mathbb{R}}(G)$ but each different choice of π leads to an isomorphic Gelfand triple as the theorem below shows.

Theorem 2.2 *Let* $\mathrm{SI}/\mathrm{Z}_{\mathbb{R}}(G) \ni \pi \sim \ell \in \omega^\times$. *Let the* Fourier transform in π-picture, $\mathcal{F}_\pi$ *be defined by*

$$\mathcal{F}_\pi := \mathfrak{Op}_\pi \circ \wp_\ell \circ \mathcal{F}_{\mathfrak{g}}, \tag{130}$$

where $\mathfrak{Op}_\pi = P_+ \otimes \mathrm{op}_\pi + P_- \otimes \mathrm{op}_{\overline{\pi}}$ *and* $P_+ = 1 - P_-$ *is the projection of* $\mathcal{S}(\mathbb{R}^\times)$ *onto* $\mathcal{S}(\mathbb{R}^+)$ *along* $\mathcal{S}(\mathbb{R}^-)$*. Then* $\mathcal{F}_\pi$ *is a Gelfand triple isomorphism*

$$\mathcal{F}_\pi : \mathcal{G}_*(G) \to \mathcal{G}(\mathbb{R}^\times; \pi). \tag{131}$$

Proof The proof essentially writes itself by now and is a summary of previous statements. The Euclidean Fourier transform $\mathcal{F}_\mathfrak{g}$ is a Gelfand triple isomorphism between $\mathcal{G}_*(\mathfrak{g}, \mu)$ and $\mathcal{G}(\mathfrak{g}, \mu') = \mathcal{G}(\omega^\times, |Pf(\ell)|\, \mu_{\omega^\circ}) \otimes \mathcal{G}(\mathfrak{z}^\circ, \theta)$ by Lemma 2.2, where we choose the Haar measures μ_{ω° and θ, such that $\mu' = |Pf(\ell)|\, \mu_{\omega^\circ} \otimes \theta$ and μ_{ω° is induced by the Lebesgue measure $\mathrm{d}\lambda$ via the map $\mathbb{R} \ni \lambda \mapsto \lambda\ell \in \omega^\circ$.

By Lemma 2.6, the pull back $\wp_\ell$ is a Gelfand triple isomorphism between $\mathcal{G}(\mathfrak{g}^\times, \mu_{\mathfrak{g}'})$ and $\mathcal{G}(\mathbb{R}^\times, \kappa\, |Pf(\ell)|\, |\lambda|^{Q-1}\, \mathrm{d}\lambda) \otimes \mathcal{G}(\mathfrak{z}^\circ, \theta)$.

For the last step we just need to use that $\mathfrak{Op}_\pi = P_+ \otimes \mathrm{op}_\pi + P_- \otimes \mathrm{op}_{\overline{\pi}}$ is a Gelfand triple isomorphism between $\mathcal{G}(\mathbb{R}^\times, \kappa_0\, |Pf(\ell)|\, \mathrm{d}\lambda) \otimes \mathcal{G}(\mathfrak{z}^\circ, \mu_{\mathfrak{z}^\circ})$ and $\mathcal{G}(\mathbb{R}^\times; \pi)$ by Theorem 2.1 and Definition 2.4. □

Let us now discuss a few properties of $\mathcal{S}_*(G)$ and $\mathcal{S}(\mathbb{R}^\times; \pi)$. Their duals can be identified with quotient spaces, in particular

$$\begin{aligned} \mathcal{S}_*'(G) &\simeq \mathcal{S}'(G)/(\mathcal{P}(\mathfrak{z}) \otimes \mathcal{S}'(\omega)) \quad \text{and} \\ \mathcal{S}'(\mathbb{R}^\times; \pi) &\simeq \mathcal{S}(\mathbb{R}^\times)\, \hat{\otimes}\, \mathcal{L}(H_\pi^{-\infty}, H_\pi^\infty)/(\mathcal{E}_0'(\mathbb{R}) \otimes \mathcal{L}(H_\pi^{-\infty}, H_\pi^\infty)), \end{aligned} \tag{132}$$

by Lemma 2.3 and Corollary 2.3. By employing Proposition 2.1, we can identify a large space of distributions on G resp. $\mathbb{R}$ that are embedded into $\mathcal{S}_*'(G)$ resp. $\mathcal{S}'(\mathbb{R}^\times; \pi)$. I.e. if we define $\dot{\mathcal{B}}'(\mathfrak{z}; \mathcal{S}'(\omega))$ by using any isomorphism $\mathbb{R} \simeq \mathfrak{z}$, then

$$\dot{\mathcal{B}}'(\mathfrak{z}; \mathcal{S}'(\omega)) \hookrightarrow \mathcal{S}_*'(G) \quad \text{and} \quad \widetilde{\mathcal{B}}'(\mathbb{R}; \mathcal{L}(H_\pi^\infty, H_\pi^{-\infty})) \hookrightarrow \mathcal{S}'(\mathbb{R}^\times; \pi). \tag{133}$$

We may, for example, identify $L^p(G, \mu)$, for $p \in [1, \infty)$, and also $\mathcal{S}(G)$ as a subspaces of $\dot{\mathcal{B}}'(\mathfrak{z}; \mathcal{S}'(\omega))$ and the Bochner–Lebesgue spaces $L^p(\mathbb{R}, \mathrm{d}\lambda; \mathcal{L}(H_\pi))$, for $p \in (1, \infty]$, and also $\mathcal{S}(\mathbb{R}; \mathcal{L}(H_\pi^{-\infty}, H_\pi^\infty))$ as subspaces of $\widetilde{\mathcal{B}}'(\mathbb{R}; \mathcal{L}(H_\pi^\infty, H_\pi^{-\infty}))$.

The definition of $\mathcal{S}(\mathbb{R}^\times; \pi)$ and $\mathcal{S}'(\mathbb{R}^\times; \pi)$ enables us to define a multiplication with a large class of smooth functions via Theorem 1.2.

Proposition 2.4 *For any* $\pi \in \mathrm{SI/Z}(G)$*, the multiplications*

$$\begin{aligned} \mathcal{O}_\mathrm{M}(\mathbb{R}^\times; \mathcal{L}(H_\pi^\infty)) \times \mathcal{S}(\mathbb{R}^\times; \pi) &: (f, \varphi) \mapsto f\,\varphi \\ \mathcal{O}_\mathrm{M}(\mathbb{R}^\times; \mathcal{L}(H_\pi^{-\infty})) \times \mathcal{S}(\mathbb{R}^\times; \pi) &: (f, \varphi) \mapsto \varphi\, f, \end{aligned} \tag{134}$$

defined pointwise by composition of operators in $\mathcal{L}(H_\pi^{-\infty}, H_\pi^\infty)$*,* $\mathcal{L}(H_\pi^\infty)$ *and* $\mathcal{L}(H_\pi^{-\infty})$*, are hypocontinuous bilinear maps.*

Proof We just need show that we may apply Theorem 1.2. The compositions of operators

$$\begin{aligned} &\mathcal{L}(H_\pi^\infty) \times \mathcal{L}(H_\pi^{-\infty}, H_\pi^\infty) \colon (A, B) \mapsto A\,B \\ &\mathcal{L}(H_\pi^{-\infty}) \times \mathcal{L}(H_\pi^{-\infty}, H_\pi^\infty) \colon (A, B) \mapsto B\,A \end{aligned} \tag{135}$$

are hypocontinuous, since separately continuous maps on barrelled spaces are hypocontinuous. Since $H_\pi^\infty \simeq \mathcal{S}(\mathbb{R}^n)$ and Theorem 1.2, the above operator spaces are barrelled by [14, Corollaire 2 on page 128 of chapter 2] and [11, Corollary to 8.4 in chapter 2]. Also, the multiplication of slowly increasing functions and Schwartz functions is hypocontinuous. This follows directly from the definition and comments on pages 243 and 244 of [17]. Now we just need to remind ourselves, that $\mathcal{S}(\mathbb{R}^\times; \pi)$ is a tensor product of nuclear Fréchet spaces. □

Now, we will prove the analogous result for the multiplication with the operator valued tempered distributions $\mathcal{S}'(\mathbb{R}^\times; \pi)$. As we used in the proof above, H_π^∞ is reflexive for any $\pi \in \mathrm{SI}/\mathrm{Z}(G)$. Thus, by using the transpose, we get the two isomorphisms of topological vector spaces

$$\mathcal{L}(H_\pi^\infty) \ni A \mapsto A^{\mathrm{t}} \in \mathcal{L}(H_\pi^{-\infty}) \quad \text{and} \quad \mathcal{L}(H_\pi^{-\infty}) \ni B \mapsto B^{\mathrm{t}} \in \mathcal{L}(H_\pi^\infty). \tag{136}$$

Denote for f in $\mathcal{O}_{\mathrm{M}}(\mathbb{R}^\times; \mathcal{L}(H_\pi^\infty))$ or in $\mathcal{O}_{\mathrm{M}}(\mathbb{R}^\times; \mathcal{L}(H_\pi^{-\infty}))$ the operator valued function $f^{\mathrm{t}}(\lambda) := f(\lambda)^{\mathrm{t}}$. Then we may define multiplications on $\mathcal{S}'(\mathbb{R}^\times; \pi)$ by

$$(f\,\phi)(\varphi) := \phi(f^{\mathrm{t}}\,\varphi) \quad \text{and} \quad (\phi\,g)(\varphi) := \phi(\varphi\,g^{\mathrm{t}}), \tag{137}$$

for all $\phi \in \mathcal{S}'(\mathbb{R}^\times; \pi)$ and $\varphi \in \mathcal{S}(\mathbb{R}^\times; \pi)$, if we choose $f \in \mathcal{O}_{\mathrm{M}}(\mathbb{R}^\times; \mathcal{L}(H_\pi^{-\infty}))$ and $g \in \mathcal{O}_{\mathrm{M}}(\mathbb{R}^\times; \mathcal{L}(H_\pi^\infty))$. We get the following corollary.

Corollary 2.4 *For any $\pi \in \mathrm{SI}/\mathrm{Z}(G)$, the multiplications*

$$\begin{aligned} &\mathcal{O}_{\mathrm{M}}(\mathbb{R}^\times; \mathcal{L}(H_\pi^\infty)) \times \mathcal{S}'(\mathbb{R}^\times; \pi) \colon (f, \phi) \mapsto f\,\phi, \\ &\mathcal{O}_{\mathrm{M}}(\mathbb{R}^\times; \mathcal{L}(H_\pi^{-\infty})) \times \mathcal{S}'(\mathbb{R}^\times; \pi) \colon (f, \phi) \mapsto \phi\,f \end{aligned} \tag{138}$$

are hypocontinuous.

Proof This follows directly from the definition of the multiplication and the fact that the dual pairing is hypocontinuous. Equivalently, we could also directly employ Theorem 1.2. □

Let us now relate the Fourier transform in π picture with the group Fourier transform. Denote by j_π the map $j_\pi(\sigma) := [\mathbb{R}^\times \ni \lambda \mapsto \sigma(\pi_\lambda) \in \mathcal{L}(\mathcal{H}_\pi^{-\infty}, \mathcal{H}_\pi^\infty)]$, defined on $\mathcal{S}(\widehat{G})$.

Due to Proposition 2.2, we know that j_π has a unitary extension from $L^2(\widehat{G}_{\text{gen}}, \widehat{\mu})$ onto $L^2(\mathbb{R}^\times; \pi)$. Because the Plancherel measure $\widehat{\mu}$ is concentrated on $\widehat{G}_{\text{gen}}$ [9, Theorem 4.3.16], we can see j_π as a map defined on $L^2(\widehat{G}, \widehat{\mu})$. Proposition 2.3 implies the L^2-diagram below.

$$\begin{array}{ccc} L^2(G,\mu) & \xrightarrow[\simeq]{\mathcal{F}_G} & L^2(\widehat{G},\widehat{\mu}) \\ & \underset{\mathcal{F}_\pi}{\overset{\simeq}{\searrow}} & \simeq\big\downarrow j_\pi \\ & & L^2(\mathbb{R}^\times;\pi) \end{array} \qquad \begin{array}{ccc} \mathcal{S}(G) & \xrightarrow[\simeq]{\mathcal{F}_G} & \mathcal{S}(\widehat{G}) \\ j_*\big\uparrow\subset & & \subset\big\uparrow j_0 \\ \mathcal{S}_*(G) & \xrightarrow[\simeq]{\mathcal{F}_G} & \mathcal{S}(\widehat{G}_{\text{gen}}) \\ & \underset{\mathcal{F}_\pi^{-1}}{\overset{\simeq}{\nwarrow}} & \simeq\big\uparrow j_\pi^{-1} \\ & & \mathcal{S}(\mathbb{R}^\times;\pi) \end{array} \qquad \begin{array}{ccc} \mathcal{S}'(G) & \xrightarrow[\simeq]{\mathcal{F}_G} & \mathcal{S}'(\widehat{G}) \\ j'_*\big\downarrow & & \big\downarrow j'_0 \\ \mathcal{S}'_*(G) & \xrightarrow[\simeq]{\mathcal{F}_G} & \mathcal{S}'(\widehat{G}_{\text{gen}}) \\ & \underset{\mathcal{F}_\pi}{\overset{\simeq}{\searrow}} & \simeq\big\downarrow (j'_\pi)^{-1} \\ & & \mathcal{S}'(\mathbb{R}^\times;\pi) \end{array}$$

For the Schwartz spaces and spaces of tempered distributions we get a very similar diagram. Denote also by $\mathcal{S}(\widehat{G}_{\text{gen}})$ the image of $\mathcal{S}(\mathbb{R}^\times; \pi)$ under j_π^{-1}. The commutative diagram for the L^2-spaces implies the commutative diagram for the Schwartz spaces above. Then, by duality, we get the commutative diagram for the tempered distributions. However, the group Fourier transformations $\mathcal{F}_G$ are defined by duality on $\mathcal{S}'(G)$ resp. $\mathcal{S}'_*(G)$ and are not the same map, even though we use the same symbol.

By Corollary 2.3 the map j'_* can be seen as the quotient map

$$\mathcal{S}'(G) \to \mathcal{S}'(G)/(\mathcal{P}(\mathfrak{z}) \otimes \mathcal{S}'(\omega)) \simeq \mathcal{S}'_*(G), \tag{139}$$

which is an open map. This also implies that j'_0 is surjective and open.

Since $w_\ell \colon \mathbb{R}^\times \times \mathfrak{z}^\circ \to \mathfrak{g}^\times$, $w_\ell(\lambda, \xi) = \delta_\lambda(\ell + \xi)$ is a tempered diffeomorphism, we can also see $\wp_\ell$ as an isomorphism between $\mathcal{O}_{\mathrm{M}}(\mathfrak{g}^\times)$ and $\mathcal{O}_{\mathrm{M}}(\mathbb{R}^\times \times \mathfrak{z}^\circ)$ resp. between $\mathcal{S}(\mathfrak{g}^\times)$ and $\mathcal{S}(\mathbb{R}^\times \times \mathfrak{z}^\circ)$. However, in order to examine the Fourier image of $\mathcal{S}(G)$, it is even better to consider mixed spaces. We equip $\omega^\times = \mathbb{R}^\times \cdot \ell$ with the polynomial structure transported from $\mathbb{R}^\times$. The space $\mathcal{O}_{\mathrm{M}}(\omega^\times) \hat{\otimes} \mathcal{S}(\mathfrak{z}^\circ)$ can be seen as a subspace of $\mathcal{O}_{\mathrm{M}}(\mathfrak{g}^\times)$. In this manner we define $\wp_\ell$ on $\mathcal{O}_{\mathrm{M}}(\omega^\times) \hat{\otimes} \mathcal{S}(\mathfrak{z}^\circ)$.

Lemma 2.7 *The Gelfand-Triple isomorphism $\wp_\ell$ restricts to an isomorphism*

$$\wp_\ell \colon \mathcal{O}_{\mathrm{M}}(\omega^\times) \hat{\otimes} \mathcal{S}(\mathfrak{z}^\circ) \to \mathcal{O}_{\mathrm{M}}(\mathbb{R}^\times) \hat{\otimes} \mathcal{S}(\mathfrak{z}^\circ). \tag{140}$$

Proof We identify $\omega^\times \simeq \mathbb{R}^\times$ and $\mathfrak{z}^\circ \simeq \mathbb{R}^{2n}$ and $\mathfrak{z}^\circ \simeq \mathbb{R}^{2n}$ via our basis of eigenvectors to the dilations. As usual, it is enough to consider the $\mathbb{R}^+$-part, since $\mathcal{O}_{\mathrm{M}}(\mathbb{R}^\times) = \mathcal{O}_{\mathrm{M}}(\mathbb{R}^+) \oplus \mathcal{O}_{\mathrm{M}}(\mathbb{R}^-)$. With these adjustments, we need to exchange $\wp_\ell$ by the map $\wp$, where

$$\wp g(\lambda, x) = g(\lambda^{\kappa_0}, (\lambda^{\kappa_j} x_j)_{j=1}^{2n}). \tag{141}$$

First of all, we realize that $\lambda \mapsto \lambda^{\kappa_0}$ is a tempered diffeomorphism. Hence $T \in \mathcal{L}(\mathcal{O}_{\mathrm{M}}(\mathbb{R}_+))$, where $T\psi(\lambda) := \psi(\lambda^{\kappa_0})$, is an isomorphism.

Now let us define linear isomorphisms $f_\lambda(x) = (\lambda^{\kappa_j/\kappa_0} x_j)_{j=1}^{2n}$ on $\mathbb{R}^{2n}$. Then it is easy to see, that both $\lambda \mapsto f_\lambda$ and $\lambda \mapsto f_\lambda^{-1}$ define functions in $\mathcal{O}_{\mathrm{M}}(\mathbb{R}^+; \mathcal{L}(\mathbb{R}^{2n}))$ with values in $\mathrm{Gl}(\mathbb{R}^{2n})$. Now we denote by F_λ the corresponding operator $F_\lambda \varphi := \varphi \circ f_\lambda$ and set $F\colon \lambda \mapsto F_\lambda$ resp. $F^{-1}\colon \lambda \mapsto F_\lambda^{-1}$. A standard calculation shows that for any continuous seminorm p on $\mathcal{L}(\mathcal{S}(\mathbb{R}^{2n}))$ and any $k \in \mathbb{N}_0$, there is a polynomial q on $\mathcal{L}(\mathbb{R}^{2n})^{k+2}$, such that

$$p(\partial_\lambda^k F_\lambda) \le q(f_\lambda^{-1}, f_\lambda, \partial_\lambda f_\lambda, \dots, \partial_\lambda^k f_\lambda). \tag{142}$$

Of course, an analogous inequality is valid for F^{-1}. Hence, we may conclude

$$F, F^{-1} \in \mathcal{O}_{\mathrm{M}}(\mathbb{R}^+; \mathcal{L}(\mathcal{S}(\mathbb{R}^{2n}))). \tag{143}$$

Here, F^{-1} is indeed the inverse of F in the algebra $\mathcal{O}_{\mathrm{M}}(\mathbb{R}^+; \mathcal{L}(\mathcal{S}(\mathbb{R}^{2n})))$. Due to Theorem 1.2, we know that the multiplication

$$\begin{aligned} &\mathcal{O}_{\mathrm{M}}(\mathbb{R}^+; \mathcal{S}(\mathbb{R}^{2n})) \ni g \mapsto F\,g \in \mathcal{O}_{\mathrm{M}}(\mathbb{R}^+; \mathcal{S}(\mathbb{R}^{2n})),\\ &\text{with } (F\,g)(\lambda, x) = H_\lambda(g(\lambda,\cdot))(x), \end{aligned} \tag{144}$$

is continuous and in fact an isomorphism.

Because $\wp g = (T \otimes 1)(F\,g)$, we can conclude that $\wp$ is an isomorphism. □

Using the above lemma, we may now prove the following continuity property for the Fourier transform in π-picture on $\mathcal{S}(G)$.

Proposition 2.5 *The Fourier transform in π-picture restricts to a continuous map*

$$\mathcal{F}_\pi : \mathcal{S}(G) \to \mathcal{O}_{\mathrm{M}}(\mathbb{R}^\times) \,\hat{\otimes}\, \mathcal{L}(\mathcal{H}_\pi^{-\infty}, \mathcal{H}_\pi^{\infty}). \tag{145}$$

Proof This statement follows from the continuity of the maps

$$\mathcal{S}(G) \xrightarrow{\mathcal{F}_{\mathfrak{g}}} \mathcal{S}(\mathfrak{g}') \hookrightarrow \mathcal{O}_{\mathrm{M}}(\omega^\times) \,\hat{\otimes}\, \mathcal{S}(\mathfrak{z}^\circ) \xrightarrow{\wp_\ell} \mathcal{O}_{\mathrm{M}}(\mathbb{R}^\times) \,\hat{\otimes}\, \mathcal{S}(\mathfrak{z}^\circ), \tag{146}$$

in which we use the continuous inclusion $\mathcal{S}(\omega^\circ) \subset \mathcal{O}_{\mathrm{M}}(\omega^\times)$, and also from the continuity of $\mathfrak{Op}_\pi = P_+ \otimes \mathfrak{op}_\pi + P_- \otimes \mathfrak{op}_{\overline{\pi}}$ as map

$$\mathfrak{Op}_\pi : \mathcal{O}_{\mathrm{M}}(\mathbb{R}^\times) \,\hat{\otimes}\, \mathcal{S}(\mathfrak{z}^\circ) \to \mathcal{O}_{\mathrm{M}}(\mathbb{R}^\times) \,\hat{\otimes}\, \mathcal{L}(\mathcal{H}_\pi^{-\infty}, \mathcal{H}_\pi^{\infty}). \tag{147}$$

□

3 Gelfand Triples for the Kohn–Nirenberg Quantization

In [7] a pseudo-differential calculus resp. a Kohn–Nirenberg quantization for graded nilpotent Lie groups was developed. We will embed this definition into our context and derive an integral formulation for the Kohn–Nirenberg quantization for a general class of symbols. First, consider the map

$$\mathcal{T}\colon \mathcal{S}(G \times G) \to \mathcal{S}(G \times G),\ \mathcal{T}f(x, y) := f(x, xy^{-1}). \tag{148}$$

Then, it is easy to see that $\mathcal{T}$ extends to a Gelfand triple isomorphism

$$\mathcal{T}\colon \mathcal{G}(G, \mu) \otimes \mathcal{G}(G, \mu) \to \mathcal{G}(G, \mu) \otimes \mathcal{G}(G, \mu). \tag{149}$$

Denote by $\mathcal{K}$ the kernel map

$$\mathcal{K}\colon \mathcal{L}(\mathcal{G}(G, \mu), \mathcal{G}(G, \mu)) \to \mathcal{G}(G, \mu) \otimes \mathcal{G}(G, \mu) \tag{150}$$

from Lemma 1.2. We may define the Kohn–Nirenberg quantization as the Gelfand triple isomorphism

$$\mathrm{Op} := \mathcal{K}^{-1}\mathcal{T}^{-1}(1 \otimes \mathcal{F}_G^{-1})\colon \mathcal{G}(G, \mu) \otimes \mathcal{G}(\widehat{G}, \widehat{\mu}) \to \mathcal{L}(\mathcal{G}(G, \mu), \mathcal{G}(G, \mu)). \tag{151}$$

That means for $a \in L^2(G \times G, \mu \otimes \mu)$, we have $\mathrm{Op}(a) \in \mathcal{HS}(L^2(G))$ and

$$\begin{aligned}&(\mathrm{Op}(a)f, g)_{L^2(G,\mu)}\\ &\quad = \int_{\widehat{G}}\int_G \mathrm{Tr}[a(x, \pi)\,((1 \otimes \mathcal{F}_G\,\mathrm{inv})\mathcal{T}g \otimes \overline{f})(x, \pi)^*]\,\mathrm{d}\mu(x)\,\mathrm{d}\widehat{\mu}([\pi])\end{aligned} \tag{152}$$

for all $f, g \in L^2(G, \mu)$, where $\mathrm{inv}\, f(x) := f(-x)$ and $(\cdot, \cdot)_{L^2(G,\mu)}$ is the inner product in $L^2(G, \mu)$. We denote the right translation of functions f by $R_x f(y) := f(yx)$. Because

$$\begin{aligned}\mathrm{Tr}&[a(x, \pi)\,((1 \otimes \mathcal{F}_G\,\mathrm{inv})\mathcal{T}g \otimes \overline{f})(x, \pi)^*]\\ &= \mathrm{Tr}\left[a(x, \pi)\left(g(x)\,(\mathcal{F}_G R_x^{-1}\,\mathrm{inv}\,\overline{f})(\pi)\right)^*\right]\\ &= \overline{g(x)}\,\mathrm{Tr}\left[a(x, \pi)\,(\mathcal{F}_G\,\mathrm{inv}\,\overline{f})(\pi)^*\,\pi(x)\right]\\ &= \overline{g(x)}\,\mathrm{Tr}[a(x, \pi)\,\mathcal{F}_G f(\pi)\,\pi(x)]\end{aligned} \tag{153}$$

for almost all $(x, [\pi]) \in G \times \widehat{G}$, we may write the operator $\mathrm{Op}(a)$ as

$$\mathrm{Op}(a)\varphi = \int_{\widehat{G}} \mathrm{Tr}[\pi(\cdot)\, a(\cdot, \pi)\, \mathcal{F}_G f(\pi)]\, \mathrm{d}\widehat{\mu}([\pi]), \quad \text{for } f \in L^2(G), \tag{154}$$

where the integral converges in $L^2(G, \mu)$.

We will now define the Kohn–Nirenberg quantization in the context of the Gelfand triples $\mathcal{G}_*(G, \mu)$ and $\mathcal{G}(\mathbb{R}^\times; \pi)$ for $\pi \in \mathrm{SI}/\mathrm{Z}_{\mathbb{R}}(G)$. Subsequently, we will discuss an integral formula similar to the L^2-case above, but for a different class of symbols.

3.1 The Kohn–Nirenberg Quantization for Operators Defined on $\mathcal{S}_*(G)$

We still take G to be a homogeneous Lie group with $\dim \mathfrak{z} = 1$ and $\pi \in \mathrm{SI}/\mathrm{Z}_{\mathbb{R}}(G)$. We already saw that $\mathcal{F}_G$ is a Gelfand triple isomorphism from $\mathcal{G}(G, \mu)$ to $\mathcal{G}(\widehat{G})$, by the definition of $\mathcal{G}(\widehat{G}, \widehat{\mu})$. Now we want to use the corresponding statement for the $\mathcal{S}_*(G)$ test functions. Again we need to show that $\mathcal{T}$ is a Gelfand triple isomorphism in this context. Although the map $\mathcal{T}$ is not well defined on $\mathcal{G}_*(G, \mu) \otimes \mathcal{G}_*(G, \mu)$, it is well defined on $\mathcal{G}(G, \mu) \otimes \mathcal{G}_*(G, \mu)$.

Lemma 3.1 *The map $\mathcal{T} \restriction_{\mathcal{S}(G\times G)}$ extends to a Gelfand triple from $\mathcal{G}(G, \mu) \otimes \mathcal{G}_*(G, \mu)$ onto itself, which we will also call $\mathcal{T}$ by a slight abuse of notation.*

Proof Suppose $\varphi \in \mathcal{S}(G) \hat{\otimes} \mathcal{S}_*(G)$ and $q \in \mathcal{P}(G)$, then for all $x \in G$ and $y \in \omega$

$$\int_{\mathfrak{z}} q(z)\varphi(x, x(-z-y))\, \mathrm{d}\mu_{\mathfrak{z}}(Z) = \int_{\mathfrak{z}} q((-x)(-z-y))\varphi(x, z)\, \mathrm{d}\mu_{\mathfrak{z}}(z) = 0 \tag{155}$$

Because $[z \mapsto q((-x)(-z-y))] \in \mathcal{P}(\mathfrak{z})$. Hence $\mathcal{T}\varphi \in \mathcal{S}(G) \hat{\otimes} \mathcal{S}_*(G)$. Analogously we may prove that $\mathcal{T}^{-1}$ maps $\mathcal{S}(G) \hat{\otimes} \mathcal{S}_*(G)$ onto itself. Because $\mathcal{S}(G) \hat{\otimes} \mathcal{S}_*(G)$ carries the subspace topology in $\mathcal{S}(G) \hat{\otimes} \mathcal{S}(G)$, the continuity of $\mathcal{T}$ and $\mathcal{T}^{-1}$ on $\mathcal{S}(G) \hat{\otimes} \mathcal{S}_*(G)$ is evident. Since also

$$\int_{G\times G} \psi\, \mathcal{T}\varphi\, \mathrm{d}(\mu \otimes \mu) = \int_{G\times G} \varphi\, \mathcal{T}^{-1}\psi\, \mathrm{d}(\mu \otimes \mu), \tag{156}$$

for all $\varphi, \psi \in \mathcal{S}(G \times G)$, we may extend $\mathcal{T} \restriction_{\mathcal{S}(G\times G)}$ to a Gelfand triple isomorphism. □

Now a direct conclusion is the formulation of the Kohn–Nirenberg quantization as a Gelfand triple isomorphism that incorporates the new Gelfand triples $\mathcal{G}_*(G, \mu)$ and $\mathcal{G}(\mathbb{R}^\times; \pi)$.

Proposition 3.1 *The* Kohn–Nirenberg quantization in π-picture

$$\mathrm{Op}_\pi := \mathcal{K}^{-1}\mathcal{T}^{-1}(1 \otimes \mathcal{F}_\pi^{-1}) \colon \mathcal{G}(G,\mu) \otimes \mathcal{G}(\mathbb{R}^\times;\pi) \to \mathcal{L}(\mathcal{G}_*(G,\mu), \mathcal{G}(G,\mu)), \tag{157}$$

where $\mathcal{K}$ is the kernel map between $\mathcal{G}(G,\mu) \otimes \mathcal{G}_*(G,\mu)$ and $\mathcal{L}(\mathcal{G}_*(G,\mu), \mathcal{G}(G,\mu))$, is a Gelfand triple isomorphism.

As for the Fourier transformation in π-picture, we may relate Op_π to the original Kohn–Nirenberg quantization Op via the diagrams on page 85.

3.2 The Integral Formula

Representation in $\mathrm{SI}/\mathrm{Z}_\mathbb{R}(\pi)$ can also be seen as slowly increasing functions. This is integral to our approach and will be proven in the proposition following the next lemma.

Lemma 3.2 *Suppose E is a complete locally convex space and $f \in \mathcal{O}_\mathrm{M}(G; E)$ and let $F(\lambda, x) := f(\delta_\lambda x)$. Then $F \in \mathcal{O}_\mathrm{M}(\mathbb{R}^\pm \times G; E)$.*

Proof It is enough to show that for each continuous seminorm p on E, each $k \in \mathbb{N}_0$ each $P \in \mathrm{Diff}_\mathcal{P}(G)$ there is a polynomial $q \in \mathcal{P}(G)$ and $l > 0$, for which

$$p(\partial_\lambda^k P_x F(\lambda, x)) \leq (1 + |\lambda|^l + |\lambda|^{-l})q(x). \tag{158}$$

We realize that there are polynomial differential operators P_v, such that

$$\partial_\lambda^k P_x F(\lambda, x) = \sum_{v \in \mathbb{R}} \lambda^v (P_v f)(\delta_\lambda x), \tag{159}$$

as a finite linear combination. Since each $p(P_v f)$ is bounded by a polynomial $\widetilde{q}_v$, we may find polynomials q_v such that

$$p(\partial_\lambda^k P_x F(\lambda, x)) \leq \sum_{v \in \mathbb{R}} |\lambda|^v \widetilde{q}_v(\delta_\lambda x) = \sum_{v \in \mathbb{R}} |\lambda|^v q_v(x). \tag{160}$$

This concludes the proof. □

Proposition 3.2 *Suppose $\pi \in \mathrm{SI}/\mathrm{Z}_\mathbb{R}(G)$, then the operator valued function $(x, \lambda) \mapsto \pi_\lambda(x)$ is both in $\mathcal{O}_\mathrm{M}(\mathbb{R}^\times \times G; \mathcal{L}(\mathcal{H}_\pi^\infty))$ and in $\mathcal{O}_\mathrm{M}(\mathbb{R}^\times \times G; \mathcal{L}(\mathcal{H}_\pi^{-\infty}))$*

Proof By Lemma 3.2 it is enough to show, that $x \mapsto \pi(x)$ is slowly increasing. For this purpose we choose an equivalent representation, that is more easily understood. There is a representation $\sigma \sim \pi$ on $\mathcal{H}_\sigma = L^2(\mathbb{R}^n)$, such that $\mathcal{H}_\sigma^\infty = \mathcal{S}(\mathbb{R}^n)$ and

$$\sigma(x)f(t) = \mathrm{e}^{2\pi \mathrm{i}\xi(a(x,t))} f(x^{-1} \cdot t) \tag{161}$$

where ξ is a linear functional on a subalgebra $\mathfrak{m}$ of $\mathfrak{g}$, $a\colon G\times\mathbb{R}^n \to \mathfrak{m}$ is polynomial and $G\times\mathbb{R}^n \ni (x,t) \mapsto x\cdot t \in \mathbb{R}^n$ is a polynomial action of G on $\mathbb{R}^n$ by Pedersen [6] and [9, Corollary 4.1.2]. Because $(x,t) \mapsto x^{-1}\cdot t$ is polynomial, we may represent the action of G on $\mathbb{R}^n$ by a linear combination

$$x\cdot t = \sum_{j,k} s_{k,j}(x)\, u_{k,j}(t)\, e_j, \tag{162}$$

where $(e_j)_j$ is the standard basis on $\mathbb{R}^n$ and $s_{k,j}$, $u_{k,j}$ are polynomials. Thus, we also have

$$t_j\, \sigma(x) f(t) = \sum_k s_{k,j}(x)\, \sigma(x)(u_{k,j}\, f)(t). \tag{163}$$

For the same reason, there are polynomials $q_{j,k}, \widetilde{q}_{j,k}$ on G, $r_{j,k}, \widetilde{r}_{j,k}$ on $\mathbb{R}^n$ such that

$$\begin{aligned} \partial_{t_j} f(x^{-1}\cdot t) &= \sum_k \widetilde{q}_{j,k}(x)\, \widetilde{r}_{j,k}(t)\, (\partial_k f)(x^{-1}\cdot t) \\ &= \sum_k q_{j,k}(x)\, r_{j,k}(x^{-1}\cdot t)\, (\partial_k f)(x^{-1}\cdot t). \end{aligned} \tag{164}$$

Hence, for all $\alpha, \beta \in \mathbb{N}_0$ we find operators $A_k \in \mathcal{L}(\mathcal{S}(\mathbb{R}^n))$ and polynomials $v_k \in \mathcal{P}(G)$, such that

$$t^\beta \partial_t^\alpha \sigma(x) f(t) = \sum_k v_k(x)\, \sigma(x)(A_k f)(t), \tag{165}$$

as a finite linear combination.

The topology on $\mathcal{L}(\mathcal{S}(\mathbb{R}^n))$ is induced by the seminorms

$$p\colon A \mapsto \sup_{f\in B}\sup_{t\in\mathbb{R}^n} |t^\beta \partial_t^\alpha A f(t)|, \quad B \subset \mathcal{S}(\mathbb{R}^n) \text{ bounded}, \alpha, \beta \in \mathbb{N}_0^n. \tag{166}$$

Now if $L \in \mathfrak{u}(\mathfrak{g})_{\mathrm{L}}$ is any left invariant differential operator on G and p is a seminorm as above, we get

$$\begin{aligned} p(L_x\sigma(x)) &\le \sum_k v_k(x) \sup_{f\in B}\sup_{t\in\mathbb{R}^n} |\sigma(x)(A_k\sigma(L) f)(t)| \\ &= \sum_k v_k(x) \sup_{f\in B}\sup_{t\in\mathbb{R}^n} |(A_k\sigma(L) f)(t)|. \end{aligned} \tag{167}$$

The right-hand side of the above inequality is a sum of continuous seminorms times polynomials, since $\sigma(L) \in \mathcal{L}(\mathcal{S}(\mathbb{R}^n))$. Thus $x \mapsto \sigma(x)$ is slowly increasing. Due to $\pi \sim \sigma$ the map $x \mapsto \pi(x)$ is slowly increasing, too. Now $(x,\lambda) \mapsto \pi_\lambda$ is slowly increasing with values in $\mathcal{L}(H_\pi^\infty)$ due to Lemma 3.2. We finish the proof by remarking that $\mathcal{L}(H_\pi^\infty)$ and $\mathcal{L}(H_\pi^{-\infty})$ are isomorphic by the transposition and $\pi_\lambda(x)^{\mathrm{t}} = C_\pi \pi_\lambda(-x) C_\pi$. This implies that π is also slowly increasing with values in $\mathcal{L}(H_\pi^{-\infty})$. □

With the help of the above proposition, we want to write the inverse Fourier transform as an integral, which converges in $\mathcal{O}_{\mathrm{M}}(G)$. For this purpose, we need to explain a small fact about the dual space $\mathcal{O}'_{\mathrm{M}}(G)$. Denote by $\partial_1, \partial_2, \ldots$ the directional derivative to any basis $v_1, v_2, \ldots$ of $\mathfrak{g}$. Each continuous linear functional on $\mathcal{O}_{\mathrm{M}}(\mathfrak{g})$ can be represented by the set

$$\mathcal{O}'_{\mathrm{M}}(G) = \operatorname{span}_{\mathbb{C}}\{\partial^\alpha f \mid \alpha \in \mathbb{N}_0^{\dim(G)} \text{ and } f \in C(G) \text{ is rapidly decreasing}\}, \tag{168}$$

where we used the standard multi-index notation, see [14, page 130 of chapter 2], if we use the dual pairing

$$\langle \partial^\alpha f, g \rangle := \int_G f\, (-\partial)^\alpha g \, \mathrm{d}\mu. \tag{169}$$

Here we say $f \colon G \to \mathbb{C}$ is rapidly decreasing, iff qf is a bounded function for any $q \in \mathcal{P}(G)$. The differential operators ∂^α, $\alpha \in \mathbb{N}_0^{\dim(G)}$ span the $\mathcal{P}(G)$-module $\mathrm{Diff}_{\mathcal{P}}(G)$. Since the multiplication of Schwartz functions with polynomials is continuous, we may exchange ∂^α with arbitrary $P \in \mathrm{Diff}_{\mathcal{P}}(G)$ in the pairing above. By [9, Lemma A.2.2] the $\mathcal{P}(G)$-span of the left invariant differential operators $\mathfrak{u}(\mathfrak{g}_{\mathrm{L}})$ is equal to $\mathrm{Diff}_{\mathcal{P}}(G)$. Now let $w^1, w^2, \ldots$ be the dual basis to $v_1, v_2, \ldots$ and let $X_1, X_2, \ldots$ be the left invariant vector fields associated to $v_1, v_2, \ldots$. A quick calculation shows that for all $\phi \in \mathcal{S}'(G)$ and all j, k there exists a polynomial $q \in \mathcal{P}(G)$ with

$$w^j X_k \phi = q\, \phi + X_k (w^j \phi). \tag{170}$$

Of course, the set of rapidly decreasing continuous functions is invariant under the multiplication with polynomials. In conclusion, we may represent the dual to $\mathcal{O}_{\mathrm{M}}(G)$ by

$$\mathcal{O}'_{\mathrm{M}}(G) = \operatorname{span}_{\mathbb{C}}\{Pf \mid P \in \mathfrak{u}(\mathfrak{g}_{\mathrm{L}}) \text{ and } f \in C(G) \text{ is rapidly decreasing}\}. \tag{171}$$

Lemma 3.3 *If $\varphi \in \mathcal{S}(\mathfrak{g})$ and $\omega^\times \ni \ell \sim \pi \in \mathrm{SI/Z}(G)$, then the integral*

$$\varphi = \int_{\mathbb{R}^\times} \mathrm{Tr}[\pi_\lambda \mathcal{F}_\pi \varphi(\lambda)] \, \mathrm{d}\lambda_\pi \tag{172}$$

exists in $\mathcal{O}_{\mathrm{M}}(G)$, where $\mathrm{d}\lambda_\pi := \kappa \, |Pf(\ell)| \, |\lambda|^{Q-1} \, \mathrm{d}\lambda$.

Proof Let $f\colon G \to \mathbb{C}$ be continuous and rapidly decreasing, let $P \in \mathfrak{u}(\mathfrak{g}_{\mathrm{L}})$ and let $\varphi \in \mathcal{S}(G)$. Then, f and $P^{\mathrm{t}}\varphi$ are L^2 functions and we may apply Plancherel for $\mathcal{F}_\pi$. Hence

$$\langle P^{\mathrm{t}}\overline{f}, \varphi\rangle = \int_G \overline{f}\, P\varphi \,\mathrm{d}\mu = \int_{\mathbb{R}^\times} \mathrm{Tr}[\widehat{f}(\lambda)^* \mathcal{F}_\pi(P\varphi)(\lambda)]\,\mathrm{d}\lambda_\pi, \tag{173}$$

in which we used the shorthand $\widehat{g}(\lambda) = \mathcal{F}_\pi g(\lambda)$ for functions g. Since $f \in L^1(G,\mu)$, we know that the integral that evaluates the Fourier transform in π-picture converges in $\mathcal{L}(H_\pi)$ with respect to the weak operator topology. That means for each pair $u, v \in H_\pi$ we have

$$(\widehat{f}(\lambda)^* u, v)_{H_\pi} = \int_G \overline{f(x)}\,(\pi_\lambda(x)u, v)_{H_\pi}\,\mathrm{d}\mu(x). \tag{174}$$

Because $P\varphi \in \mathcal{S}(G)$, we have $\mathcal{F}_\pi(P\varphi)(\lambda) = \pi_\lambda(P)\,\widehat{\varphi}(\lambda) \in \mathcal{L}(H_\pi^{-\infty}, H_\pi^{\infty})$, which is a nuclear operator for each $\lambda \in \mathbb{R}^\times$. Hence for each orthonormal basis $(e_k)_{k\in\mathbb{N}} \subset H_\pi$

$$\int_G \sum_{k\in\mathbb{N}} |\overline{f(x)}\,(\pi_\lambda(x)\,\pi_\lambda(P)\,\widehat{\varphi}(\lambda)e_k, e_k)_{H_\pi}|\,\mathrm{d}\mu(x)$$
$$\le \|f\|_{L^1(G,\mu)}\,\|\pi_\lambda(P)\,\widehat{\varphi}(\lambda)\|_{\mathcal{N}(H_\pi)} < \infty, \tag{175}$$

where $\|\cdot\|_{\mathcal{N}(H_\pi)}$ is the trace-norm on the space of nuclear operators on H_π. Using Fubini with respect to the counting measure and μ results in

$$\mathrm{Tr}[\widehat{f}(\lambda)^* \mathcal{F}_\pi(P\varphi)(\lambda)] = \int_G \overline{f(x)}\,\mathrm{Tr}[\pi_\lambda(x)\,\pi_\lambda(P)\,\widehat{\varphi}(\lambda)]\,\mathrm{d}\mu(x), \tag{176}$$

since $f \in L^1(G,\mu)$. Naturally, we have $\pi_\lambda(x)\,\pi_\lambda(P) = P_x\pi_\lambda(x)$. By the embedding of $\mathcal{L}(H_\pi^{-\infty}, H_\pi^{\infty})$ into the nuclear operators $\mathcal{N}(H_\pi)$, we may see Tr as a continuous functional on $\mathcal{L}(H_\pi^{-\infty}, H_\pi^{\infty})$. Because the operator valued function $\pi_\lambda\,\widehat{\varphi}(\lambda)$ is a slowly increasing map from G to $\mathcal{L}(H_\pi^{-\infty}, H_\pi^{\infty})$, we get

$$\begin{aligned}\mathrm{Tr}[\pi_\lambda(x)\,\pi_\lambda(P)\,\widehat{\varphi}(\lambda)] &= P_x\,\mathrm{Tr}[\pi_\lambda(x)\,\widehat{\varphi}(\lambda)],\\ \mathrm{Tr}[\pi_\lambda\,\widehat{\varphi}(\lambda)] &\in \mathcal{O}_{\mathrm{M}}(G).\end{aligned} \tag{177}$$

Finally we get

$$\begin{aligned}\langle P^{\mathrm{t}}\overline{f}, \varphi\rangle &= \int_{\mathbb{R}^\times}\int_G \overline{f(x)}\, P_x \operatorname{Tr}[\pi_\lambda(x)\,\widehat{\varphi}(\lambda)]\,\mathrm{d}\mu(x)\mathrm{d}\lambda_\pi \\ &= \int_{\mathbb{R}^\times}\langle P^{\mathrm{t}}\overline{f}, \operatorname{Tr}[\pi_\lambda(\cdot)\,\widehat{\varphi}(\lambda)]\rangle\,\mathrm{d}\lambda_\pi,\end{aligned} \tag{178}$$

which concludes the proof. □

Let us write $\rho(x,\lambda) := \pi_\lambda(x)$ and $\rho^*(\lambda, x) := \pi_\lambda(-x)$ for some $\pi \in \mathrm{SI/Z}(G)$. With Lemma 1.1, we already proved the continuity of the map

$$\mathcal{L}(\mathcal{O}_\mathrm{M}(G)) \to \mathcal{L}(\mathcal{O}_\mathrm{M}(G\times\mathbb{R}^\times; \mathcal{L}(H_\pi^\infty))),\ A \mapsto A\otimes 1. \tag{179}$$

Of course the evaluation map

$$\mathcal{L}(\mathcal{O}_\mathrm{M}(G\times\mathbb{R}^\times; \mathcal{L}(H_\pi^\infty))) \to \mathcal{O}_\mathrm{M}(G\times\mathbb{R}^\times; \mathcal{L}(H_\pi^\infty)),\ \Phi \mapsto \Phi(\rho) \tag{180}$$

is continuous, as well. Finally, since the multiplication in $\mathcal{O}_\mathrm{M}(G\times\mathbb{R}^\times)$ is continuous [17, page 248] and because of Theorem 1.2, the map S defined by

$$S\colon \mathcal{L}(\mathcal{O}_\mathrm{M}(G)) \to \mathcal{O}_\mathrm{M}(G\times\mathbb{R}^\times; \mathcal{L}(H_\pi^\infty)),\ A \mapsto \rho^*\cdot(A\otimes 1)(\rho) \tag{181}$$

is continuous. Now this map looks exactly like the inverse Kohn–Nirenberg quantization on compact Lie groups H from [8]. Namely, for any $B \in \mathcal{L}(\mathcal{D}(H))$ the unique Kohn–Nirenberg symbol b with $B = \mathrm{Op}(b)$, evaluated at the irreducible unitary representation ξ, is given by $\xi^*\cdot(A\otimes 1)(\xi) \in \mathcal{D}(H; \mathcal{L}(H_\xi))$.

Lemma 3.4 *The embedding $\mathcal{S}_*(G) \hookrightarrow \mathcal{O}_\mathrm{M}(G)$ is continuous and has dense range.*

Proof The multiplication on $\mathcal{S}(G)$ is a continuous bilinear map. This implies the continuity of the canonical embedding $\tau\colon \mathcal{S}_*(G) \hookrightarrow \mathcal{O}_\mathrm{M}(G)$, since $\mathcal{S}_*(G)$ carries the subspace topology in $\mathcal{S}(G)$. Now consider the dual map

$$\tau'\colon \mathcal{O}_\mathrm{M}'(G) \to \mathcal{S}_*'(G), \quad \text{where} \quad \langle\tau'\phi, \varphi\rangle = \langle\phi,\varphi\rangle, \quad \text{for all } \varphi \in \mathcal{S}_*(G). \tag{182}$$

That this is indeed an embedding, can be seen from Proposition 2.1 and the representation (168) of the dual space $\mathcal{O}_\mathrm{M}'(G)$. By the Hahn-Banach theorem, the operator τ has dense image. □

In the Lemma above we saw that $\mathcal{S}_*(G) \hookrightarrow \mathcal{O}_\mathrm{M}(G)$ has dense range. Naturally we also have $\mathcal{O}_\mathrm{M}(G) \hookrightarrow \mathcal{S}'(G)$ and $\mathcal{O}_\mathrm{M}(\mathbb{R}^\times) \hookrightarrow \mathcal{S}'(\mathbb{R}^\times)$, thus we get embeddings

$$\mathcal{L}(\mathcal{O}_\mathrm{M}(G)) \hookrightarrow \mathcal{L}(\mathcal{S}_*(G), \mathcal{S}'(G)), \tag{183}$$

$$\mathcal{O}_\mathrm{M}(G\times\mathbb{R}^\times; \mathcal{L}(H_\pi^\infty)) \hookrightarrow \mathcal{S}'(G; \mathcal{S}'(\mathbb{R}^\times; \pi)). \tag{184}$$

Notice that we can exchange $\mathcal{L}(H_\pi^\infty)$ with $\mathcal{L}(H_\pi^{-\infty})$, in the paragraph above. By using the embeddings above, we will see that the map S does indeed reproduce the Kohn–Nirenberg symbol. We can even go one step further. Of course for $A \in \mathcal{L}(\mathcal{O}_\mathrm{M}(G), \mathcal{S}(G))$, we can still define the map S, since $\mathcal{S}(G) \hookrightarrow \mathcal{O}_\mathrm{M}(G)$. However, we are lacking tools to check whether $S(A) \in \mathcal{S}(G) \hat{\otimes} \mathcal{O}_\mathrm{M}(\mathbb{R}^\times; \mathcal{L}(H_\pi^\infty))$ or not, since we cannot apply Theorem 1.2. We run into the same problem if we try to define S for operators $A \in \mathcal{L}(\mathcal{O}_\mathrm{M}(G); \mathcal{S}'(G))$.

Before we prove that the definition of S gives us Kohn–Nirenberg symbols, we need two final lemmata.

Lemma 3.5 *Suppose* $a \in \mathcal{S}'(G) \hat{\otimes} \mathcal{S}'(\mathbb{R}^\times; \pi)$, *then*

$$\rho \cdot a = (1 \otimes \mathcal{F}_\pi \operatorname{inv})\mathcal{T}^{-1}(1 \otimes \mathcal{F}_\pi^{-1})a, \tag{185}$$

where $\operatorname{inv} f(x) = f(-x)$, $f \in \mathcal{S}(G)$, *continued to distributions.*

Proof First we take $a \in \mathcal{S}(G) \hat{\otimes} \mathcal{S}(\mathbb{R}^\times; \pi)$. Then we just have

$$(1 \otimes \mathcal{F}_\pi^{-1})(\rho \cdot a)(x, y) = (1 \otimes \mathcal{F}_\pi^{-1})a(x, yx) = (1 \otimes \operatorname{inv})\mathcal{T}^{-1}(1 \otimes \mathcal{F}_G^{-1})a(x, y), \tag{186}$$

by the integral formula for the inverse Fourier transform from Lemma 3.3. Now the rest simply follows due to the continuity of the involved maps. □

Lemma 3.6 *Define* $\chi_x(\xi) := \mathrm{e}^{2\pi \mathrm{i}\xi(x)}$ *for* $x \in \mathfrak{g}$ *and* $\xi \in \mathfrak{g}'$. *Then*

$$\mathfrak{Op}_\pi \wp_\ell(\chi_x) = \pi_\lambda(x) \tag{187}$$

for any $\lambda \in \mathbb{R}^\times$ *and* $\mathrm{SI}/\mathrm{Z}_\mathbb{R}(G) \ni \pi \sim \ell \in \omega^\times$.

Proof Let $\psi \in \mathcal{D}(\mathfrak{z}^\circ)$, such that $\psi \equiv 1$ on some neighbourhood of zero. If we define $\psi_k(x) := \psi(x/k)$ for $k \in \mathbb{N}$, then $\psi_k \chi_x \to \chi_x$ for $k \to \infty$ in $\mathcal{S}'(\mathfrak{z}^\circ)$. Due to the continuity of $\mathfrak{op}_\pi$, we may deduce for $\lambda > 0$

$$\begin{aligned}\mathfrak{Op}_\pi \wp_\ell(\chi_x) &= \lim_{k\to\infty} \mathrm{e}^{2\pi \mathrm{i}\ell(\delta_\lambda x)} \mathfrak{op}_\pi(\psi_k \cdot \chi_{\delta_\lambda x} \restriction_{\mathfrak{z}^\circ}) \\ &= \lim_{k\to\infty} \mathrm{e}^{2\pi \mathrm{i}\ell(\delta_\lambda x)} \int_\omega \pi(y)\widehat{\psi}_k(y - \delta_\lambda \widetilde{x})\, \mathrm{d}\nu(y),\end{aligned} \tag{188}$$

in which $\widehat{\psi}_k \in \mathcal{S}(\omega)$ is the Euclidean Fourier transform of ψ_k and $\widetilde{x}$ is the projection of x onto ω along $\mathfrak{z}$. If we consider the functions $\widehat{\psi}_k(\cdot - \delta_\lambda \widetilde{x})$ as distributions

$$\mathcal{S}(G) \ni \varphi \mapsto \int_\omega \widehat{\psi}_k(y - \delta_\lambda \widetilde{x})\varphi(y)\, \mathrm{d}\nu(y) \tag{189}$$

in $\mathcal{S}'(G)$, then the sequence of functions $\widehat{\psi}_k(\cdot - \delta_\lambda \widetilde{x})$ converges to the Dirac distribution supported on $\delta_\lambda x$ in $\mathcal{S}'(G)$. By Proposition 2.1 and the continuity of

$\mathcal{F}_\pi$, we arrive at

$$\mathfrak{Op}_\pi \wp_\ell(\chi_x) = \mathrm{e}^{2\pi \mathrm{i}\ell(\delta_\lambda x)}\pi(\delta_\lambda \widetilde{x}) = \pi_\lambda(x). \tag{190}$$

For $\lambda < 0$ the calculation is analogous, we merely need to exchange π with $\overline{\pi}$. □

Theorem 3.1 *For any $A \in \mathcal{L}(\mathcal{O}_{\mathrm{M}}(G), E)$, $E \in \{\mathcal{S}(G), \mathcal{O}_{\mathrm{M}}(G)\}$, the equality $a := S(A) = \mathrm{Op}_\pi^{-1}(A)$ is valid. Furthermore,*

$$A\varphi = \int_{\mathbb{R}^\times} \mathrm{Tr}[\pi_\lambda\, a(\cdot,\lambda)\, \mathcal{F}_\pi\varphi(\lambda)]\, \mathrm{d}\lambda_\pi, \quad \textit{for} \quad \varphi \in \mathcal{S}(G), \tag{191}$$

where the integral exists in E.

Proof First we will prove the integral formula for $A \in \mathcal{L}(\mathcal{O}_{\mathrm{M}}(G), E)$. From Lemma 3.3 we know, that for $\varphi \in \mathcal{S}(G)$

$$A\varphi = A\int_G \mathrm{Tr}[\pi_\lambda\, \widehat{\varphi}(\lambda)]\, \mathrm{d}\lambda_\pi = \int_G A\big(\mathrm{Tr}[\pi_\lambda\, \widehat{\varphi}(\lambda)]\big)\, \mathrm{d}\lambda_\pi, \tag{192}$$

where the integral converges in E and where we used the shorthand $\mathcal{F}_\pi\varphi(\lambda) = \widehat{\varphi}(\lambda)$. Due to Propositions 2.5 and 3.2, we know that

$$(\pi_\lambda\, \widehat{\varphi}(\lambda)) \in \mathcal{O}_{\mathrm{M}}(G)\, \hat{\otimes}\, \mathcal{L}(H_\pi^{-\infty}, H_\pi^\infty). \tag{193}$$

The trace operator Tr, restricted from the nuclear operators on H_π, is a continuous functional on $\mathcal{L}(H_\pi^{-\infty}, H_\pi^\infty)$, so we may use the tensor product structure of the above expression to get

$$A\big(\mathrm{Tr}[\pi_\lambda\, \widehat{\varphi}(\lambda)]\big) = (A \otimes \mathrm{Tr})\big(\pi_\lambda\, \widehat{\varphi}(\lambda)\big) = (1 \otimes \mathrm{Tr})(A \otimes 1)\big(\pi_\lambda\, \widehat{\varphi}(\lambda)\big), \tag{194}$$

for each $\lambda \in \mathbb{R}^\times$. Furthermore,

$$(A \otimes 1)(\pi_\lambda\, \widehat{\varphi}(\lambda)) = \pi_\lambda \cdot \pi_\lambda^* \cdot (A \otimes 1)(\pi_\lambda) \cdot \widehat{\varphi}(\lambda), \tag{195}$$

in which the multiplication is defined pointwise by the multiplication in $\mathcal{L}(H_\pi^\infty)$. Hence, we can represent $A\varphi$ by the integral

$$A\varphi = \int_{\mathbb{R}^\times} \mathrm{Tr}[\pi_\lambda\, a(\cdot,\lambda)\, \widehat{\varphi}(\lambda)]\, \mathrm{d}\lambda_\pi, \tag{196}$$

with $a := S(A)$.

Now it is left to check that indeed $A = \mathrm{Op}_\pi(a)$. First of all, due to Lemma 3.5

$$\mathcal{T}^{-1}(1 \otimes \mathcal{F}_\pi^{-1})a = (1 \otimes \mathrm{inv}\,\mathcal{F}_\pi^{-1})(\rho \cdot a). \tag{197}$$

We define the function $\chi(x,\xi) := \mathrm{e}^{2\pi \mathrm{i}\xi(x)}$ for $\xi \in \mathfrak{g}'$, $x \in \mathfrak{g}$, then $\chi \in \mathcal{O}_{\mathrm{M}}(\mathfrak{g}\times\mathfrak{g}^{\times})$. Because $(1 \otimes \mathfrak{Op}_\pi \wp\ell)\chi(x,\lambda) = \pi_\lambda(x) = \rho(x,\lambda)$, due to Lemma 3.6, and $\mathcal{F}_\pi = \mathfrak{Op}_\pi \wp\ell \mathcal{F}_{\mathfrak{g}}$, we know that

$$(1 \otimes \operatorname{inv} \mathcal{F}_\pi^{-1})(A \otimes 1)(\rho) = (A \otimes \operatorname{inv} \mathcal{F}_{\mathfrak{g}}^{-1})(\chi) = (A \otimes \mathcal{F}_{\mathfrak{g}'})(\chi). \tag{198}$$

We choose arbitrary $\varphi \in \mathcal{S}(\mathfrak{g})$ and $\psi \in \mathcal{S}_*(G)$. The integral

$$\varphi = \int_{\mathfrak{g}'} \chi(\cdot,\xi)\mathcal{F}_{\mathfrak{g}}\varphi(\xi)\,\mathrm{d}\mu'(\xi) \tag{199}$$

converges in $\mathcal{O}_{\mathrm{M}}(\mathfrak{g})$. Hence

$$\begin{aligned} \langle (A\otimes \mathcal{F}_{\mathfrak{g}'})\chi, \psi\otimes\varphi\rangle &= \langle (A\otimes 1)\chi, \psi \otimes \mathcal{F}_{\mathfrak{g}}\varphi\rangle \\ &= \int_{\mathfrak{g}'} \langle A(\chi(\cdot,\xi)), \psi\rangle \mathcal{F}_{\mathfrak{g}}\varphi(\xi)\,\mathrm{d}\mu'(\xi) \\ &= \langle A\varphi, \psi\rangle. \end{aligned} \tag{200}$$

Combining the calculations above implies

$$\mathcal{K}A = \mathcal{T}^{-1}(1\otimes\mathcal{F}_\pi^{-1})a, \tag{201}$$

for the kernel map $\mathcal{K}$ for $\mathcal{G}(G)\otimes\mathcal{G}_*(G)$. I.e. $\mathrm{Op}_\pi^{-1}(A) = S(A)$. □

Acknowledgments Jonas Brinker was supported by ISAAC as a young scientist for his participation at the 12th ISAAC Congress in Aveiro.

We thank David Rottensteiner for discussing the possibility of our approach in this setting (homogeneous Lie groups with one-dimensional center and flat orbits) and we thank Christian Bargetz for a very helpful discussion and for pointing out the relevant literature to the multiplication of vector valued distributions with vector valued functions.

We especially thank the reviewer for many helpful comments and suggestions, which allowed to improve the paper.

References

1. Geller, D.: Fourier analysis on the Heisenberg group. I. Schwartz space. J. Funct. Anal. **36**(2), 205–254 (1980)
2. Astengo, F., Di Blasio, B., Ricci, F.: Fourier transform of Schwartz functions on the Heisenberg group. Stud. Math. **214**, 201–222 (2013)
3. Măntoiu, M., Ruzhansky, M.: Quantizations on nilpotent lie groups and algebras having flat coadjoint orbits. J. Geom. Anal. **29**(2), 2823–2861 (2018)
4. Gröchenig, K., Rottensteiner, D.: Orthonormal bases in the orbit of square-integrable representations of nilpotent Lie groups. J. Funct. Anal. **275**(12), 3338–3379 (2018)

5. Bargetz, C., Ortner, N.: Characterization of L. Schwartz' convolutor and multiplier spaces $\mathcal{O}'_C$ and $\mathcal{O}_M$ by the short-time Fourier transform. RACSAM **108**(2), 833–847 (2014)
6. Pedersen, N.V.: Geometric quantization and the universal enveloping algebra of a nilpotent Lie group. Trans. Am. Math. Soc. **315**(2), 511–563 (1989)
7. Fischer, V., Ruzhansky M.: Quantization on Nilpotent Lie Groups. Birkäuser, Boston (2016)
8. Ruzhansky, M., Turunen, V.: Pseudo-Differential Operators and Symmetries. Birkäuser, Boston (2010)
9. Corwin, L., Greenleaf, F.P.: Representations of Nilpotent Lie Groups and Their Applications. Cambridge University Press, Cambridge (1990)
10. Treves, F.: Topological Vectors Spaces, Distributions and Kernels. Academic Press, San Diego (1967)
11. Schaefer, H.H.: Topological Vector Spaces, Springer, New York (1971)
12. Schwartz, L.: Théorie des distributions à valeurs vectorielles. II. Ann. Inst. Fourier **8**, 1–209 (1958)
13. Kaballo, W.: Aufbaukurs Funktionalanalysis und Operatortheorie. Springer, Berlin (2014)
14. Grothendieck, A.: Produit tensoriels topologiques et espace nucléaire. Memoirs of the American Mathematical Society, vol. 16. American Mathematical Society, Providence (1955)
15. Bargetz, C: Explicit representations of spaces of smooth functions and distributions. J. Math. Anal. Appl. **424**(2), 1491–1505 (2015)
16. Gel'fand, I.M., Vilenkin, N.Y.: Generalized Functions, vol. 4, Academic Press, New York (1964)
17. Schwartz, L.: Théorie des Distributions. Hermann, Paris (1978)
18. Grafakos, L.: Modern Fourier Analysis. Springer, New York (2009)
19. Komatsu, H.: Projective and injective limits of weakly compact sequences of locally convex spaces. J. Math. Soc. Jpn. **19**(3), 366–383 (1967)
20. Pedersen, N.V.: Matrix coefficients and a Weyl correspondence for nilpotent Lie groups. Invent. Math. **118**, 1–36 (1994)
21. Cartier, P.: Vecteurs différentiables dans les représentations unitaires des groupes de Lie. Sémin. Bourbaki, Exp. No. **454**, 20–34 (1976)
22. Moore, C., Wolf, J.: Square integrable representations of nilpotent groups. Trans. Am. Math. Soc. **185**, 445–445 (1973)

A Multiplicity Result for a Non-Homogeneous Subelliptic Problem with Sobolev Exponent

Annunziata Loiudice

Abstract We prove a multiplicity result for the inhomogeneous subelliptic problem

$$-\Delta_{\mathbb{G}} u = |u|^{2^*-2} u + f \quad \text{in } \Omega, \qquad u = 0 \text{ on } \partial\Omega,$$

where $\Delta_{\mathbb{G}}$ is a sub-Laplacian on a Carnot group $\mathbb{G}$, $2^* = 2Q/(Q-2)$ is the critical Sobolev exponent in this context, Ω is a bounded domain of $\mathbb{G}$ and f is small in a suitable sense. Precisely, we prove the existence of two distinct solutions, that are positive if f is. We adapt to the present subelliptic setting the well-known technique developed by Tarantello (Ann Inst H Poincarè Anal Non Linéaire 9:281–309, 1992).

Keywords Non-homogeneous problem · Critical Sobolev exponent · Carnot groups · Multiplicity result

1 Introduction

In this paper we provide existence results for the following non-homogeneous subelliptic problem

$$\begin{cases} -\Delta_{\mathbb{G}} u = |u|^{2^*-2} u + f & \text{in } \Omega, \\ \quad u = 0 & \text{on } \partial\Omega. \end{cases} \tag{1}$$

Here, $\Delta_{\mathbb{G}}$ is a sub-Laplacian on a Carnot group $\mathbb{G}$, $2^* = 2Q/(Q-2)$ is the critical Sobolev exponent in this context, Ω is a bounded domain of $\mathbb{G}$ and the inhomogeneous term f satisfies suitable summability assumptions.

A. Loiudice (✉)
Department of Mathematics, University of Bari, Bari, Italy
e-mail: annunziata.loiudice@uniba.it

© Springer Nature Switzerland AG 2020

V. Georgiev et al. (eds.), *Advances in Harmonic Analysis and Partial Differential Equations*, Trends in Mathematics, https://doi.org/10.1007/978-3-030-58215-9_4

We recall that a great deal of interest has been paid in the literature to subelliptic equations with critical exponent on Stratified Lie groups, starting from the papers [19, 21] (See e.g. [2, 3, 6, 12, 13, 27, 29–34, 36, 38, 45]).

The present results further explore the analysis performed by the author in [29, 31], where the homogeneous case $f = \lambda u$, $\lambda \in \mathbb{R}$, has been investigated for problem (1). We recall that, for $f = 0$, problem (1) does not admit any nonnegative non trivial solution, sufficiently regular up to the boundary, on starshaped domains of $\mathbb{G}$ (see [19, 21]; see also [31] for a Hardy–Sobolev version of the problem). In [29] the author proves that, when f is a homogeneous linear perturbation $f = \lambda u$, problem (1) admits at least one positive solution for any dimension $Q \geq 4$, when $0 < \lambda < \lambda_1$, λ_1 being the first eigenvalue of the sub-Laplacian $\Delta_{\mathbb{G}}$ with Dirichlet boundary conditions, extending the celebrated results by Brezis–Nirenberg [7] to the subelliptic Carnot context.

In the present paper, we investigate the problem with a non-homogeneous term f and, under suitable assumptions, we prove the existence of at least two solutions, which are positive if f is. We recall that multiplicity results for critical problems in the subelliptic context can be found e.g. in [10, 13, 18, 29, 35, 37]. In particular, Garagnani–Uguzzoni [18] and Maalaoui–Martino [35] obtain existence of multiple solutions depending on the topology of the domain when $\mathbb{G}$ is the Heisenberg group $\mathbb{H}^n$.

Our multiplicity result depends on the variational structure of the problem, it is obtained by using Ekeland's variational principle and the Nehari manifold and generalizes to the present context a celebrated Euclidean result by Tarantello [44].

Let us introduce our results. We denote by $S_0^1(\Omega)$ the Sobolev–Stein space defined as the completion of $C_0^\infty(\Omega)$ with respect to the norm

$$\|u\| := \left(\int_\Omega |\nabla_{\mathbb{G}} u|^2\right)^{1/2}. \tag{2}$$

We are interested in weak solutions of problem (1), i.e. $u \in S_0^1(\Omega)$ such that

$$\int_\Omega \nabla_{\mathbb{G}} u \nabla_{\mathbb{G}} \phi = \int_\Omega |u|^{2^*-2} u\phi + \int_\Omega f\phi, \quad \forall \phi \in S_0^1(\Omega). \tag{3}$$

We assume that the non-homogeneous term f belongs to $S^{-1}(\Omega)$, that is the dual space of $S_0^1(\Omega)$, and that its S^{-1}-norm is appropriately small. We look for solutions to (1) as critical points of the functional

$$J(u) = \frac{1}{2}\int_\Omega |\nabla_{\mathbb{G}} u|^2 - \frac{1}{2^*}\int_\Omega |u|^{2^*} - \int_\Omega fu, \quad u \in S_0^1(\Omega). \tag{4}$$

Let us set

$$C_Q := \frac{4}{Q-2}\left(\frac{Q-2}{Q+2}\right)^{(Q+2)/4}. \tag{5}$$

Our main results can be stated as follows.

Theorem 1.1 *Let $f \neq 0$ satisfy the following inequality*

$$\int_\Omega fu \leq C_Q \left(\|\nabla_{\mathbb{G}} u\|_2\right)^{(Q+2)/2}, \tag{6}$$

for any $u \in S_0^1(\Omega)$, $\int_\Omega |u|^{2^*} = 1$, where C_Q is the constant defined in (5)*. Then, problem* (1) *has at least one weak solution $u_0 \in S_0^1(\Omega)$. Moreover, $u_0 \geq 0$ if $f \geq 0$.*

If, moreover, the estimate (6) holds with the strict sign, the existence of a second solution can be proved. We have the following result.

Theorem 1.2 *Let $f \neq 0$ satisfy the following strict inequality*

$$\int_\Omega fu < C_Q \left(\|\nabla_{\mathbb{G}} u\|_2\right)^{(Q+2)/2}, \tag{7}$$

for any $u \in S_0^1(\Omega)$, $\int_\Omega |u|^{2^*} = 1$, where C_Q is the constant defined in (5)*. Then, problem* (1) *has at least two weak solutions $u_0, u_1 \in S_0^1(\Omega)$. Moreover, $u_0 \geq 0$, $u_1 \geq 0$ if $f \geq 0$.*

We also note that, for $f \geq 0$, by means of Bony's maximum principle [5], such solutions turn out to be strictly positive.

Remark 1.1 We observe that condition (6) naturally holds if

$$\|f\|_{S^{-1}} \leq C_Q S^{Q/4}, \tag{8}$$

where S is the best Sobolev constant on $\mathbb{G}$.

We point out that the technique introduced in [44] has been applied in many different contexts: see e.g. some applications to singular elliptic problems in Chen–Chen [9], Kang–Deng [26], a generalization to the fractional Laplacian in Shang–Zhang-Yang [43], an application to a critical problem on the Heisenberg group with indefinite nonlinearity in Chen–Huang–Rocha in [10].

The main difficulty one encounters when adapting the Euclidean technique to the general context of Carnot groups is the lack of knowledge of the related Sobolev-type extremals, whose explicit form is not known in the abstract Carnot setting. Indeed, the explicit form of Sobolev-type minimizers on Carnot groups is known only when $\mathbb{G}$ is a group of Iwasawa type (see Jerison–Lee [25] and Frank–Lieb [17] for the Sobolev minimizers in the Heisenberg group $\mathbb{H}^n$, Ivanov–Minchev–Vassilev [23] and Christ–Liu–Zhang [11] for the remaining cases).

However, as observed by the author in [29], this difficulty can be overcome since some qualitative properties of the extremal functions, namely their existence and their exact asymptotic behavior at infinity, are sufficient to perform the required Brezis–Nirenberg type asymptotic expansions [7] and achieve the existence results by a mountain pass procedure (see e.g. [24, 29, 31]). We also recall that the sharp asymptotic behavior at infinity of Sobolev-type minimizers has been studied by the author in [29, 30, 32].

Finally, we note that obviously the preceding theorems hold for the subcritical case, i.e. when the power $p = 2^*$ is replaced by $q \in (2, 2^*)$: in such a case standard compactness arguments apply (see [20] for compact Sobolev-type embeddings in the subelliptic setting).

The paper is organized as follows: in Sect. 2 we recall the main notation and definitions related to the Carnot setting; in Sect. 3 we prove the existence of the first solution; in Sect. 4 we establish the existence of the second solution.

2 The Functional Setting of Carnot Groups

We recall the main definitions and notations related to the Carnot groups setting. For a complete treatment, we refer to the monograph [4] and the classical papers [15, 16]. We also quote [14, 40] for an overview on general homogeneous Lie groups.

A *Carnot group* (or *Stratified group*) $(\mathbb{G}, \circ)$ is a connected, simply connected nilpotent Lie group, whose Lie algebra g admits a stratification, namely a decomposition $g = \bigoplus_{j=1}^{r} V_j$, such that $[V_1, V_j] = V_{j+1}$ for $1 \leq j < r$, and $[V_1, V_r] = \{0\}$. The number r is called the *step* of the group $\mathbb{G}$. The integer $Q = \sum_{i=1}^{r} i \dim V_i$ is called the *homogeneous dimension* of $\mathbb{G}$. We shall assume throughout that $Q \geq 3$.

By means of the natural identification of $\mathbb{G}$ with its Lie algebra via the exponential map (which we shall assume throughout), it is not restrictive to suppose that $\mathbb{G}$ is a homogeneous Lie group on $\mathbb{R}^N = \mathbb{R}^{N_1} \times \mathbb{R}^{N_2} \times \ldots \times \mathbb{R}^{N_r}$, with $N_i = \dim V_i$, equipped with a family of group-automorphisms (called *dilations*) δ_λ of the form

$$\delta_\lambda(\xi) = (\lambda\, \xi^{(1)}, \lambda^2\, \xi^{(2)}, \cdots, \lambda^r\, \xi^{(r)}), \tag{9}$$

where $\xi^{(j)} \in \mathbb{R}^{N_j}$ for $j = 1, \ldots, r$. Let $m := N_1$ and let $X_1, \ldots, X_m$ be the set of left invariant vector fields of V_1 that coincide at the origin with the first m partial derivatives. The second order differential operator

$$\Delta_{\mathbb{G}} = \sum_{i=1}^{m} X_i^2 \tag{10}$$

is called the canonical *sub-Laplacian* on $\mathbb{G}$. We shall denote by

$$\nabla_{\mathbb{G}} = (X_1, \ldots, X_m) \tag{11}$$

the related horizontal gradient. The operator $\Delta_{\mathbb{G}}$ is left-translation invariant with respect to the group action $\tau_\xi(\eta) = \xi \circ \eta$ and δ_λ-homogeneous of degree two. In other words, $\Delta_{\mathbb{G}}(u \circ \tau_\xi) = \Delta_{\mathbb{G}} u \circ \tau_\xi$, $\Delta_{\mathbb{G}}(u \circ \delta_\lambda) = \lambda^2 \Delta_{\mathbb{G}} u \circ \delta_\lambda$. Moreover, due to the stratification condition, the Lie algebra generated by $X_1, \dots, X_m$ is the whole g, and therefore it is everywhere of rank N; therefore, the sub-Laplacian operator $\Delta_{\mathbb{G}}$ satisfies the well-known Hörmander's hypoellipticity condition.

The simplest example of Carnot group is the additive group $\mathbb{G} = (\mathbb{R}^N, +)$. In this case $Q = N$ and the sub-Laplacians are exactly the constant coefficients elliptic operators on $\mathbb{R}^N$. Moreover, if $\mathbb{G}$ is a Carnot group of homogeneous dimension $Q \leq 3$, then necessarily $\mathbb{G}$ is the ordinary Euclidean space. The simplest non-abelian Carnot group is the Heisenberg group $\mathbb{H}^n = (\mathbb{R}^{2n+1}, \circ)$ with homogeneous dimension $Q = 2n + 2$ and composition law given by $\xi \circ \xi' = (x + x', y + y', t + t' + 2(\langle x', y\rangle - \langle x, y'\rangle))$, for every $\xi = (x, y, t)$, $\xi' = (x', y', t') \in R^{2n+1}$, where $x, y \in \mathbb{R}^n$ and $t \in \mathbb{R}$.

When $Q \geq 3$, Carnot groups possess the following property: there exists a suitable homogeneous norm d on $\mathbb{G}$ such that the function

$$\Gamma(\xi) = \frac{C}{d(\xi)^{Q-2}} \tag{12}$$

is a fundamental solution of $-\Delta_{\mathbb{G}}$ with pole at 0, for a suitable constant $C > 0$ (see [15]). By definition, a *homogeneous norm* on $\mathbb{G}$ is a continuous function $d : \mathbb{G} \to [0, +\infty)$, smooth away from the origin, such that $d(\delta_\lambda(\xi)) = \lambda d(\xi)$, for every $\lambda > 0$ and $\xi \in \mathbb{G}$, $d(\xi^{-1}) = d(\xi)$ and $d(\xi) = 0$ if and only if $\xi = 0$. Moreover, if we define $d(\xi, \eta) := d(\eta^{-1} \circ \xi)$, then d is a pseudo-distance on $\mathbb{G}$. Throughout the paper, we shall denote by d the homogeneous norm associated with the fundamental solution of the sub-Laplacian operator by (12). We shall denote by $B_r(\xi) = B_d(\xi, r)$ the d-ball with center at ξ and radius r.

A fundamental rôle in the functional analysis on Carnot groups is played by the following Sobolev-type inequality due to Folland and Stein [15]: if $\mathbb{G}$ is a Carnot group of homogeneous dimension $Q \geq 3$, there exists a positive constant $S = S(\mathbb{G})$ such that

$$\int_{\mathbb{G}} |\nabla_{\mathbb{G}} u|^2 \, d\xi \geq S \left(\int_{\mathbb{G}} |u|^{2^*} \, d\xi \right)^{\frac{2}{2^*}} \quad \forall u \in C_0^\infty(\mathbb{G}), \tag{13}$$

where $2^* = 2Q/(Q-2)$ denotes the critical exponent in this context.

It is known that the best constant in (13) is attained (see [21, 45]), but the explicit form of the minimizers is known only for the class of Iwasawa-type groups (see [11, 17, 23, 25]). Qualitative properties of Sobolev minimizers on Carnot groups, such as their sharp summability in L^p-weak spaces and their asymptotic behavior at infinity, have been studied by the author in [29] for the pure Sobolev case, in [30] for the Hardy–Sobolev case, in [33] for the general Sobolev inequality with exponent $1 < p < Q$. We also recall that a quantitative form of Sobolev inequality (13) on the Heisenberg group $\mathbb{H}^n$ has been obtained by the author in [28].

Finally, we point out that a large variety of functional inequalities can be obtained on Carnot groups, and more generally on graded and homogeneous Lie groups. We quote [39–42] and the references therein for recent developments on this topic.

3 Existence of the First Solution

In this section, we prove the existence of the first solution for problem (1). Since J is bounded from below on the so-called Nehari manifold

$$\begin{aligned}\mathcal{N} &:= \{u \in S_0^1(\Omega) \mid \, < J'(u), u >= 0\} \\ &= \left\{u \in S_0^1(\Omega) \mid \int_\Omega |\nabla_{\mathbb{G}} u|^2 - \int_\Omega |u|^{2^*} - \int_\Omega fu = 0\right\},\end{aligned} \tag{14}$$

we shall consider the following minimization problem

$$c_0 := \inf_{\mathcal{N}} J. \tag{15}$$

Let us split $\mathcal{N}$ into the following sets

$$\mathcal{N}^+ := \left\{u \in \mathcal{N} \mid \int_\Omega |\nabla_{\mathbb{G}} u|^2 - (2^* - 1)\int_\Omega |u|^{2^*} > 0\right\}, \tag{16}$$

$$\mathcal{N}^0 := \left\{u \in \mathcal{N} \mid \int_\Omega |\nabla_{\mathbb{G}} u|^2 - (2^* - 1)\int_\Omega |u|^{2^*} = 0\right\}, \tag{17}$$

$$\mathcal{N}^- := \left\{u \in \mathcal{N} \mid \int_\Omega |\nabla_{\mathbb{G}} u|^2 - (2^* - 1)\int_\Omega |u|^{2^*} < 0\right\}. \tag{18}$$

We shall obtain that, under the assumption (6), c_0 is attained at a function $u_0 \in \mathcal{N}$ and, if the more restrictive assumption (7) holds, u_0 is a local minimum for J and belongs to $\mathcal{N}^+$, i.e.

$$c_0 = \inf_{\mathcal{N}} J = \inf_{\mathcal{N}^+} J. \tag{19}$$

We begin with some preliminary lemmas. The first lemma clarifies the purpose of the assumption (7) on the non-homogeneous term f.

Lemma 3.1 *Let $f \neq 0$ satisfy condition (7). Then, for any $u \in S_0^1(\Omega)$, $u \neq 0$, there exists a unique $t^+ = t^+(u) > 0$ such that $t^+ u \in \mathcal{N}^-$,*

$$t^+ > \left\{\frac{\int_\Omega |\nabla_{\mathbb{G}} u|^2}{(2^* - 1)\int_\Omega |u|^{2^*}}\right\}^{1/(2^*-2)} := t_{\max} \tag{20}$$

and

$$J(t^+u) = \max_{t \geq t_{\max}} J(tu). \tag{21}$$

Moreover, if $\int_\Omega fu > 0$, *there exists a unique* $t^- = t^-(u) > 0$ *such that* $t^-u \in \mathcal{N}^+$,

$$t^- < \left\{ \frac{\int_\Omega |\nabla_{\mathbb{G}} u|^2}{(2^* - 1) \int_\Omega |u|^{2^*}} \right\}^{1/(2^*-2)} \tag{22}$$

and

$$J(t^-u) = \min_{0 \leq t \leq t^+} J(tu). \tag{23}$$

Proof Consider the function

$$\Phi(t) := t \int_\Omega |\nabla_{\mathbb{G}} u|^2 - t^{2^*-1} \int_\Omega |u|^{2^*}. \tag{24}$$

Since $2^* > 2$, Φ is concave and achieves its maximum in

$$t_{\max} = \left\{ \frac{\int_\Omega |\nabla_{\mathbb{G}} u|^2}{(2^* - 1) \int_\Omega |u|^{2^*}} \right\}^{1/(2^*-2)} \tag{25}$$

and

$$\Phi(t_{\max}) = C_Q \frac{\left(\int_\Omega |\nabla_{\mathbb{G}} u|^2\right)^{(Q+2)/4}}{\left(\int_\Omega |u|^{2^*}\right)^{(Q-2)/4}} \tag{26}$$

where C_Q is the constant defined in (5).

Now, if $\int_\Omega fu \leq 0$, there exists a unique $t^+ > t_{\max}$ such that $\Phi(t^+) = \int_\Omega fu$ and $\Phi'(t^+) < 0$, that is, $t^+u \in \mathcal{N}^-$ and $J(t^+u) \geq J(tu)$ for any $t \geq t_{\max}$.

If, instead, $\int_\Omega fu > 0$, by the assumption (7) on f we get that

$$\int_\Omega fu < \Phi(t_{\max}). \tag{27}$$

Hence, there exists a unique $0 < t^- < t_{\max} < t^+$ such that

$$\Phi(t^+) = \int_\Omega fu = \Phi(t^-) \tag{28}$$

and

$$\Phi'(t^+) < 0 < \Phi'(t^-), \tag{29}$$

that is, $t^+u \in \mathcal{N}^-$ and $t^-u \in \mathcal{N}^+$.

Moreover, $J(t^+u) \geq J(tu), \forall t \geq t^-$ and $J(t^-u) \leq J(tu), \forall 0 \leq t \leq t^+$. □

Lemma 3.2 *For $f \neq 0$, the infimum*

$$\inf_{\int_\Omega |u|^{2^*}=1} \left(C_Q \left(\int_\Omega |\nabla_{\mathbb{G}} u|^2 \right)^{(Q+2)/4} - \int_\Omega fu \right) := \mu_0 \tag{30}$$

is attained. In particular, if f satisfies (7)*, then $\mu_0 > 0$.*

Proof The proof is rather technical and it straightforwardly follows the Euclidean outline. We omit it, referring to the Appendix in [44]. □

Lemma 3.3 *Assume that condition* (7) *holds. Then, for any $u \in \mathcal{N}$, $u \neq 0$, we have*

$$\int_\Omega |\nabla_{\mathbb{G}} u|^2 - (2^* - 1) \int_\Omega |u|^{2^*} \neq 0, \tag{31}$$

i.e. $\mathcal{N}_0 = \{0\}$.

Proof For any $u \in \mathcal{N}$, we have

$$\int_\Omega |\nabla_{\mathbb{G}} u|^2 = \int_\Omega |u|^{2^*} + \int_\Omega fu. \tag{32}$$

By contradiction, assume that (31) does not hold. Then, there exists $u_0 \in \mathcal{N}$, $u_0 \neq 0$ such that

$$\int_\Omega |\nabla_{\mathbb{G}} u_0|^2 - (2^* - 1) \int_\Omega |u_0|^{2^*} = 0. \tag{33}$$

From (32) and (33) we get

$$\int_\Omega |u_0|^{2^*} = \frac{1}{2^* - 2} \int_\Omega fu_0 \tag{34}$$

and

$$\int_\Omega |\nabla_{\mathbb{G}} u_0|^2 = \frac{2^* - 1}{2^* - 2} \int_\Omega fu_0. \tag{35}$$

Set

$$v := u_0 / \|u_0\|_{2^*}^{2^*}. \tag{36}$$

From the preceding equalities, we easily get

$$\frac{1}{2^*-1}(2^*-2)^{(2^*-2)/(2^*-1)}\int_\Omega |\nabla_{\mathbb{G}} v|^2 = \left(\int_\Omega f v\right)^{(2^*-2)/(2^*-1)}, \tag{37}$$

that is

$$\int_\Omega f v = \frac{4}{Q-2}\left(\frac{Q-2}{Q+2}\right)^{(Q+2)/4}\left(\int_\Omega |\nabla_{\mathbb{G}} v|^2\right)^{(Q+2)/4}, \tag{38}$$

which contradicts assumption (7). □

Lemma 3.4 *Assume that $f \neq 0$ satisfy (7). Then, for any $u \in \mathcal{N}^\pm$, there exist $\varepsilon > 0$ and a differentiable function $t = t(v) > 0$, $v \in S_0^1(\Omega)$, $\|v\| < \varepsilon$, such that*

$$t(0) = 1, \quad t(v)(u-v) \in \mathcal{N}^\pm, \quad \text{for } \|v\| < \varepsilon, \tag{39}$$

and

$$\langle t'(0), v\rangle = \frac{2\int_\Omega \nabla_{\mathbb{G}} u \nabla_{\mathbb{G}} v - 2^*\int_\Omega |u|^{2^*-2} u v - \int_\Omega f v}{\|u\|^2 - (2^*-1)\int_\Omega |u|^{2^*}}. \tag{40}$$

Proof Define the map $F : \mathbb{R} \times S_0^1(\Omega) \to \mathbb{R}$ as

$$F(t, v) = t\|u-v\|^2 - t^{2^*-1}\int_\Omega |u-v|^{2^*} - \int_\Omega f(u-v). \tag{41}$$

Since

$$F(1,0) = 0, \quad \frac{\partial F}{\partial t}(1,0) = \|u\|^2 - (2^*-1)\int_\Omega |u|^{2^*} \neq 0, \tag{42}$$

by applying the implicit function Theorem at the point $(1,0)$ we get the result. □

The following lemma provides an upper bound for c_0.

Lemma 3.5 *Let $f \in S^{-1}(\Omega)$ satisfy (7). Then, there exists a negative constant θ_0, depending on f, such that $c_0 < \theta_0 < 0$.*

Proof Let $v \in S_0^1(\Omega)$ be the unique solution of

$$\begin{cases} -\Delta_{\mathbb{G}} v = f & \text{in } \Omega, \\ v = 0 & \text{on } \partial\Omega. \end{cases} \tag{43}$$

Since $f \neq 0$, we have that $\|v\|^2 > 0$ and $\|v\|^2 = \|f\|^2_{S^{-1}}$. Moreover, consider $t^- = t^-(v) > 0$ provided by Lemma 3.1. Hence $t^- v \in \mathcal{N}^+$ and consequently

$$\begin{aligned} c_0 &\leq J(t^- v) \\ &= -\frac{(t^-)^2}{2}\|v\|^2 + \frac{2^* - 1}{2^*}(t^-)^{2^*}\int_\Omega |v|^{2^*} \\ &< -\frac{(t^-)^2}{2}\|v\|^2 + \frac{(t^-)^2}{2^*}\|v\|^2 = -\frac{(t^-)^2}{Q}\|f\|^2_{S^{-1}}. \end{aligned} \tag{44}$$

Hence, $c_0 < \theta_0 < 0$, where $\theta_0 = -\frac{(t^-)^2}{Q}\|f\|^2_{S^{-1}}$. □

We are now ready to prove Theorem 1.1.

Proof *(Theorem 1.1)* It is easy to prove that J is bounded from below on $\mathcal{N}$. Then, by Ekeland's variational principle (see [1, Coroll. 5.3.2]), there exists a minimizing sequence for the minimization problem $c_0 = \inf_{\mathcal{N}} J$ such that

$$J(u_n) < c_0 + 1/n, \tag{45a}$$

$$J(w) \geq J(u_n) - (1/n)\|w - u_n\|, \quad \forall w \in \mathcal{N}. \tag{45b}$$

We first obtain our result for f satisfying the strict inequality (7). By the preceding lemma, we know that $c_0 \leq \theta_0$. Therefore, by (45a), for n large we get

$$\left(\frac{1}{2} - \frac{1}{2^*}\right)\|u_n\|^2 < \theta_0 + \frac{2^* - 1}{2^*}\int_\Omega f u_n, \tag{46}$$

and so

$$-\frac{2^*}{2^* - 1}\theta_0 \leq \int_\Omega f u_n \leq \|f\|_{S^{-1}}\|u_n\|. \tag{47}$$

Putting together (46) and (47), we obtain that

$$C_1 \leq \|u_n\| \leq C_2, \tag{48}$$

where $C_1 = -\frac{2^*}{2^*-1}\theta_0\|f\|^{-1}_{S^{-1}}$ and $C_2 = \frac{2(2^*-1)}{2^*-2}\|f\|_{S^{-1}}$.

Our aim is to prove that

$$J'(u_n) \to 0 \text{ in } S^{-1}. \tag{49}$$

Assume that $\|J'(u_n)\|_{S^{-1}} > 0$ for n large (otherwise we are done). By Lemma 3.1 applied to $w = \delta J'(u_n)/\|J'(u_n)\|$ with $\delta > 0$ small enough, we can get

$t_n(\delta) = t(w)$ such that $w_\delta := t_n(\delta)(u_n - w) \in \mathcal{N}$. From condition (45b) we have

$$\begin{aligned}\frac{1}{n}\|w_\delta - u_n\| & \\ &\geq J(u_n) - J(w_\delta) \qquad (50)\\ &= (1 - t_n(\delta))\langle J'(w_\delta), u_n\rangle + \delta t_n(\delta)\langle J'(w_\delta), \frac{J'(u_n)}{\|J'(u_n)\|}\rangle + o(\delta).\end{aligned}$$

Dividing by $\delta > 0$ and passing to the limit as $\delta \to 0$, we obtain

$$\frac{1}{n}(1 + |t_n'(0)|\,\|u_n\|) \geq -t_n'(0)\langle J'(u_n), u_n\rangle + \|J'(u_n)\| = \|J'(u_n)\|, \qquad (51)$$

where we have set

$$t_n'(0) := \langle t'(0), \frac{J'(u_n)}{\|J'(u_n)\|}\rangle. \qquad (52)$$

Hence, taking into account (48), we get

$$\|J'(u_n)\| \leq \frac{C}{n}\left(1 + |t_n'(0)|\right), \qquad (53)$$

for a suitable constant $C > 0$. The thesis then follows by showing that $|t_n'(0)|$ is uniformly bounded with respect to n. Indeed, from (40) and (48), it follows that

$$|t_n'(0)| \leq \frac{C_1}{|\,\|u_n\|^2 - (2^* - 1)\|u_n\|_{2^*}^{2^*}|} \qquad (54)$$

for a suitable constant $C_1 > 0$, and it can be verified, by using Lemma 3.2, that the denominator in the r.h.s. of (54) is bounded away from 0. We refer to [44] for further details.

Now, let $u_0 \in S_0^1(\Omega)$ be the weak limit, up to a subsequence, of u_n. From (47) we get that

$$\int_\Omega f u_0 > 0, \qquad (55)$$

and from the fact that $\|J'(u_n)\| \to 0$ as $n \to \infty$, we have that

$$< J'(u_0), w >= 0, \quad \forall\, w \in S_0^1(\Omega), \qquad (56)$$

i.e. u_0 is a weak solution for problem (1) and $u_0 \in \mathcal{N}$. Hence

$$c_0 \le J(u_0) = \frac{1}{Q}\|u_0\|^2 - \left(1 - \frac{1}{2^*}\right)\int_\Omega f u_0 \le \lim_{n\to\infty} J(u_n) = c_0. \tag{57}$$

So we deduce that $u_n \to u_0$ strongly in $S_0^1(\Omega)$ and $J(u_0) = c_0$. We can also deduce by Lemma 3.1 that $u_0 \in \mathcal{N}^+$ (see e.g. [43, page 575]).

Finally, if $f \ge 0$, take $t_0 = t^-(|u_0|) > 0$ with $t_0|u_0| \in \mathcal{N}^+$. Necessarily, $t_0 \ge 1$ and

$$J(t_0|u_0|) \le J(|u_0|) \le J(u_0). \tag{58}$$

So we can always take $u_0 \ge 0$.

The proof for the case when f satisfies the weaker assumption (6) is achieved by an approximation argument, reasoning with $f_\varepsilon = (1-\varepsilon)f$, $0 < \varepsilon < 1$, for which the strict condition (7) holds and the preceding results apply. See the details in [44, page 292]. □

4 Existence of the Second Solution

In this section, we prove that, under the strict assumption (7), the existence of a second solution to our problem (1) can be proved. Precisely, we show that, under the assumption (7), the following infimum

$$c_1 := \inf_{u\in\mathcal{N}^-} J(u) \tag{59}$$

is achieved at a point $u_1 \in \mathcal{N}^-$ which defines a critical point of J.

We need the following preliminary results. In what follows, S will denote the best Sobolev constant on $\mathbb{G}$.

Lemma 4.1 *Assume that condition* (7) *holds. Then, the solution u_0 obtained in Theorem 1.1 is a local minimum for J.*

Proof From Lemma 3.1, for any $u \in S_0^1(\Omega)$ with $\int_\Omega fu > 0$ we know that

$$J(su) \ge J(t^- u), \quad \text{for any } 0 < s < \left\{\frac{\int_\Omega |\nabla_{\mathbb{G}} u|^2}{(2^*-1)\int_\Omega |u|^{2^*}}\right\}^{1/(2^*-2)}. \tag{60}$$

In particular, for $u = u_0 \in \mathcal{N}^+$ we have

$$t^-(u_0) = 1 < \left\{\frac{\int_\Omega |\nabla_{\mathbb{G}} u_0|^2}{(2^*-1)\int_\Omega |u_0|^{2^*}}\right\}^{1/(2^*-2)}. \tag{61}$$

Choose $\varepsilon > 0$ small enough so that the following inequality holds

$$1 < \left\{ \frac{\int_\Omega |\nabla_{\mathbb{G}}(u_0 - w)|^2}{(2^* - 1)\int_\Omega |u_0 - w|^{2^*}} \right\}^{1/(2^*-2)}, \tag{62}$$

for any w such that $\|w\| < \varepsilon$.

By Lemma 3.4, let $t(w) > 0$ satisfy $t(w)(u_0 - w) \in \mathcal{N}$ for any $\|w\| < \varepsilon$. Since $t(w) \to 1$ as $\|w\| \to 0$, we can assume that the following inequality holds

$$t(w) < \left\{ \frac{\int_\Omega |\nabla_{\mathbb{G}}(u_0 - w)|^2}{(2^* - 1)\int_\Omega |u_0 - w|^{2^*}} \right\}^{1/(2^*-2)}, \tag{63}$$

for any w such that $\|w\| < \varepsilon$. Namely,

$$t(w)(u_0 - w) \in \mathcal{N}^+ \text{ and, for } 0 < s < \left\{ \frac{\int_\Omega |\nabla_{\mathbb{G}}(u_0 - w)|^2}{(2^* - 1)\int_\Omega |u_0 - w|^{2^*}} \right\}^{1/(2^*-2)}, \tag{64}$$

we have

$$J(s(u_0 - w)) \geq J(t(w)(u_0 - w)) \geq J(u_0). \tag{65}$$

From (61) we can take $s = 1$ and then we obtain that

$$J(u_0 - w) \geq J(u_0), \quad \forall w : \|w\| < \varepsilon. \tag{66}$$

The proof is therefore complete. □

The following lemma gives a range where the Palais–Smale condition, briefly denoted by (PS), holds for J.

Lemma 4.2 *The functional J satisfies the (PS)-condition at level c, for all c satisfying*

$$c < c_0 + \frac{1}{Q} S^{Q/2}. \tag{67}$$

Proof The proof is standard and it straightforwardly follows the Euclidean outline (See [44]). □

Now, following [29], let $U > 0$ be a fixed minimizer for the Sobolev inequality (13) and consider, for $\varepsilon > 0$ and $\eta \in \mathbb{G}$, the family of rescaled and translated functions

$$U_{\varepsilon,\eta}(\xi) = \varepsilon^{(2-Q)/2} U(\delta_{1/\varepsilon}(\eta^{-1} \circ \xi)), \quad \xi \in \mathbb{G}. \tag{68}$$

The functions $U_{\varepsilon,\eta}$ are solutions, up to multiplicative constants, of the equation $-\Delta_{\mathbb{G}} u = u^{2^*-1}$ in $\mathbb{G}$. Moreover, they satisfy

$$\int_{\mathbb{G}} |\nabla_{\mathbb{G}} U_{\varepsilon,\eta}|^2 \mathrm{d}\xi = \int_{\mathbb{G}} U_{\varepsilon,\eta}^{2^*} \mathrm{d}\xi = S^{Q/2}, \quad \text{for all } \varepsilon > 0 \text{ and } \eta \in \mathbb{G}. \tag{69}$$

Now, recall that $u_0 \neq 0$. We set $\Sigma \subset \Omega$ to be a set of positive measure such that $u_0 > 0$ on Σ (replace u_0 with $-u_0$ and f with $-f$ if necessary).

Let $\eta \in \Sigma$ and let $R > 0$ be such that $B_d(\eta, R) \subset \Omega$. Let $\varphi_\eta \in C_0^\infty(B_d(\eta, R))$ be a cut-off function, $0 \leq \varphi_\eta \leq 1$, $\varphi_\eta \equiv 1$ in $B_d(\eta, R/2)$. Define

$$u_{\varepsilon,\eta} := \varphi_\eta U_{\varepsilon,\eta}, \quad \varepsilon > 0, \eta \in \Sigma. \tag{70}$$

We get the following crucial lemma.

Lemma 4.3 *For any $t > 0$ and for a.e. $\eta \in \Sigma$, there exists $\varepsilon_0 = \varepsilon_0(t, \eta) > 0$ such that*

$$J(u_0 + t u_{\varepsilon,\eta}) < c_0 + \frac{1}{Q} S^{Q/2}, \tag{71}$$

for every $0 < \varepsilon < \varepsilon_0$.

Proof We have

$$\begin{aligned} J(u_o + t u_{\varepsilon,\eta}) = &\frac{1}{2} \int_\Omega |\nabla_{\mathbb{G}} u_0|^2 + t \int_\Omega \nabla_{\mathbb{G}} u_0 \nabla_{\mathbb{G}} u_{\varepsilon,\eta} + \frac{t^2}{2} \int_\Omega |\nabla_{\mathbb{G}} u_{\varepsilon,\eta}|^2 \\ &- \frac{1}{2^*} \int_\Omega |u_0 + t u_{\varepsilon,\eta}|^{2^*} - \int_\Omega f u_0 - t \int_\Omega f u_{\varepsilon,\eta}. \end{aligned} \tag{72}$$

We claim that, for a.e. $\eta \in \Sigma$,

$$\begin{aligned} \int_\Omega |u_0 + t u_{\varepsilon,\eta}|^{2^*} = &\int_\Omega |u_0|^{2^*} + t^{2^*} \int_\Omega u_{\varepsilon,\eta}^{2^*} + 2^* t \int_\Omega |u_0|^{2^*-2} u_0 u_{\varepsilon,\eta} \\ &+ 2^* t^{2^*-1} \int_\Omega u_{\varepsilon,\eta}^{2^*-1} u_0 + o(\varepsilon^{(Q-2)/2}). \end{aligned} \tag{73}$$

To verify (73), following Brezis–Nirenberg [8], we use the following formulas: for any $1 < q \leq 3$, there exists a positive constant C depending on q such that for $\alpha, \beta \in \mathbb{R}$ we have

$$\begin{aligned} &||\alpha + \beta|^q - |\alpha|^q - |\beta|^q - q\alpha\beta(|\alpha|^{q-2} + |\beta|^{q-2})| \\ &\qquad \leq C \begin{cases} |\alpha||\beta|^{q-1} & \text{if } |\alpha| \geq |\beta|, \\ |\alpha|^{q-1}|\beta| & \text{if } |\alpha| \leq |\beta|, \end{cases} \end{aligned} \tag{74}$$

and for $q \geq 3$, there exists a positive constant C depending on q such that for $\alpha, \beta \in \mathbb{R}$ we have

$$||\alpha+\beta|^q - |\alpha|^q - |\beta|^q - q\alpha\beta(|\alpha|^{q-2}+|\beta|^{q-2})| \leq C(|\alpha|^{q-2}\beta^2+\alpha^2|\beta|^{q-2}). \tag{75}$$

By the above formulas, we have that

$$\begin{aligned}\int_\Omega |u_0 + tu_{\varepsilon,\eta}|^{2^*} &= \int_\Omega |u_0|^{2^*} + t^{2^*}\int_\Omega u_{\varepsilon,\eta}^{2^*} + 2^*t\int_\Omega |u_0|^{2^*-2}u_0u_{\varepsilon,\eta} \\ &\quad + 2^*t^{2^*-1}\int_\Omega u_{\varepsilon,\eta}^{2^*-1}u_0 + R_\varepsilon^{(1)} + R_\varepsilon^{(2)},\end{aligned} \tag{76}$$

where $R_\varepsilon^{(1)}$ and $R_\varepsilon^{(2)}$ denote the terms corresponding, respectively, to the two terms in the right hand side of (74), if $Q \geq 6$, and the two terms in the right hand side of (75) if $3 \leq Q \leq 5$.

Now, if $Q \geq 6$, by (74) we have

$$|R_\varepsilon^{(1)}| \leq C\int_{\{|u_0|\geq tu_{\varepsilon,\eta}\}} |u_0|(tu_{\varepsilon,\eta})^{2^*-1}. \tag{77}$$

We split the integration into $d(\xi,\eta) < R/2$ and $d(\xi,\eta) \geq R/2$. Taking into account that the following decay estimate holds for Sobolev minimizers on Carnot groups

$$U(\xi) \leq C(1+d(\xi)^{Q-2})^{-1}, \quad \forall \xi \in \mathbb{G}, \tag{78}$$

(see [3, 29]), we have that $U_{\eta,\varepsilon}^{2^*-1} \leq C\varepsilon^{(Q+2)/2}$ on $\Omega \setminus B_d(\eta, R/2)$. Hence, letting u_0 to be zero outside Ω, from (77) we get

$$|R_\varepsilon^{(1)}| \leq C_t'\varepsilon^{\gamma_2(Q-2)/2}\int_{\mathbb{G}} \frac{|u_0(\xi)|^{1+\gamma_1}}{d(\eta^{-1}\circ\xi)^{\gamma_2(Q-2)}}\,\mathrm{d}\xi + C_t'\varepsilon^{(Q+2)/2}, \tag{79}$$

where $\gamma_1, \gamma_2 > 0$ are chosen to satisfy $\gamma_1+\gamma_2 = 2^*-1$ and $1 < \gamma_2 < Q/(Q-2)$. Note that

$$|u_0|^{1+\gamma_1} \in L^{2^*/(1+\gamma_1)} = L^{2^*/(2^*-\gamma_2)} \text{ and } d^{-\gamma_2(Q-2)} \in L^{Q/[(Q-2)\gamma_2],\infty}. \tag{80}$$

Hence, their convolution on $\mathbb{G}$ belongs to $L^{r,\infty}$, where $r = 2Q/[(Q-2)\gamma_2]$ (see [22, Theorem 1.2.13]). In particular

$$\int_{\mathbb{G}} \frac{|u_0(\xi)|^{1+\gamma_1}}{d(\eta^{-1}\circ\xi)^{\gamma_2(Q-2)}}\,\mathrm{d}\xi < +\infty, \quad \text{for a.e. } \eta \in \Sigma. \tag{81}$$

It follows that $R_\varepsilon^{(1)} = o(\varepsilon^{(Q-2)/2})$. In a similar fashion, estimating the right hand side of

$$|R_\varepsilon^{(2)}| \le C \int_{\{|u_0|<tu_{\varepsilon,\eta}\}} |u_0|^{2^*-1} tu_{\varepsilon,\eta}, \tag{82}$$

we get that $R_\varepsilon^{(2)} = o(\varepsilon^{(Q-2)/2})$, and so (73) follows for $Q \ge 6$.

Let us now consider the remaining cases $Q = 4, 5$: we omit the case $Q = 3$ since in this case $\mathbb{G}$ is necessarily the ordinary Euclidean group $(\mathbb{R}^3, +)$ and the result is proved in [8]. If $Q = 4$, in which case $2^* = 4$, by formula (75) we have

$$\begin{aligned} |R_\varepsilon^{(1)}|, |R_\varepsilon^{(2)}| &\le C_t' \int_{\mathbb{G}} |u_0(\xi)|^2 u_{\varepsilon,\eta}^2(\xi)\,\mathrm{d}\xi \\ &\le C_t' \int_{\mathbb{G}} |u_0(\xi)|^2 \frac{\varepsilon^{-2}}{(1+d(\delta_{1/\varepsilon}(\eta^{-1}\circ\xi))^2)^2}\,\mathrm{d}\xi \\ &\le C_t' \int_{\mathbb{G}} |u_0(\xi)|^2 \frac{\varepsilon^2}{(\varepsilon^2+d(\eta^{-1}\circ\xi)^2)^2}\,\mathrm{d}\xi, \end{aligned} \tag{83}$$

where we have used again (78). Let $\gamma_1, \gamma_2 > 0$ satisfy $\gamma_1 + \gamma_2 = 2$, with $\gamma_2 > 1$. Then

$$\frac{1}{(\varepsilon^2+d(\eta^{-1}\circ\xi)^2)^2} \le \frac{1}{\varepsilon^{2\gamma_1}} \frac{1}{d(\eta^{-1}\circ\xi)^{2\gamma_2}}, \tag{84}$$

which implies

$$|R_\varepsilon^{(1)}|, |R_\varepsilon^{(2)}| \le C_t' \varepsilon^{2-2\gamma_1} \int_{\mathbb{G}} |u_0(\xi)|^2 \frac{1}{d(\eta^{-1}\circ\xi)^{2\gamma_2}}\,\mathrm{d}\xi, \tag{85}$$

where the convolution integral in the right hand side is finite for a.e. $\eta \in \Sigma$, being $u_0 \in L^{2^*}$ (see [22]). Therefore, $R_\varepsilon^{(1)} = R_\varepsilon^{(2)} = O(\varepsilon^{2\theta})$, for any $\theta < 1$, and so (73) follows for $Q = 4$.

Finally, if $Q = 5$, by (75) we get

$$\begin{aligned} |R_\varepsilon^{(1)}| &\le C_t' \int_{\mathbb{G}} |u_0(\xi)|^{4/3} u_{\varepsilon,\eta}^2(\xi)\,\mathrm{d}\xi \\ &\le C_t' \varepsilon^2 \int_{\mathbb{G}} |u_0(\xi)|^{4/3} \frac{1}{\varepsilon^Q} U^2(\delta_{1/\varepsilon}(\eta^{-1}\circ\xi))\,\mathrm{d}\xi \end{aligned} \tag{86}$$

and

$$|R_\epsilon^{(2)}| \le C_t' \varepsilon^2 \int_{\mathbb{G}} |u_0(\xi)|^2 \frac{1}{d(\eta^{-1}\circ\xi)^4}\,\mathrm{d}\xi. \tag{87}$$

Note that, since $U \in L^{2^*/2,\infty}(\mathbb{G}) \cap L^\infty(\mathbb{G})$ (see the regularity results in [29, 30]), we have in particular that $U \in L^2(\mathbb{G})$ for $Q > 4$. Therefore, the kernel

$$K(\xi) := \frac{1}{\varepsilon^Q} U^2(\delta_{1/\varepsilon}(\eta^{-1} \circ \xi)) \tag{88}$$

appearing in (86), is a mollifier (up to a constant) and so

$$\int_{\mathbb{G}} |u_0(\xi)|^{4/3} \frac{1}{\varepsilon^Q} U^2(\delta_{1/\varepsilon}(\eta^{-1} \circ \xi))\, d\xi \to |u_0(\eta)|^{4/3} \int_{\mathbb{G}} U^2(\xi)\, d\xi \tag{89}$$

as $\varepsilon \to 0$. Moreover, concerning (87), it holds that

$$\int_{\mathbb{G}} |u_0(\xi)|^2 \frac{1}{d(\eta^{-1} \circ \xi)^4}\, d\xi < +\infty, \tag{90}$$

for a.e. $\eta \in \Sigma$.

Henceforth, we obtain that $R_\varepsilon^{(1)} = R_\varepsilon^{(2)} = O(\varepsilon^2)$, and so (73) holds also in this case.

Now, from [29] we know that, as $\varepsilon \to 0$

$$\int_\Omega |\nabla_{\mathbb{G}} u_{\varepsilon,\eta}|^2 = S^{Q/2} + O(\varepsilon^{Q-2}), \tag{91a}$$

$$\int_\Omega u_{\varepsilon,\eta}^{2^*} = S^{Q/2} + O(\varepsilon^Q). \tag{91b}$$

Therefore, by (72), (73), (91a), and (91b), and taking into account that u_0 is a weak solution of Eq. (1), we get

$$\begin{aligned} J(u_0 + t u_{\varepsilon,\eta}) \leq & J(u_0) + \frac{t^2}{2} S^{Q/2} - \frac{t^{2^*}}{2^*} S^{Q/2} - t^{2^*-1} \int_\Omega u_{\varepsilon,\eta}^{2^*-1} u_0 \\ & + o(\varepsilon^{(Q-2)/2}), \end{aligned} \tag{92}$$

for a.e. $\eta \in \Sigma$. Setting $u_0 = 0$ outside Ω, we have

$$\begin{aligned} \int_\Omega u_{\varepsilon,\eta}^{2^*-1} u_0 &= \varepsilon^{-(Q+2)/2} \int_{\mathbb{G}} u_0(\xi) \varphi_\eta^{2^*-1}(\xi) U^{2^*-1}(\delta_{1/\varepsilon}(\eta^{-1} \circ \xi))\, d\xi \\ &= \varepsilon^{(Q-2)/2} \int_{\mathbb{G}} u_0(\xi) \varphi_\eta^{2^*-1}(\xi) \frac{1}{\varepsilon^Q} U^{2^*-1}(\delta_{1/\varepsilon}(\eta^{-1} \circ \xi))\, d\xi. \end{aligned} \tag{93}$$

So, letting $\varepsilon \to 0$, as before we get

$$\int_{\mathbb{G}} u_0 \varphi_\eta^{2^*-1} \frac{1}{\varepsilon^Q} U^{2^*-1}(\delta_{1/\varepsilon}(\eta^{-1} \circ \xi))\, d\xi \to u_0(\eta) \int_{\mathbb{G}} U^{2^*-1}\, d\xi, \tag{94}$$

for a.e. $\eta \in \Sigma$. In other words, setting $\mathrm{D} = \int_{\mathbb{G}} U^{2^*-1}$, it holds

$$\int_\Omega u_{\varepsilon,\eta}^{2^*-1} u_0 = \varepsilon^{(Q-2)/2} u_0(\eta)\mathrm{D} + o(\varepsilon^{(Q-2)/2}). \tag{95}$$

Consequently, by (92) and (95) we get

$$\begin{aligned} J(u_0 + tu_{\varepsilon,\eta}) \leq & c_0 + (\frac{t^2}{2} - \frac{t^{2^*}}{2^*})S^{Q/2} - t^{2^*-1}u_0(\eta)\mathrm{D}\varepsilon^{(Q-2)/2} \\ & + o(\varepsilon^{(Q-2)/2}). \end{aligned} \tag{96}$$

Define

$$h(t) := (\frac{t^2}{2} - \frac{t^{2^*}}{2^*})S^{Q/2} - t^{2^*-1}u_0(\eta)\mathrm{D}\varepsilon^{(Q-2)/2}, \quad \text{for } t \geq 0. \tag{97}$$

Since $h(t)$ goes to $-\infty$ as $t \to \infty$, $\sup_{t\geq 0} h(t)$ is achieved at some $t_\varepsilon > 0$. Since t_ε satisfies

$$(t_\varepsilon - t_\varepsilon^{2^*-1})S^{Q/2} = (2^* - 1)t_\varepsilon^{2^*-2}u_0(\eta)\mathrm{D}\varepsilon^{(Q-2)/2}, \tag{98}$$

then necessarily $t_\varepsilon < 1$ and $t_\varepsilon \to 1$ as $\varepsilon \to 0$.

Set $t_\varepsilon = 1 - \delta_\varepsilon$. From (98), we obtain

$$(1 - \delta_\varepsilon - (1 - \delta_\varepsilon)^{2^*-1})S^{Q/2} = (2^* - 1)(1 - \delta_\varepsilon)^{2^*-2}u_0(\eta)\mathrm{D}\varepsilon^{(Q-2)/2}, \tag{99}$$

and so

$$(2^* - 2)S^{Q/2}\delta_\varepsilon = (2^* - 1)u_0(\eta)\mathrm{D}\varepsilon^{(Q-2)/2} + o(\varepsilon^{(Q-2)/2}). \tag{100}$$

Therefore

$$\begin{aligned} J(u_0 + tu_{\varepsilon,\eta}) &\leq c_0 + (\frac{t_\varepsilon^2}{2} - \frac{t_\varepsilon^{2^*}}{2^*})S^{Q/2} - t_\varepsilon^{2^*-1}u_0(\eta)\mathrm{D}\varepsilon^{(Q-2)/2} + o(\varepsilon^{(Q-2)/2}) \\ &= c_0 + (\frac{1}{2} - \frac{1}{2^*})S^{Q/2} - u_0(\eta)\mathrm{D}\varepsilon^{(Q-2)/2} + o(\varepsilon^{(Q-2)/2}) \\ &= c_0 + \frac{1}{Q}S^{Q/2} - u_0(\eta)\mathrm{D}\varepsilon^{(Q-2)/2} + o(\varepsilon^{(Q-2)/2}). \end{aligned} \tag{101}$$

Hence, for $\varepsilon_0 = \varepsilon_0(t, \eta) > 0$ sufficiently small, we conclude

$$J(u_0 + tu_{\varepsilon,\eta}) < c_0 + \frac{1}{Q}S^{Q/2}, \tag{102}$$

for any $0 < \varepsilon < \varepsilon_0$. □

Now, by means of the preceding lemma, we are able to prove that the value c_1 stands below the (PS)-threshold.

Lemma 4.4 *Let $f \neq 0$ satisfy* (7). *Then,*

$$c_1 < c_0 + \frac{1}{Q} S^{Q/2}. \tag{103}$$

Proof Following [44], to prove estimate (103) we provide a mountain pass critical level that is below the threshold $c_0 + \frac{1}{Q} S^{Q/2}$ and compares with c_1.

To this aim, observe that, under the assumption (7), the manifold $\mathcal{N}^-$ disconnects $S_0^1(\Omega)$ in two connected components E_1 and E_2.

Indeed, notice that, for every $u \in S_0^1(\Omega)$, $\|u\| = 1$, by Lemma 3.1, there exists a unique $t^+(u) > 0$ such that

$$t^+(u)u \in \mathcal{N}^- \quad \text{and} \quad J(t^+(u)u) = \max_{t \geq t_{\max}} J(tu). \tag{104}$$

Set

$$E_1 = \{0\} \cup \left\{u : \|u\| < t^+\left(\frac{u}{\|u\|}\right)\right\}, \quad E_2 = \left\{u : \|u\| > t^+\left(\frac{u}{\|u\|}\right)\right\}. \tag{105}$$

We have that $S_0^1 \setminus \mathcal{N}^- = E_1 \cup E_2$ and $\mathcal{N}^+ \subset E_1$.

Moreover, observe that there exists a suitable constant $C_1 > 0$ such that $0 < t^+(u) < C_1$, for all u, $\|u\| = 1$. Then, set $T := (S^{-Q/2}|C_1^2 - \|u_0\|^2|)^{1/2} + 1$, and fix $\eta \in \Sigma$ such that Lemma 4.3 applies. It is possible to prove that

$$u_0 + T u_{\varepsilon,\eta} \in E_2 \tag{106}$$

for $\varepsilon > 0$ sufficiently small (see [44]). So, for such a choice of T and $\eta \in \Sigma$, fix $\varepsilon > 0$ such that both (71) and (106) hold, and define:

$$\Gamma := \{h : [0,1] \to S_0^1(\Omega) \text{ continuous, } h(0) = u_0, h(1) = u_0 + T u_{\varepsilon,\eta}\}. \tag{107}$$

Clearly, $\widetilde{h} : [0,1] \to S_0^1(\Omega)$ defined by $\widetilde{h}(t) := u_0 + tTu_{\varepsilon,\eta}$ belongs to Γ. So, by Lemma 4.3, we conclude that

$$c := \inf_{h \in \Gamma} \max_{t \in [0,1]} J(h(t)) < c_0 + \frac{1}{Q} S^{Q/2}. \tag{108}$$

Moreover, since the range of any $h \in \Gamma$ intersects $\mathcal{N}^-$, it holds that

$$c_1 = \inf_{u \in \mathcal{N}^-} J(u) \leq c, \tag{109}$$

which concludes the proof. □

We are now able to complete the proof of Theorem 1.2, by providing the existence of the second solution.

Proof *(Theorem 1.2)* Analogously to the proof of Theorem 1.1, by Ekeland's variational principle, there exists a sequence $\{u_n\} \subset \mathcal{N}^-$ such that

$$J(u_n) \to c_1 \text{ and } \|J'(u_n)\| \to 0. \tag{110}$$

From Lemma 4.2 and (103), there exists $u_1 \in S_0^1(\Omega)$ such that, up to a subsequence

$$u_n \to u_1 \text{ strongly in } S_0^1(\Omega). \tag{111}$$

Hence, u_1 is a critical point for J, $u_1 \in \mathcal{N}^-$ (since $\mathcal{N}^-$ is closed) and $J(u_1) = c_1$. In particular, u_1 is a solution to problem (1), $u_1 \neq u_0$.

Moreover, if $f \geq 0$, let $t^+ > 0$ satisfy $t^+|u_1| \in \mathcal{N}^-$. From Lemma 3.1, we get

$$J(u_1) = \max_{t \geq t_{\max}} J(tu_1) \geq J(t^+u_1) \geq J(t^+|u_1|). \tag{112}$$

So, we can always take $u_1 \geq 0$. □

Acknowledgments The author thanks Professors Georgiev, Ozawa, Ruzhansky and Wirth, organizers of the Session "Harmonic Analysis and Partial Differential Equations" at ISAAC Congress 2019 in Aveiro (Portugal), where some of the present results were presented.

References

1. Aubin, J.P., Ekeland, I.: Applied Nonlinear Analysis, Pure and Applied Mathematics. Wiley Interscience Publications, New York (1984)
2. Balogh, Z.M., Kristály, A.: Lions-type compactness and Rubik actions on the Heisenberg group. Calc. Var. Partial Diff. Equ. **48**(1–2), 89–109 (2013)
3. Bonfiglioli, A., Uguzzoni, F.: Nonlinear Liouville theorems for some critical problems on H-type groups. J. Funct. Anal. **207**, 161–215 (2004)
4. Bonfiglioli, A., Lanconelli, E., Uguzzoni, F.: Stratified Lie groups and potential theory for their Sub-Laplacians. In: Springer Monographs in Mathematics. Springer, Berlin (2007)
5. Bony J.M.: Principe du maximum, inégalité de Harnack et unicité du problème de Cauchy pour les opérateurs elliptiques dégénérés. Ann. Inst. Fourier Grenobles **19**, 277–304 (1969)
6. Bordoni, S., Filippucci, R., Pucci, P.: Existence problems on Heisenberg groups involving hardy and critical terms. J. Geom. Anal. **30**, 1887–1917 (2020). Special Issue: Perspectives of Geometric Analysis in PDEs (2019)
7. Brezis, H., Nirenberg L.: Positive solutions of nonlinear elliptic equations involving critical exponents. Comm. Pure Appl. Math. **36**, 437–477 (1983)
8. Brezis, H., Nirenberg L.: A minimization problem with critical exponent and nonzero data. Symmetry in Nature Scuola Norm. Sup. Pisa **1**, 129–140 (1989)
9. Chen, Y., Chen, J.: Multiple positive solutions for a semilinear equation with critical exponent and prescribed singularity. Nonlinear Anal. **130**, 121–137 (2016)

10. Chen, J., Huang, L., Rocha, E.: Multiple non-negative solutions to a semilinear equation on Heisenberg group with indefinite nonlinearity. Boundary Value Problems **2015**(1), Article number: 165 (2015)
11. Christ, M., Liu, H., Zhang, A.: Sharp Hardy–Littlewood–Sobolev inequalities on the octonionic Heisenberg group. Calc. Var. Partial Differ. Equ. **55**(1), Art. 11, 18 pp. (2016)
12. Citti, G.: Semilinear Dirichlet problem involving critical exponent for the Kohn Laplacian. Ann. Mat. Pura Appl. **169**, 375–392 (1995)
13. Ferrara, M., Molica Bisci G.: Subelliptic and parametric equations on Carnot groups. Proc. Am. Math. Soc. **144**(7), 3035–3045 (2016)
14. Fischer, V., Ruzhansky, M.: Quantization on nilpotent Lie groups. In: Progress in Mathematics, vol. 314. Birkhäuser/Springer, Cham (2016)
15. Folland, G.B.: Subelliptic estimates and function spaces on nilpotent Lie groups. Ark. Mat. **13**, 161–207 (1975)
16. Folland, G.B., Stein, E.: Hardy spaces on homogeneous groups. In: Mathematical Notes, vol. 28. University Press, Princeton (1982)
17. Frank, R.L., Lieb E.H.: Sharp constants in several inequalities on the Heisenberg group. Ann. of Math. (2) **176**(1), 349–381 (2012)
18. Garagnani, E., Uguzzoni, F.: A multiplicity result for a degenerate-elliptic equation with critical growth on noncontractible domains. Topol. Methods Nonlinear Anal. **22**, 53–68 (2003)
19. Garofalo, N., Lanconelli, E.: Existence and nonexistence results for semilinear equations on the Heisenberg group. Indiana Univ. Math. J. **41**, 71–98 (1992)
20. Garofalo, N., Nhieu, D.M.: Isoperimetric and Sobolev inequalities for Carnot-Carathéodory spaces and the existence of minimal surfaces. Comm. Pure Appl. Math. **49**, 1081–1144 (1996)
21. Garofalo, N., Vassilev, D.: Regularity near the characteristic set in the non-linear Dirichlet problem and conformal geometry of sub-Laplacians on Carnot Groups. Math. Ann. **318**, 453–516 (2000)
22. Grafakos, L.: Classical Fourier Analysis, 2nd edn. Graduate Text in Mathematics, voi. 249. Springer, New York (2008)
23. Ivanov, S., Minchev, I., Vassilev, D.: The optimal constant in the L^2 Folland-Stein inequality on the quaternionic Heisenberg group. Ann. Sc. Norm. Super. Pisa Cl. Sci. (5) **11**(3), 635–652 (2012)
24. Jannelli, E., Loiudice, A.: Critical polyharmonic problems with singular nonlinearities. Nonlinear Anal. **110**, 77–96 (2014)
25. Jerison, D., Lee, J.: Extremals for the Sobolev inequality on the Heisenberg group and the CR Yamabe problem. J. Am. Math. Soc. **1**, 1–13 (1988)
26. Kang, D., Deng, Y.: Multiple solutions for inhomogeneous elliptic problems involving critical Sobolev-Hardy exponents. Nonlinear Anal. **60**, 729–753 (2005)
27. Lanconelli, E., Uguzzoni, F.: Non-existence results for semilinear Kohn-Laplace equations in unbounded domains. Comm. Partial Differ. Equations **25**, 1703–1739 (2000)
28. Loiudice, A.: Improved Sobolev inequalities on the Heisenberg group. Nonlinear Anal. **62**(5), 953–962 (2005)
29. Loiudice, A.: Semilinear subelliptic problems with critical growth on Carnot groups. Manuscripta Math. **124**, 247–259 (2007)
30. Loiudice, A.: L^p-weak regularity and asymptotic behavior of solutions for critical equations with singular potentials on Carnot groups. NoDEA Nonlinear Differ. Equ. Appl. **17**, 575–589 (2010)
31. Loiudice, A.: Critical growth problems with singular nonlinearities on Carnot groups. Nonlinear Anal. **126**, 415–436 (2015)
32. Loiudice, A.: Local behavior of solutions to subelliptic problems with Hardy potential on Carnot groups. Mediterr. J. Math. **15**(3), Art. 81, 20 (2018)
33. Loiudice, A.: Optimal decay of p-Sobolev extremals on Carnot groups. J. Math. Anal. Appl. **470**(1), 619–631 (2019)
34. Loiudice, A.: Asymptotic estimates and non-existence results for critical problems with Hardy term involving Grushin-type operators. Ann. Mat. Pura. Appl. **198**, 1909–1930 (2019)

35. Maalaoui, A., Martino, V.: Multiplicity result for a non-homogeneous Yamabe type equation involving the Kohn Laplacian. J. Math. Anal. Appl. **399**, 333–339 (2013)
36. Molica Bisci, G., Repovŝ, D.: Yamabe-type equations on Carnot groups. Potential Anal. **46**(2), 369–383 (2017)
37. Pinamonti, A.: Multiple solutions for possibly degenerate equations in divergence form. Electron. J. Differ. Equ. **220**, 1–10 (2016)
38. Pucci, P.: Critical Schrödinger-Hardy systems in the Heisenberg group. Discrete Contin. Dyn. Syst. Ser. S **12**(2), 375–400 (2019)
39. Ruzhansky, M., Suragan, D.: Hardy and Rellich inequalities, identities, and sharp remainders on homogeneous groups. Adv. Math. **308**, 483–528 (2017)
40. Ruzhansky, M., Suragan, D.: Hardy inequalities on homogeneous groups. 100 years of Hardy inequalities. In: Progress in Mathematics, vol. 327. Birkhäuser/Springer, Cham (2019)
41. Ruzhansky, M., Suragan, D., Yessirkegenov, N.: Caffarelli-Kohn-Nirenberg and Sobolev-type inequalities on stratified Lie groups. NoDEA Nonlinear Differ. Equ. Appl. Art. **56**, 12 (2017)
42. Ruzhansky, M., Suragan, D., Yessirkegenov, N.: Caffarelli-Kohn-Nirenberg inequalities, and remainders, stability, and superweights for L_p-weighted Hardy inequalities. Trans. Am. Math. Soc. B **5**, 32–62 (2018)
43. Shang, X., Zhang, J., Yang, Y.: Positive solutions of non-homogeneous fractional Laplacian problem with critical exponent. Commun. Pure Appl. Anal. **13**, 567–584 (2014)
44. Tarantello, G.: On non-homogeneous elliptic equations involving critical Sobolev exponent. Ann. Inst. H. Poincarè Anal. Non Linéaire **9**, 281–309 (1992)
45. Vassilev, V.: Existence of solutions and regularity near the characteristic boundary for sub-laplacian equations on Carnot Groups. Pacific J. Math. **227**(2) 361–397 (2006)

The Dixmier Trace and the Noncommutative Residue for Multipliers on Compact Manifolds

Duván Cardona and César Del Corral

Abstract In this paper we investigate the Dixmier trace and the noncommutative residue (also called Wodzicki's residue) of pseudo-differential operators by using the notion of a global symbol. We consider both cases, compact manifolds with or without boundary. Our analysis on the Dixmier trace of invariant pseudo-differential operators on closed manifolds will be based on the Fourier analysis associated with every elliptic and positive operator and the quantization process developed by Delgado and Ruzhansky. In particular, for compact Lie groups, this can be done by using the representation theory of the group in view of the Peter–Weyl Theorem and the Ruzhansky–Turunen symbolic calculus. The analysis of invariant pseudo-differential operators on compact manifolds with boundary will be based on the global calculus of pseudo-differential operators developed by Ruzhansky and Tokmagambetov.

Keywords Dixmier trace · Non-commutative residue · Global operators · Representation theory · Closed manifolds · Manifolds with boundary · Hörmander classes · Boutet de Monvel's algebra

1 Introduction

The Dixmier trace arises in functional analysis, from the problem if there exists a non-trivial trace—which does not coincide with the spectral trace—for an ideal containing the set of trace class operators on a Hilbert space H. In 1966, J. Dixmier in [15] showed that there exists a trace which vanishes in the ideal of trace class operators, but it is non-trivial in the so-called Dixmier ideal $\mathcal{L}^{1,\infty}(H)$. As it was

D. Cardona (✉)
Department of Analysis, Logic and Discrete Mathematics, Ghent University, Ghent, Belgium
e-mail: duvan.cardonasanchez@ugent.be

C. Del Corral
Department of Mathematics, Edificio Franco, Universidad de Los Andes, Bogota, Colombia
e-mail: ce-del@uniandes.edu.co

© Springer Nature Switzerland AG 2020
V. Georgiev et al. (eds.), *Advances in Harmonic Analysis and Partial Differential Equations*, Trends in Mathematics, https://doi.org/10.1007/978-3-030-58215-9_5

pointed out in [40] and references therein, the Dixmier trace is important due to its applications in fractal theory, foliation theory, spaces of non-commuting coordinates, perturbation theory, noncommutative geometry and quantum field theory.

In this paper, we study the Dixmier trace of invariant pseudo-differential operators (multipliers) on compact manifolds with or without boundary. By applying the Connes trace theorem (see [4]) we derive a formula for the noncommutative residue of Fourier multipliers on compact Lie groups, and manifolds with boundary. This is possible based on the recent quantizations of pseudo-differential operators associated with global symbols introduced by M. Ruzhansky and V. Turunen [32] (for compact Lie groups), J. Delgado and M. Ruzhansky [10] (for compact manifolds without boundary), and finally, by M. Ruzhansky and N. Tokmagambetov [14] (for compact manifolds with boundary). One of the advantages of our approach is that we obtain results on the Dixmier trace and the noncommutative residue for Fourier multipliers in terms of global (full) symbols, where we give a new viewpoint, instead of the classical results where the problem was treated by using local symbols (see e.g. [4, 15, 16, 18, 21, 35, 41] and references therein). We summarize our research in the following results:

- We provide necessary and sufficient conditions in order to guarantee that global invariant pseudo-differential operators on a manifold M with or without boundary lie in the *Dixmier class* $\mathcal{L}^{(1,\infty)}(H)$, $H = L^2(M)$, (see Eq. (2)).
- We provide sufficient and necessary conditions in order to obtain Dixmier traceability for a type of pseudo-differential operators with global symbols in Hörmander classes on compact Lie groups and we express our results by using the representation theory of these groups. Also, we use Connes' trace theorem in order to show formulae for the noncommutative residue of classical pseudo-differential operators on compact Lie groups.
- On a compact Lie group, we provide formulae in terms of global symbols for the noncommutative residue of a type of classical pseudo-differential operators in terms of the representation theory of the group (see Proposition 3.1).
- For a compact manifold M with or without boundary, we find criteria for global symbols in order that the corresponding operators belong to the Marcinkiewicz ideal $\mathcal{L}^{(p,\infty)}(H)$, $1 < p < \infty$, where $H = L^2(M)$, (see Eq. (4)).

In order to present our main results we precise some definitions. By following Connes [4], if H is a Hilbert space, the class $\mathcal{L}^{(1,\infty)}(H)$ consists of those bounded linear operators $A \in \mathcal{L}(H)$ satisfying

$$\sum_{1\leq n\leq N} s_n(A) = O(\log(N)), \quad N \to \infty, \tag{1}$$

where $\{s_n(A)\}$ denotes the sequence of singular values of A, i.e. the square roots of the eigenvalues of the non-negative self-adjoint operator A^*A. So, $\mathcal{L}^{(1,\infty)}(H)$ is endowed with the norm

$$\|A\|_{\mathcal{L}^{(1,\infty)}(H)} = \sup_{N\geq 2} \frac{1}{\log(N)} \sum_{1\leq n\leq N} s_n(A). \tag{2}$$

The Dixmier trace Tr_ω on positive operators in $\mathcal{L}^{(1,\infty)}(H)$ can be formulated by using an average function ω of the following form

$$\lim_{\omega} \frac{1}{\log(N)} \sum_{1\leq n\leq N} s_n(A) := \omega\big(\frac{1}{\log(N)} \sum_{1\leq n\leq N} s_n(A)\big), \tag{3}$$

where $\lim_\omega$ is a positive linear functional on $l^\infty(\mathbb{N})$, the set of bounded sequences, satisfying:

- $\lim_\omega(\alpha_n) \geq 0$ if all $\alpha_n \geq 0$;
- $\lim_\omega(\alpha_n) = \lim(\alpha_n)$ whenever the ordinary limit exists;
- $\lim_\omega(\alpha_1, \alpha_1, \alpha_2, \alpha_2, \alpha_3, \alpha_3, \dots) = \lim_w(\alpha_n)$.

The functional Tr_ω can be defined for positive operators, and later it can be extended to the whole ideal $\mathcal{L}^{(1,\infty)}(H)$ by linearity (in this case Tr_ω is not necessarily a positive functional). The functional Tr_ω is trivial on the ideal $\mathcal{L}^1(H)$ of trace class operators, (see [4] or [15]). The subset of $\mathcal{L}^{(1,\infty)}(H)$ with Dixmier trace independent of the average function ω defines the class of Dixmier measurable operators.

In noncommutative geometry, a remarkable result due to A. Connes shows that classical pseudo-differential operators acting on $L^2(M)$ with order $-\dim(M)$, belong to the Dixmier class, and the Dixmier trace coincides with the noncommutative residue (see [41]), which is the unique trace (up to by factors) in the algebra of pseudo-differential operators with classical symbols on a closed manifold, (see [4, pag. 305] for this fact). R. Nest, E. Schrohe in [26], show that in general Connes' result does not hold for the Boutet de Monvel algebra on a manifold with boundary (cf. (62) and (63)).

More generally, the ideal $\mathcal{L}^{(p,\infty)}(H)$ consists of those linear bounded operators A on H satisfying the condition:

$$\sum_{1\leq n\leq N} s_n(A) = O(N^{(1-1/p)}),\ \ N \to \infty, \tag{4}$$

for $1 < p < \infty$. On $\mathcal{L}^{(p,\infty)}(H)$ the usual norm is given by

$$\|A\|_{\mathcal{L}^{(p,\infty)}(H)} := \sup_{N\geq 1} N^{(\frac{1}{p})-1} \sum_{1\leq n\leq N} s_n(A). \tag{5}$$

In order to present our results we introduce briefly the theory of pseudo-differential operators that we use here [9, 23, 32]. The notion of global symbol on $\mathbb{R}^n$ is natural because a pseudo-differential operator A is an integral operator defined by

$$Af(x) = \int_{\mathbb{R}^n} e^{i2\pi x\xi} \sigma_A(x,\xi) \hat{f}(\xi)\, d\xi, \tag{6}$$

for a suitable smooth function $\sigma_A(x,\xi)$—called the symbol of A—and satisfying some bounded conditions on its derivatives (see [23]). Here $\hat{f}$ denotes the euclidean Fourier transform of the function f. If we consider a closed manifold M (i.e a compact manifold without boundary) and $H = L^2(M)$, there exists a (global) Fourier analysis associated to every positive elliptic pseudo-differential operator E on M which gives for certain pseudo-differential operators A—called E-*invariants*—a discrete Fourier representation of the form

$$Af(x) = \sum_{l=0}^{\infty} \langle \sigma_{A,E}(l)\widehat{f}(l), e_l(x)\rangle_{\mathbb{C}^{d_l}}, \tag{7}$$

where $e_l(x) := (e_l^m)_{1\leq m\leq d_l}$ and $\{e_l^m : l \in \mathbb{N}, 1 \leq m \leq d_l\}$ provides a basis of $L^2(M)$ consisting of eigenfunctions e_l^m, associated to certain eigenvalues λ_l, $l \in \mathbb{N}_0$, of the operator E. The function $\sigma_{A,E}$ is called the *matrix valued symbol* of A with respect to E. The process associating to every operator A acting in $C^\infty(M)$ a such matrix valued symbol $\sigma_{A,E}$ was introduced by J. Delgado and M. Ruzhansky, [13]. Also, If $M = G$ is a compact Lie group and $E = -\mathcal{L}_G$ is minus the Laplace–Beltrami operator on G, Ruzhansky and Turunen [32] give a discrete representation to every pseudo-differential operator A on $C^\infty(G)$ in terms of the representation theory of the group G, in the following way

$$Af(x) = \sum_{[\xi]\in\widehat{G}} d_\xi \operatorname{Tr}[\xi(x)\sigma_A(x,\xi)\widehat{f}(\xi)], \tag{8}$$

here $\widehat{G}$ denotes the unitary dual of the group G. Ruzhansky–Turunen's calculus gives a characterization of the Hörmander classes $\Psi^m_{\rho,\delta}(G)$ on a compact Lie group G by using the notion of global symbols. In fact, $A \in \Psi^m_{\rho,\delta}(G)$ if and only if its global symbol $\sigma_A(x,\xi)$ as in (8) satisfies

$$\|\Delta_\xi^\alpha \partial_x^\beta \sigma_A(x,\xi)\|_{op} \leq C_{\alpha,\beta}\langle\xi\rangle^{m-\rho|\alpha|+\delta|\beta|}, \quad x \in G, [\xi] \in \widehat{G}, \tag{9}$$

where $\langle\xi\rangle := (1+\lambda_{[\xi]})^{\frac{1}{2}}$ and $\{\lambda_{[\xi]} : [\xi] \in \widehat{G}\}$ is the spectrum of $-\mathcal{L}_G$. In Remark 2.1 we will recall the relationship between the symbols $\sigma_{A,-\mathcal{L}(G)}$ and σ_A associated to the quantizations (7) and (8), respectively.

Now, we recall that for a compact manifold M (without boundary) of dimension $\varkappa$, a classical pseudo-differential operator A on M of order m, can be defined by local symbols. This means that for any local chart U, the operator A has the form

$$Au(x) = \int_{T_x^*U} e^{ix\xi}\sigma^A(x,\xi)\widehat{u}(\xi)\,d\xi \tag{10}$$

where $\sigma^A(x,\xi)$ is a smooth function on $T^*U \cong U \times \mathbb{R}^n$, $T_x^*U = \mathbb{R}^n$, admitting an asymptotic expansion

$$\sigma^A(x,\xi) \sim \sum_{j=0}^{\infty} \sigma^{A_{m-j}}(x,\xi) \tag{11}$$

where $\sigma_{m-j}(x,\xi)$ are homogeneous functions in $\xi \neq 0$, of degree $m-j$ for ξ far from zero. The set of classical pseudo-differential operators of order m is denoted by $\Psi_{cl}^m(M)$. If $A \in \Psi_{cl}(M)$, for all $x \in M$, $\int_{|\xi|=1} \sigma_{-\varkappa}(x,\xi)\, d\xi\, dx$ defines a local density which can be glued over M. So, the noncommutative residue, is the functional defined on classical operators by

$$\mathrm{res}\,(A) = \frac{1}{\varkappa(2\pi)^{\varkappa}} \int_M \int_{|\xi|=1} \sigma^{A_{-\varkappa}}(x,\xi)\, d\xi\, dx. \tag{12}$$

An important feature of the noncommutative residue is that it vanishes on non-integer order classical operators. A complementary trace to the noncommutative residue is the canonical trace, $A \mapsto \mathrm{TR}(A)$, on the set of classical pseudo-differential operators of non-integer order. The term complementary is justified because the noncommutative residue is defined on the whole algebra of pseudo-differential operators, but does not extend the usual L^2-trace. However the canonical trace is not well defined on the whole algebra of classical pseudo-differential operators, but it extends the usual L^2-trace on a suitable set, see, e.g. [7, 22, 28, 29].

Now we present our main results. Our formulae for Dixmier traces will be presented for positive operators in $\mathcal{L}^{(1,\infty)}(L^2(M))$ because such formulae can be extended to the whole ideal $\mathcal{L}^{(1,\infty)}(L^2(M))$ by linearity. We start with the case of manifolds without boundary. Here, $\varkappa$ denotes the dimension of a compact manifold without boundary M. For every compact Lie group G, we define $\mathcal{L}^{(1,\infty)}(\widehat{G})$ to be the class of vector-valued matrices $M : \widehat{G} \to \bigcup_{[\xi]\in\widehat{G}} \mathbb{C}^{d_\xi \times d_\xi}$, such that, for every $[\xi] \in \widehat{G}$, $M([\xi]) \equiv M(\xi) \in \mathbb{C}^{d_\xi \times d_\xi}$, and

$$\|M\|_{\mathcal{L}^{(1,\infty)}(\widehat{G})} := \lim_{N\to\infty} \frac{1}{\log N} \sum_{[\xi]:\langle\xi\rangle\leq N} d_\xi \mathrm{Tr}(|M(\xi)|) < \infty.^1 \tag{13}$$

In a similar way, we define

$$\|M\|_{\mathcal{L}^{(p,\infty)}(\widehat{G})} := \sup_{N\geq 1} N^{(\frac{1}{p}-1)\dim G} \sum_{[\xi]:\langle\xi\rangle\leq N} d_\xi \mathrm{Tr}(|M(\xi)|) < \infty, \tag{14}$$

for every $1 < p < \infty$.

[1] As usual, $|M(\xi)|$ is defined via the spectral calculus of matrices by $|M(\xi)| = \sqrt{M(\xi)^*M(\xi)}$.

Theorem 1.1 *Let M be a $\varkappa$-dimensional compact manifold without boundary and let $E \in \Psi^{\nu}_{+e}(M)$ be a positive elliptic pseudo-differential operator on M. If $A : L^2(M) \to L^2(M)$ is a E-invariant bounded operator with matrix-valued symbol $(\sigma_{A,E}(l))_l$, then we have,*

- *A is Dixmier traceable, i.e. $A \in \mathcal{L}^{(1,\infty)}(L^2(M))$ if and only if*

$$\tau(A) := \frac{1}{\dim(M)} \lim_{N\to\infty} \frac{1}{\log N} \sum_{l:(1+\lambda_l)^{\frac{1}{\nu}} \leq N} \mathrm{Tr}(|\sigma_{A,E}(l)|) < \infty, \tag{15}$$

where $\sigma_{A,E}$ are as in (7). Moreover, if A is positive, $\tau(A) = \mathrm{Tr}_w(A)$.

- *The E-invariant operator $A \in \mathcal{L}^{(p,\infty)}(L^2(M))$ if and only if*

$$\gamma_p(A) := \sup_{N\geq 1} N^{\dim M(\frac{1}{p}-1)} \cdot \sum_{l:(1+\lambda_l)^{\frac{1}{\nu}} \leq N} \mathrm{Tr}(|\sigma_{A,E}(l)|) < \infty, \tag{16}$$

for all $1 < p < \infty$. In this case, the norms $\|A\|_{\mathcal{L}^{(p,\infty)}(L^2(M))}$ and $\gamma_p(A)$ are equivalents, which we denote by $\|A\|_{\mathcal{L}^{(p,\infty)}(L^2(M))} \asymp \gamma_p(A)$. If $M = G$ is a compact Lie group then

$$\|A\|_{\mathcal{L}^{(p,\infty)}(L^2(G))} \asymp \|\sigma_A\|_{\mathcal{L}^{(p,\infty)}(\widehat{G})}. \tag{17}$$

- *Let $M = G$ be a compact Lie group of dimension $\varkappa$ and let $\widehat{G}$ be the unitary dual of G. If we denote by $\sigma_A(x,\xi)$ the matrix valued symbol associated to A, then under the condition*

$$\|\Delta_\xi^\alpha \partial_x^\beta \sigma_A(x,\xi)\|_{op} \leq C\langle\xi\rangle^{-\varkappa-|\alpha|}, \quad x \in G, [\xi] \in \widehat{G}, \quad \forall\alpha \in \mathbb{R}^n, \tag{18}$$

the operator A is Dixmier measurable. Moreover, if A is left-invariant, then A is Dixmier traceable if and only if

$$\tau(A) := \frac{1}{\dim(G)} \times \|\sigma_A\|_{\mathcal{L}^{(1,\infty)}(\widehat{G})} < \infty. \tag{19}$$

In this case, if A is positive, $\tau(A) = \mathrm{Tr}_w(A)$.

- *If $A \in \Psi^{-\varkappa}_{cl}(G)$ is a classical, positive and left-invariant pseudo-differential operator, the noncommutative residue of A (cf. (12)), is given in terms of representations on G by*

$$\mathrm{res}(A) = \tau(A) \equiv \frac{1}{\dim(G)} \times \|\sigma_A\|_{\mathcal{L}^{(1,\infty)}(\widehat{G})}. \tag{20}$$

Moreover, if A is as above and its symbol admits an asymptotic expansion in homogeneous components of the form:

$$\sigma^A(x,\xi) \sim \sum_{j=0}^{\infty} a_{m-j}(x)\sigma^{A_{m-j}}(\xi), \tag{21}$$

then

$$\mathrm{res}(A) = \frac{1}{\dim(G)} \int_G a_{-\varkappa}(x)dx \times \|\sigma_{A_{-\varkappa}}\|_{\mathcal{L}^{(1,\infty)}(\widehat{G})}. \tag{22}$$

It is important to mention that our formula (15) for the Dixmier trace of a global operator A on a closed manifold M has two components: one is, the inverse of the dimension of M, which has a geometric nature. The other one has an analytic nature, which is defined by the global symbol $\sigma_{A,E}(\cdot)$ of A, determined by the Fourier analysis associated to E.

We observe that the approach in the preceding result can be used in order to analyze the Dixmier traceability of Fourier multipliers on compact homogeneous manifolds; this analysis in terms of the Fourier analysis and the representation theory of such manifolds has been considered in Theorem 4.1. Equation (22) provides a formula of the noncommutative residue for a class of global pseudo-differential operators on compact Lie groups. In particular, our approach, provide formulae for the noncommutative residue of operators on the torus, in a different way to the work of Pietsch [30].

Now we present our result concerning to global operators on a manifold with boundary. In the formulation of our result we use the *global quantization* of pseudo-differential operators on compact manifolds with boundary due to J. Delgado, M. Ruzhansky and N. Tokmagambetov [31] which we briefly describe as follows. If M denotes a compact manifold M with boundary ∂M and L is a pseudo-differential operator on M satisfying some boundary conditions on ∂M and with a merely discrete spectrum $\{\lambda_\xi \mid \xi \in \mathcal{I}\}$ (see Sect. 2.3), then every continuous linear operator A acting on a suitable domain $C_L^\infty(M)$, has associated a function $\sigma_{A,L} : M \times \mathcal{I} \to \mathbb{C}$—called the *$L$-global symbol* of A—satisfying

$$\sigma_{A,L}(x,\xi) = u_\xi(x)^{-1} A(u_\xi)(x), \quad \xi \in \mathcal{I},\ x \in M, \tag{23}$$

where u_ξ is the eigenvalue corresponding to λ_ξ. In this case, the operator A can be written in terms of the global symbol $\sigma_{A,L}(x,\xi)$ as

$$Af(x) = \sum_{\xi \in \mathcal{I}} u_\xi(x)\sigma_{A,L}(x,\xi)\widehat{f}(\xi), \quad \text{for } f \in C_L^\infty(M), \tag{24}$$

where $\widehat{f} := \mathcal{F}_L(f)$ denotes the L-Fourier transform of f introduced in [31], (see also (79)).

An *L-Fourier multiplier* is a bounded linear operator $A : C_L^\infty(M) \to C_L^\infty(M)$ if it satisfies

$$\mathcal{F}_L(Af)(\xi) = \sigma_{A,L}(\xi)\mathcal{F}_L(f)(\xi), \quad \text{for } f \in C_L^\infty(M), \tag{25}$$

for some function $\sigma_{A,L} : \mathcal{I} \to \mathbb{C}$ which depends only on the Fourier variable $\xi \in \mathcal{I}$, in this case $\sigma_{A,L}(\xi)$ corresponds with the L-symbol of A.

With the notation above our result on the Dixmier traceability of operators on manifolds with boundary can be enunciated in the following way. We write $\mathcal{I} = \{\xi_l : l \in \mathbb{N}_0\}$.

Theorem 1.2 *Let M be a compact manifold with boundary ∂M. If $A : L^2(M) \to L^2(M)$ is a bounded Fourier multiplier and L is a self-adjoint operator on $L^2(M)$ as in Sect. 2.3, then we have the following assertions,*

- *A is Dixmier traceable if and only if*

$$\tau'(A) := \lim_{N\to\infty} \frac{1}{\log N} \sum_{l\le N} |\sigma_{A,L}(\xi_l)| < \infty. \tag{26}$$

In this case, if A is positive, $\tau'(A) = \mathrm{Tr}_\omega(A)$.

- *Moreover, if L is an operator of order m satisfying the Weyl Counting Eigenvalue Formula, that is,*

$$N_L(\lambda) := \#\{l : |\lambda_{\xi_l}|^{\frac{1}{m}} \le \lambda\} = C_0\lambda^{\dim M} + O(\lambda^{\dim M-1}), \ \lambda \to \infty, \tag{27}$$

then A is Dixmier traceable if only if (cf. (26))

$$\tau'(A) = \frac{1}{\dim M} \lim_{N\to\infty} \frac{1}{\log N} \sum_{l:|\lambda_{\xi_l}|^{\frac{1}{m}}\le N} |\sigma_{A,L}(\xi_l)| < \infty. \tag{28}$$

In this case, if A is positive, $\tau'(A) = \mathrm{Tr}_\omega(A)$.

- *$A \in \mathcal{L}^{(p,\infty)}(L^2(M))$ if and only if*

$$\gamma_p'(A) := \sup_{N\ge 1} N^{(\frac{1}{p}-1)} \sum_{l\le N} |\sigma_{A,L}(\xi_l)| < \infty, \tag{29}$$

for all $1 < p < \infty$. In this case, $\|A\|_{\mathcal{L}^{(p,\infty)}(L^2(M))} \asymp \gamma_p'(A)$. Moreover, if L satisfies (27), we have

$$\gamma_p'(A) \asymp \sup_{N\ge 1} N^{\dim M(\frac{1}{p}-1)} \sum_{l:|\lambda_{\xi_l}|^{\frac{1}{m}}\le N} |\sigma_{A,L}(\xi_l)|. \tag{30}$$

Moreover, if we assume the existence of L as before, and $A = \begin{bmatrix} P_+ \\ T \end{bmatrix}$ *is an elliptic operator and positive in the Boutet de Monvel algebra of order n, then the operator* P_T *(realization P with respect to T) can be regarded as an operator in Ruzhansky–Tokmagambetov's calculus (see Remark* (5.2)*), and any parametrix R of* P_T *is Dixmier's traceable with*

$$\mathrm{Tr}_\omega(R) = \mathrm{res}\,(R) = \lim_{N\to\infty} \frac{1}{\ln N} \sum_{l\leq N} |(\sigma_{P,L}(\xi_l))^{-1}|, \tag{31}$$

here $(\sigma_{P,L}(\xi_l))_{l\in\mathbb{N}_0}$ *denotes the global symbol of the operator P with respect to* L_M*. The expression is the same for all parametrices and independent of the choice of the boundary condition and independent of the average function* ω*.*

As it was pointed out in [25, 39], the condition (27) holds true for several classes of elliptic operators on manifolds with boundary and without boundary, which are not necessarily compact manifolds. For our purposes, (27) is not restrictive because such condition implies that for some $s_0 \in \mathbb{R}$,

$$\tau(s_0, L) := \sum_{\xi\in I} \langle\xi\rangle^{-s_0} < \infty, \tag{32}$$

which is an important condition in the Ruzhansky–Tokmagambetov quantization (See Sect. 2.3) for manifolds with boundary. For an exhaustive historical perspective on this subject, we refer to [25]. Now, we present some references on the Dixmier traceability of pseudo-differential operators and related topics. The problem of classifying global pseudo-differential operators on certain ideals of bounded operators—as Schatten von Neumann classes and Gröthendieck nuclear operators—in terms of global symbols has been considered of a very satisfactory way in the references [9–13]. Nevertheless, the problem of finding sufficient conditions on the symbol σ of a pseudo-differential operator in order that the corresponding operator A will be trace class, Hilbert Schmidt or Dixmier traceable is classical. If $H = L^2(\mathbb{R}^n)$ and we consider the pseudo-differential operator A defined by the Weyl quantization,

$$Af(x) = \int_{\mathbb{R}^n}\int_{\mathbb{R}^n} e^{i2\pi(x,\xi)}\sigma_A\left(\frac{1}{2}(x+y),\xi\right) f(y)dyd\xi \tag{33}$$

is well known that $\sigma \in L^2(\mathbb{R}^{2n})$ implies that A is Hilbert–Schmidt. In the framework of the Weyl–Hörmander calculus of operators A associated to symbols σ in the $S(m, g)$-classes (see the definition of these classes in [23]), there exist two remarkable results. The first, due to Hörmander, which asserts that if $\sigma \in S(m, g)$ and $\sigma \in L^1(\mathbb{R}^{2n})$ then A is a trace class operator. The second, due to L. Rodino and

F. Nicola expresses that if $\sigma \in S(m, g)$ and $m \in L^1_w$, (the weak-L^1 space), then A is Dixmier traceable [27]. Moreover, an open conjecture by Rodino and Nicola (see [27]) says that $\sigma \in L^1_w(\mathbb{R}^{2n})$ gives an operator A with finite Dixmier trace. More general regularized traces for pseudo-differential operators (see S. Scoot [38] or S. Paycha [28] for complete a description) have been studied in different contexts, in first instance, G. Grubb, [19, 20] has obtained results about the zeta-regularized trace by using a resolvent approach; S. Paycha [29] has obtained important results about traces on meromorphic families of pseudo-differential operator; B. Fedosov, F. Golse, E. Leichtnam, E. Schrohe, [16] have defined the noncommutative residue and its properties on the algebra of pseudo-differential boundary value problems called Boutet de Monvel's algebra, in the case of manifolds with boundary; G. Grubb and E. Schrohe in [22] have studied defect formulas for regularized traces on the Boutet de Monvel algebra, whose operators are choosing to satisfy the so-called Hörmander transmission property. Recently, one of the authors has studied in [8] the traceability property and uniqueness of the canonical trace for a class of operators not necessarily satisfying the transmission property.

Finally, we explain the structure of our paper.

- In Sect. 2, we present some preliminaries on the quantization of global pseudo-differential operators on compact Lie groups and general compact manifolds with or without boundary and we briefly describe the Boutet de Monvel calculus. Also, we present some notions and known results about the Dixmier trace as well as the noncommutative residue.
- In Sect. 3 we prove our main results for operators acting on functions defined on compact manifolds without boundary. The case of compact manifolds with boundary will be addressed in Section 5.
- In Sect. 4 we study the Dixmier traceability of multipliers and we investigate the noncommutative residue in this setting.
- We end our paper with some examples in Sect. 6.

2 Preliminaries

In this section, we present some basics about the Fourier analysis used in this paper. Also, we present some definitions on the noncommutative residue (sometimes called Wodzicki's residue) and the Dixmier trace for pseudo-differential operators on compact manifolds with or without boundary. In this article, $\mathcal{L}^{(1,\infty)}(H)$ denotes the Dixmier ideal of the algebra of bounded operators on an Hilbert space H, $\mathcal{L}(H)$, $\mathrm{Spec}_p(A)$ denotes the pointwise spectrum of a (not necessarily bounded) linear operator A on H, and $\mathrm{Tr}_\omega(\cdot)$ is the Dixmier trace functional on the ideal $\mathcal{L}^{(1,\infty)}(H)$.

2.1 *Pseudo-Differential Operators on Compact Manifolds Without Boundary*

We recall that for every $m \in \mathbb{R}$ and for every open subset $U \subset \mathbb{R}^n$, the Hörmander class $S^m(U \times \mathbb{R}^n)$, (for a detailed description see [23]), is defined by functions satisfying the usual estimates

$$|\partial_x^\alpha \partial_\xi^\beta \sigma(x,\xi)| \leq C_{\alpha,\beta}\langle\xi\rangle^{m-|\beta|}, \tag{34}$$

for all $(x,\xi) \in T^*U \cong U \times \mathbb{R}^n$ and $\alpha, \beta \in \mathbb{N}^n$. Then, a pseudo-differential operator associated to a function $\sigma \in S^m(U \times \mathbb{R}^n)$ is a operator of the form

$$\sigma(x,D)f(x) = \int_{\mathbb{R}^n} e^{i2\pi(x,\xi)}\sigma(x,\xi)\widehat{f}(\xi)d\xi, \quad \text{for } f \in C_0^\infty(U), \tag{35}$$

where $C_0^\infty(U)$ denotes the set of smooth compact supported functions on U. The set of such operators is denoted by $\Psi^m(U \times \mathbb{R}^n)$. On a smooth manifold M without boundary of dimension n, for every $m \in \mathbb{R}$, the Hörmander class of order m, $S^m(M)$ is defined by smooth functions on the cotangent T^*M which in local coordinates coincides with symbols in some open sets U of $\mathbb{R}^n$ satisfying inequalities as in (34). We observe that the notion of symbol on arbitrary manifolds is of local nature. However, it is possible to define a notion of global symbol if we restrict our attention to the case of compact Lie groups. Such a notion was developed by M. Ruzhansky and V. Turunen in [32]. In this theory, every operator A mapping $C^\infty(G)$ itself, where G is a compact Lie group, can be described in terms of representations of G as follows.

Let $\widehat{G}$ be the unitary dual of G (i.e, the set of equivalence classes of continuous irreducible unitary representations on G), the Ruzhansky–Turunen approach establishs that A has associated a *matrix-valued global (or full) symbol* $\sigma_A(x,\xi) \in \mathbb{C}^{d_\xi \times d_\xi}$, $[\xi] \in \widehat{G}$, on the noncommutative phase space $G \times \widehat{G}$ satisfying

$$\sigma_A(x,\xi) = \xi(x)^*(A\xi)(x). \tag{36}$$

Then it can be shown that the operator A can be expressed in terms of such a symbol as [32],

$$Af(x) = \sum_{[\xi]\in\widehat{G}} d_\xi \mathrm{Tr}(\xi(x)\sigma_A(x,\xi)\widehat{f}(\xi)). \tag{37}$$

An important feature in this setting is that the Hörmander classes $\Psi^m(G)$, $m \in \mathbb{R}$ where characterized in [32, 34] by the condition: $A \in \Psi^m(G)$ if only if its matrix-valued symbol $\sigma_A(x,\xi)$ as before satisfies

$$\|\partial_x^\alpha \mathbb{D}^\beta \sigma_A(x,\xi)\|_{op} \le C_{\alpha,\beta}\langle\xi\rangle^{m-|\beta|}, \tag{38}$$

for every $\alpha, \beta \in \mathbb{N}^n$. For a rather comprehensive treatment of this quantization process we refer to [32].

The notion of global symbol on arbitrary compact manifolds is more delicate. This problem has been considered by J. Delgado and M. Ruzhansky in [11, 12]. Now, we explain this notion. If M is a compact manifold without boundary with a volume element dx, and E is a positive elliptic classical pseudo-differential operator of order $\nu > 0$, we say that $A : C^\infty(M) \to C^\infty(M)$ is E-invariant if A and E commutes. In this case we can associate a full symbol to A in terms of the spectrum of E as follows: the eigenvalues of E form a positive sequence λ_j, which we label as

$$0 \le \lambda_0 \le \lambda_1 \le \lambda_2 \le \cdots \le \lambda_n \le \cdots. \tag{39}$$

For every eigenvalue λ_j the corresponding eigenspace $H_j = \mathrm{Ker}(E - \lambda_j I)$. If $\lambda_0 = 0$ then $d_0 = \dim H_0 = \dim \mathrm{Ker}(E)$. By the spectral theorem we can write

$$L^2(M) = \bigoplus_{j=0}^{\infty} H_j. \tag{40}$$

So, if we denote by $\{e_{jk}\}_{k=1}^{d_j}$ a basis of H_j for $j \in \mathbb{N}$, where $d_j = \dim H_j$, and for every $f \in L^2(M)$, we define the E-Fourier transform of f (relative to the operator E and the basis $\{e_{jk}\}_{k,j}^{d_j}$) at $j \in \mathbb{N}$ by

$$(\mathcal{F}_E f)(j) := \widehat{f}(j) := \begin{bmatrix} \langle f, e_{j1}\rangle_{L^2(M)} \\ \langle f, e_{j2}\rangle_{L^2(M)} \\ \vdots \\ \langle f, e_{jd_j}\rangle_{L^2(M)} \end{bmatrix}_{d_j\times 1}. \tag{41}$$

If e_j denotes the column vector with entries $e_{j1}, e_{j2}, \cdots e_{jd_j}$ then, for every $f \in L^2(M)$, it follows from the Parseval Theorem that f has a Fourier series representation of the type,

$$f(\cdot) = \sum_{l=0}^{\infty} \langle \widehat{f}(l), e_l(\cdot)\rangle_{d_l}. \tag{42}$$

where $\langle\cdot,\cdot\rangle_{d_l}$ is the usual inner product on $\mathbb{C}^{d_l}$. If $\{\sigma(l)\}_{l\in\mathbb{N}_0}$ is a sequence of matrices such that for every j, $\sigma(j)$ is a square matrix of order d_j, the E-invariant operator A associated to $\{\sigma(l)\}_{l\in\mathbb{N}_0}$ is the operator on $C^\infty(M)$ defined by

$$(Af)(\cdot) = \sum_{l=0}^{\infty} \langle \sigma(l)\widehat{f}(l), e_l(\cdot)\rangle_{d_l}. \tag{43}$$

Since $\widehat{Af}(l) = \sigma(l)\widehat{f}(l)$ we refer to E-invariant operators A as *E-Fourier multipliers*, or simply by Fourier multipliers, multipliers or E-multipliers. An important feature in the Delgado–Ruzhansky's approach is that there exist formulae between the symbol $\sigma(l)$ of a E-multiplier A from $\mathscr{D}'(M)$ into $\mathscr{D}'(M)$ (we use the notation $\mathscr{D}'(M)$ for the space of distributions in M) and its Schwartz kernel $K(x, y)$ (see [23]). In fact we have

$$\sigma(l) = \int_M \int_M K(x, y) Q_l(x, y)^* dx dy \tag{44}$$

and

$$K(x, y) = \sum_{l=0}^{\infty} \mathrm{Tr}(\sigma(l) Q_l(x, y)) \tag{45}$$

where $Q_l(x, y) = \overline{e_l(y)} e_l(x)^t$. We end this section with the following remark of global symbols on compact Lie groups which will be useful in our analysis related with the Dixmier trace for Fourier multipliers.

Remark 2.1 If A is left-invariant (with respect to the Laplacian) then we have two notions of global symbols for A. One is defined in terms of the representation theory of the group G and we will denote this symbol by $(\sigma_A(\xi))_{[\xi]\in\widehat{G}}$, and the other one is that defined when we consider the compact Lie group as a manifold, and in this case the symbol will be denoted by $(\sigma_A(l))_{l\in\mathbb{N}_0}$. The relation of this two symbols was discovered in [13, pag. 25]. Now, we describe this relation. In the setting of compact Lie groups the unitary dual being discrete, we can enumerate the unitary dual as $[\xi_j]$, for $j \in \mathbb{N}_0$. In this way we fix the orthonormal basis

$$\{e_{jk}\}_{k=1}^{d_j} = \{d_{\xi_j}^{\frac{1}{2}}(\xi_j)_{il}\}_{i,l=1}^{d_{\xi_j}} \tag{46}$$

where $d_j = d_{\xi_j}^2$. Then, we have the subspaces $H_j = \mathrm{span}\{(\xi_j)_{i,l} : i, l = 1, \cdots, d_{\xi_j}\}$. With the notation above we have

$$\sigma_A(l) = \begin{bmatrix} \sigma_A(\xi_l) & 0_{d_{\xi_l}\times d_{\xi_l}} & 0_{d_{\xi_l}\times d_{\xi_l}} & \cdots & 0_{d_{\xi_l}\times d_{\xi_l}} \\ 0_{d_{\xi_l}\times d_{\xi_l}} & \sigma_A(\xi_l) & 0_{d_{\xi_l}\times d_{\xi_l}} & \cdots & 0_{d_{\xi_l}\times d_{\xi_l}} \\ \vdots & \vdots & \vdots & \ddots & \vdots \\ 0_{d_{\xi_l}\times d_{\xi_l}} & 0_{d_{\xi_l}\times d_{\xi_l}} & 0_{d_{\xi_l}\times d_{\xi_l}} & \cdots & \sigma_A(\xi_l) \end{bmatrix}_{d_l\times d_l}. \tag{47}$$

As a consequence of this discussion we obtain the following relation:

$$\mathrm{Tr}(|\sigma_A(l)|) = d_{\xi_l}\mathrm{Tr}(|\sigma_A(\xi_l)|), \tag{48}$$

where $|A| := \sqrt{A^*A}$ denotes the absolute value of the matrix A.

2.2 Global Fourier Multipliers on Compact Homogeneous Manifolds

In order to present our theorem for multipliers on compact homogeneous spaces, we recall some definitions on the subject. Compact homogeneous manifolds can be obtained if we consider the quotient of a compact Lie groups G with one of its closed subgroups K—there exists an unique differential structure for the quotient $M := G/K$–. Examples of compact homogeneous spaces are spheres $\mathbb{S}^n \cong \mathrm{SO}(n+1)/\mathrm{SO}(n)$, real projective spaces $\mathbb{RP}^n \cong \mathrm{SO}(n+1)/\mathrm{O}(n)$, complex projective spaces $\mathbb{CP}^n \cong \mathrm{SU}(n+1)/\mathrm{SU}(1)\times\mathrm{SU}(n)$ and more generally Grassmannians $\mathrm{Gr}(r,n) \cong \mathrm{O}(n)/\mathrm{O}(n-r)\times\mathrm{O}(r)$.

Let us denote by $\widehat{G}_0$ the subset of $\widehat{G}$, representations of G, that are class I with respect to the subgroup K. This means that $\pi \in \widehat{G}_0$ if there exists at least one non trivial invariant vector a with respect to K, i.e., $\pi(h)a = a$ for every $h \in K$. Let us denote by B_π to the vector space of these invariant vectors and $k_\pi = \dim B_\pi$. Now we follow the notion of Multipliers as in [1]. Let us consider the class of symbols $\Sigma(M)$, for $M = G/K$, consisting of those matrix-valued functions

$$\sigma : \bigcup \widehat{G} \to \bigcup_{n=1}^{\infty} \mathbb{C}^{n\times n} \quad \text{such that} \quad \sigma(\pi)_{ij} = 0 \text{ for all } i, j > k_\pi. \tag{49}$$

Following [1], a Fourier multiplier A on M is a bounded operator on $L^2(M)$ such that for some $\sigma_A \in \Sigma(M)$ satisfies

$$Af(x) = \sum_{\pi\in\widehat{G}_0} d_\pi \mathrm{Tr}(\pi(x)\sigma_A(\pi)\widehat{f}(\pi)), \quad \text{for } f \in C^\infty(M), \tag{50}$$

where $\widehat{f}$ denotes the Fourier transform of the lifting $\dot{f} \in C^\infty(G)$ of f to G, given by $\dot{f}(x) := f(xK)$, $x \in G$.

Remark 2.2 For every symbols of a Fourier multipliers A on M, only the upper-left block in $\sigma_A(\pi)$ of the size $k_\pi \times k_\pi$ can be not the trivial matrix zero.

2.3 Global Pseudo-Differential Operators on Compact Manifolds with Boundary

The natural analogy of the algebra of pseudo-differential operators on a closed manifold is the algebra of Boutet de Monvel (BdM), which will be described in this section (we will use a kernel integral description of such algebra, but also we give a local symbolic description), where we follow the references [2, 3, 36]. It was defined in [16], by B. Fedosov, F. Golsche, E. Leichtmann, and E. Schrohe a linear functional trace, agreeing with the noncommutative residue whether the boundary is empty, and it can be shown that such functional is the unique trace on the BdM algebra algebra. Moreover, R. Nest and E. Schrohe showed in [26] that such functional does not coincide with the Dixmier trace (with the exception of the closed case where the boundary is empty), however, they provide a particular set of operators in Boutet de Monvel's algebra where these two traces agree.

Let M be an n-dimensional compact smooth manifold with boundary ∂M which is embedded on a compact closed manifold Ω, (we use n to denote the dimension of the manifolds underlying instead of $\varkappa$, as in the case of a closed manifold). Boutet de Monvel's approach comprises operators acting on smooth functions over M and over ∂M as follows. A matrix operator of the form

$$A = \begin{bmatrix} P_{+}+G & K \\ T & S \end{bmatrix} : C^{\infty}(\bar{M}) \oplus C^{\infty}(\partial M) \to C^{\infty}(\bar{M}) \oplus C^{\infty}(\partial M), \tag{51}$$

is said to be an operator in the Boutet de Monvel algebra of order $m \in \mathbb{Z}$ and of type $r \in \mathbb{N}$, if

- $P_{+} := r^{+}Pe^{+}$ is an operator, called a *truncated operator*, obtained from a classical pseudo-differential operator P on Ω (here r^{+} corresponds to the restriction map from functions on Ω to functions on M and the last one e^{+} refers to the extension by zero from functions on M to functions Ω), and the operator P satisfies the *Hörmander transmission property* (see [2]), namely, if $\sigma^{P}(x,\xi)$ denotes the local symbol satisfying (11), (in local coordinates $(x,\xi) = (x', x_n, \xi', \xi_n) \in \mathbb{R}^{n-1} \times \mathbb{R} \times \mathbb{R}^{n-1} \times \mathbb{R}$ on the boundary) of P, then for all α, β and any $j = 0, 1, \cdots$, holds

$$\partial_{\xi'}^{\alpha}\partial_{x_n}^{\beta}\sigma_j^{P}(x', 0, 0, +1) = (-1)^{j-|\alpha|}\partial_{\xi'}^{\alpha}\partial_{x_n}^{\beta}\sigma_j^{P}(x', 0, 0, -1). \tag{52}$$

- K is called a *Poisson operator* of order m, defined by

$$(Kv)(x) = \int_{\mathbb{R}^{n-1}} e^{ix'\xi'}\tilde{k}(x', x_n, \xi')\mathcal{F}[v](\xi')d\xi', \tag{53}$$

with $\tilde{k} \in S^{m-1}(\mathbb{R}^{n-1}, \mathcal{S}(\mathbb{R}_+))$, in the sense of (58). In this case, the symbol associated with the operator K is defined by

$$k(x', \xi', \xi_n) := \mathcal{F}_{x_n \to \xi_n}(\tilde{k}(x', x_n, \xi')) \tag{54}$$

(here $\mathcal{F}_{x \to y}(f)$ means the Fourier transform of f which transform the variable x to y) and it is assumed an expansion in homogeneous functions $k_{m-1-j}(x', \xi', \xi_n)$ with respect to ξ (for $|\xi| \geq 1$) of degree $m - 1 - j$.

- T is called a *Trace Operator* of order m, a type $r \in \mathbb{N}$, and it has the form $T = \sum\limits_{j=0}^{r-1} S_J \gamma_j + T'$, where S_j is pseudo-differential operator on $\mathbb{R}^{n-1}$ of order $m - j$, and T' can be defined by the integral operator

$$(T'u)(x') = \int_{\mathbb{R}^{n-1}} t'(x', x_n, \xi')\mathcal{F}_{x' \to \xi'}[u](\xi', x_n) dx_n d\xi', \tag{55}$$

with $\tilde{t} \in S^m(\mathbb{R}^{n-1} \times \mathbb{R}^{n-1}, \mathcal{S}(\mathbb{R}_+))$, in the sense of (58), similarly the symbol of T is defined by

$$t(x', \xi', \xi_n) := \mathcal{F}_{x_n \to \xi_n}(\tilde{t}(x', x_n, \xi')). \tag{56}$$

- G is called a singular Green operator (s.G.o.) of order m and type r; it is defined by an operator having the form $G = \sum_{j=0}^{r-1} K_j \gamma_j + G'$, where K_j are Poisson operators; $\gamma_j u(x') := D_{x_n}^j u(x', x_n)|_{x_n=0}$ (defines trace operators), and G' can be defined by the integral operator

$$(G'u)(x) := \int_{\mathbb{R}^{n-1}} e^{ix'\xi'} \Big(\int_0^\infty \tilde{g}(x', x_n, y_n, \xi')\mathcal{F}_{x' \to \xi'}[u](\xi', y_n) \Big) d\xi', \tag{57}$$

with $\tilde{g} \in S^m(\mathbb{R}^{n-1} \times \mathbb{R}^{n-1}; \mathcal{S}(\overline{\mathbb{R}}_+ \times \overline{\mathbb{R}}_+))$, i.e. $\tilde{g} \in C^\infty(\mathbb{R}^{n-1} \times \mathbb{R}^{n-1}; \mathcal{S}(\overline{\mathbb{R}}_+ \times \overline{\mathbb{R}}_+))$ such that for all $\alpha, \beta, l, l', k, k'$

$$\|x_n^k D_{x_n}^{k'} y_n^l D_{y_n}^{l'} D_{\xi'}^\alpha D_{x'}^\beta \tilde{g}(x', x_n, y_n, \xi')\|_{L^2(\mathbb{R}^2_{++})} \leq C(x') \langle \xi' \rangle^{m+1-k+k'-l+l'-|\alpha|} \tag{58}$$

for a suitable positive constant $C(x')$ depending of x'. The function $\tilde{g}$ has an asymptotic expansion $\tilde{g} \sim \sum_{j \geq 0} \tilde{g}_{m-j}$ where $\tilde{g}_{m-j}$ satisfy the following homogenety property: for $\lambda \geq 1$ and $|\xi'| \geq 1$,

$$\tilde{g}_{d-j}\Big(x', \frac{1}{\lambda}x_n, \frac{1}{\lambda}y_n, \xi'\Big) = \lambda^{m+2-j} \tilde{g}_{d-j}(x', x_n, y_n, \xi'). \tag{59}$$

In this case, the symbol associated with the operator G is defined by

$$g(x', \xi', \xi_n, \eta_n) := \mathcal{F}_{x_n \to \xi_n} \overline{\mathcal{F}}_{y_n \to \eta_n} (e^+_{x_n} e^+_{y_n} \tilde{g}'(x', x_n, y_n, \xi')), \tag{60}$$

here $(\overline{\mathcal{F}}\phi)(\xi) := (\mathcal{F}\phi)(-\xi)$ for any $\phi \in L^2(\mathbb{R})$, the homogeneity property (59) corresponds to

$$g_{d-j}(x', \lambda\xi', \lambda\xi_n, \lambda\eta_n) = \lambda^{d-j} g(x', \xi', \xi_n, \eta_n), \; |\xi| \geq 1, \; \lambda \geq 1. \tag{61}$$

- Finally, S is a classical pseudo-differential operator on $\mathbb{R}^{n-1}$ of order m.

We have described Boutet de Monvel's algebra acting on functions with values in $\mathbb{R}$ or $\mathbb{C}$, which our setting of interest, however, such description can be done for a general setting as vector bundles.

Now, let us mention an important result about the set of the boundary operators described above.

Theorem 2.1 **(*L. Boutet de Monvel, [3]*)** *Matrix operators as in (51) form an algebra, called the* Boutet de Monvel algebra, *more precisely, the composition AB of an operator $A \in \mathcal{B}$ with all entries of order m and type r with an operator $B \in \mathcal{B}$ with all entries of order m' and type r' yields an operator in $\mathcal{B}$ of order $m + m'$ and type* $\max\{m' + r, r'\}$.

Let us mention that Boutet de Monvel's algebra contains the parametrix operators of elements in such algebra (whether this exists). A parametrix here means that $AB - I = S_1$ and $BA - I = S_2$ are regularizing operators; their types are 0 and m, respectively.

On the one hand, in [16] it is defined a linear functional on $\mathcal{B}$, which extends the noncommutative residue to the case of a manifold with boundary,

$$\begin{aligned} \operatorname{res}(A) =& \frac{(2\pi)^{-n}}{n} \int_M \int_{|\xi|=1} p_{-n}(x, \xi)\, d\xi dx \\ &+ \frac{(2\pi)^{-n+1}}{n} \int_{\partial M} \int_{|\xi'|=1} \operatorname{tr}_n(g_{-n})(x', \xi') + s_{1-n}(x', \xi')\, d\xi' dx', \end{aligned} \tag{62}$$

where $\operatorname{tr}_n(g_{-n})(x', \xi') := \int_0^\infty \tilde{g}_{-n}(x', x_n, x_n, \xi')\, dx_n$. On the other hand, R. Nest and E. Schrohe in [26] study the noncommutative trace and the Dixmier trace for suitable sets of the Boutet de Monvel algebra. The authors obtain a first result about the Dixmier class.

Proposition 2.1 **(*[26], Proposition 2.1.*)** *A bounded operator on the zero order Sobolev space $L^2(M)$ with range in the n-order Sobolev space $H^n(M)$ is an element of $\mathcal{L}^{(1,\infty)}(M)$; moreover, if its range even is contained in $H^{n+1}(M)$ then it is trace class.*

If $\mathcal{D}^m$ denotes the set of matrix-operator A in $\mathcal{B}$, as (51), for which P is an operator of order $m \in \mathbb{Z}$; G is a. s.G.o. of order m and type zero, K is of order $m+1$ and type zero, and S is an operator on the boundary of order $m+1$. In [26] is showed that any operator A in $\mathcal{D}^{-n}$ defines a bounded map $A : L^2(M) \oplus L^2(\partial M) \to H^{-n}(M) \oplus H^{1-n}(\partial M)$. Therefore, $\mathcal{D}^{-n}$ forms a subset of the Dixmier class $\mathcal{L}^{(1,\infty)}(H)$ where $H = L^2(M; E) \oplus L^2(\partial M; F)$. Moreover, the authors show in the following result a explicit formula for the Dixmier trace in terms of the terms appearing in the noncommutative residue (cf. (62)), for operators in $\mathcal{D}^{-n}$.

Theorem 2.2 *For $A \in \mathcal{D}^{-n}$ acting on $H = L^2(M; E) \oplus L^2(\partial M; F)$, we have*

$$\begin{aligned} \mathrm{Tr}_\omega(A) =& \frac{(2\pi)^{-n}}{n} \int_M \int_{|\xi|=1} p_{-n}(x,\xi)\, d\xi dx \\ &+ \frac{(2\pi)^{-n+1}}{n-1} \int_{\partial M} \int_{|\xi'|=1} s_{-n+1}(x',\xi')\, d\xi' dx'. \end{aligned} \tag{63}$$

In particular, $\mathrm{Tr}_\omega(A)$ *is independent of the averaging procedure* ω.

Remark 2.3 The Theorem 2.2 shows that Conne's theorem does not hold for the Boutet de Monvel algebra, see (62).

Given $P : C^\infty(M) \to C^\infty(M)$ a differential operator of order $m > 0$. In general one is interested in solving either the in-homogeneous problem

$$Pu = f \ \text{ on } M; \quad Tu = g \ \text{ at } \partial M, \tag{64}$$

for f and g given, or else the homogeneous problem

$$Pu = f \ \text{ on } M; \quad Tu = 0 \ \text{ at } \partial M. \tag{65}$$

In order to treat problem (64), one consider operator of the form

$$A = \begin{bmatrix} P \\ T \end{bmatrix} : H^m(M) \to L^2(M) \oplus H^{m-m'-1/2}(\partial M), \tag{66}$$

where m' denotes de order of the trace operator T and it is such that $m' < m$. For treat (65) one studies the *realization* P_T which is defined as the unbounded operator P_T on $L^2(M)$, acting like P over

$$\mathcal{D}(P_T) = \{u \in H^m(M) \mid Tu = 0\}. \tag{67}$$

The ellipticity implies that there is a parametrix to P_T in Boutet de Monvel's calculus. It is of the form $B = (Q_+ + G \;\; K)$; the pseudo-differential part Q is a parametrix to P, while G is a singular Green operator of order $-m$ and type zero and K is a potential operator of order $-m$.

Proposition 2.2 (*[26], Theorem 3.2.*) *Let P_T be the above elliptic boundary value problem and $B = (Q_+ + G \quad K)$ its parametrix. Then there is a regularizing singular Green operator G_0 of type zero, i.e. an integral operator with smooth kernel on $M \times M$, such that $R = (Q_+ + G + G_0)$ has the following properties:*

(i) R maps $L^2(M)$ to $\mathcal{D}(P_T)$ and
(ii) $RP_T - I$ and $P_T R - I$ are finite rank operators whose range consists of smooth functions.

It shown in Corollary 3.2. in [26] that the parametrix R in the above theorem is unique up to a regularizing singular Green operator of type 0.

Proposition 2.3 ([26, Corollary 3.2.]) *If $m = n$ and $A = \begin{bmatrix} P_+ \\ T \end{bmatrix}$ is elliptic and positive, the Dixmier trace for an arbitrary parametrix R to the operator P_T, we have*

$$\operatorname{Tr}_\omega(R) = \frac{1}{(2\pi)^n n} \int_M \int_{|\xi|=1} (\sigma_n^P(x, \xi))^{-1} \, d\xi dx, \tag{68}$$

where $\sigma_n^P(x, \xi)$ denotes the homogeneous component for $\sigma^P(x, \xi)$ of degree n. The expression is the same for all parametrices and independent of the choice of the boundary condition. Moreover, it coincides with the noncommutative residue res (R) *for R.*

By using non-harmonic analysis, M. Ruzhansky and N. Tokmagambetov in [31] give a different approach in terms of global symbols, which we summarize here. As above, M denotes a compact manifold of dimension n with boundary ∂M, and L_M denotes the boundary value problem determined by a pseudo-differential operator L of order m on function in $L^2(M)$, on the interior of M, satisfying a suitable boundary conditions (BC) on ∂M. It is assumed the following conditions:

- The operator L associated with the (BC) has a discrete spectrum $\{\lambda_\xi \in \mathbb{C} \,|\, \xi \in \mathcal{I}\}$, where $\mathcal{I}$ is a countable subset of $\mathbb{Z}^k$ for some $k \geq 1$, and the eigenvalues are ordered with the occurring multiplicities in the increasing order $|\lambda_j| \leq |\lambda_k|$ for $|j| \leq |k|$. The authors introduce the weight

$$\langle \xi \rangle := (1 + |\lambda_\xi|^2)^{1/2m}, \tag{69}$$

  and they assume that there exists a number $s_0 \in \mathbb{R}$ such that

$$\tau(s_0, L) := \sum_{\xi \in \mathcal{I}} \langle \xi \rangle^{-s_0} < \infty. \tag{70}$$

- If u_ξ denote the eigenfunction in $L^2(M)$ of L associated with the (BC) corresponding to the eigenvalue λ_ξ for $\xi \in \mathcal{I}$, so

$$Lu_\xi = \lambda_\xi u_\xi \quad \text{for} \quad \xi \in \mathcal{I}. \tag{71}$$

The adjoint spectrum problem is

$$L^* v_\xi = \overline{\lambda}_\xi v_\xi \quad \text{for} \quad \xi \in \mathcal{I}, \tag{72}$$

which is equipped with the conjugate boundary conditions (BC)*. We can take biorthonormal system $\{u_\xi\}$ and $\{v_\xi\}$ for $\xi \in \mathcal{I}$, i.e. $\|u_\xi\| = \|v_\xi\| = 1$ for $\xi \in \mathcal{I}$ and

$$\langle u_\xi, v_\eta \rangle = 0 \quad \text{for} \quad \xi \neq \eta \quad \text{and} \ \langle u_\xi, v_\eta \rangle = 1 \quad \text{for} \quad \xi = \eta, \tag{73}$$

where $\langle f, g \rangle := \int_M f x)\overline{g(x)}\, dx$ is the usual product on the Hilbert space $L^2(M)$. In this context, we assume that the system $\{u_\xi \mid \xi \in \mathcal{I}\}$ forms a basis of $L^2(M)$, i.e. for any $f \in L^2(M)$ there exists a unique series $\sum_{\xi \in \mathcal{I}} a_\xi u_\xi(x)$, so $\{v_\xi \mid \xi \in \mathcal{I}\}$ is a basis of $L^2(M)$ too.

- Also we assume the functions u_ξ and v_ξ do not have zeros in M for all $\xi \in \mathcal{I}$ and there exits $C > 0$ and $N \geq 0$ such that as $\langle \xi \rangle \to \infty$,

$$\inf_{x \in M} |u_\xi(x)| \geq C \langle \xi \rangle^{-N}, \quad \inf_{x \in M} |u_\xi(x)| \geq C \langle \xi \rangle^{-N}. \tag{74}$$

In this case the systems $\{u_\xi \mid \xi \in \mathcal{I}\}$ is called a *WZ-system (without zeros system).*

As it was mentioned in [14, Remark 2.2], the condition given by WZ-system can be removed, however, under this situation the analysis underlying leads a matrix-valued version of the symbolic calculus, similar to Sect. 2.1, which consists in vectors of eigenfunctions so that its elements do not all vanish at the same time-typical examples of such situation is of operators on compact Lie groups, as it was described before.

Now, we describe some elements involved in Ruzhansky–Tokmagambetov's calculus which will be needed in this paper. The space

$$C_L^\infty(M) := \bigcap_{k=1}^{\infty} \mathcal{D}(L^k) \tag{75}$$

where $\mathcal{D}(L^k) := \{f \in L^2(M) \mid L^j f \in \mathcal{D}(L),\ j = 0, 1, \cdots, k\}$, so that the boundary condition (BC) are satisfied by all the operators L^j. The Fréchet topology of $C_L^\infty(M)$ is given by the family of norms

$$\|f\|_{C_L^k} := \max_{j \leq k} \|L^j f\|_{L^2(M)}, \ k \in \mathbb{N}_0, \ f \in C_L^\infty(M). \tag{76}$$

Similarly, it is defined $C_{L^*}^\infty(M)$ corresponding to the adjoint L^* by

$$C_{L^*}^\infty(M) := \bigcap_{k=1}^{\infty} \mathcal{D}((L^*)^k) \tag{77}$$

where $\mathcal{D}((L^*)^k) := \{f \in L^2(M) \mid (L^*)^j f \in \mathcal{D}(L),\ j = 0, 1, \cdots, k\}$, which also has to satisfy the adjoint boundary conditions corresponding to the operator L^*. The Fréchet topology of $C_{L^*}^{\infty}(M)$ is given by the family of norms

$$\|f\|_{C_{L^*}^k} := \max_{j \le k} \|(L^*)^j f\|_{L^2(M)},\ k \in \mathbb{N}_0,\ f \in C_L^{\infty}(M). \tag{78}$$

Since $\{u_\xi\}$ and $\{v_\xi\}$ are dense in $L^2(M)$ that $C_L^{\infty}(M)$ and $C_{L^*}^{\infty}(M)$ are dense in $L^2(M)$.

L-Fourier transform: Let $\mathcal{S}(\mathcal{I})$ be the space of rapidly decreasing functions $\phi : \mathcal{I} \to \mathbb{C}$, i.e. for any $N < \infty$, there exists a constant $C_{\phi,N}$ such that $|\phi(\xi)| \le C_{\phi,N}\langle\xi\rangle^{-N}$ for all $\xi \in \mathcal{I}$. The space $\mathcal{S}(\mathcal{I})$ forms a Fréchet space with the family of semi-norms $p_k(\phi) := \sup_{\xi\in\mathcal{I}}\langle\xi\rangle^k|\phi(\xi)|$. The L-Fourier transform is a bijective homeomorphism $\mathcal{F}_L : C_L^{\infty}(M) \to \mathcal{S}(\mathcal{I})$ defined by

$$(\mathcal{F}_L f)(\xi) := \hat{f}(\xi) := \int_M f(x)\overline{v_\xi(x)}\,dx. \tag{79}$$

The inverse operator $\mathcal{F}_L^{-1} : \mathcal{S}(\mathcal{I}) \to C_L^{\infty}(M)$ is given by

$$(\mathcal{F}^{-1}h)(x) := \sum_{\xi\in\mathcal{I}} h(\xi)u_\xi(x), \tag{80}$$

so that the Fourier inversion formula is given by

$$f(x) = \sum_{\xi\in\mathcal{I}} \hat{f}(\xi)u_\xi(x),\ \ f \in C_L^{\infty}(M). \tag{81}$$

Similarly, the L^*-Fourier transform is a bijective homeomorphism $\mathcal{F}_L : C_{L^*}^{\infty}(M) \to \mathcal{S}(\mathcal{I})$ defined by

$$(\mathcal{F}_{L^*} f)(\xi) := \hat{f}_*(\xi) := \int_M f(x)\overline{u_\xi(x)}\,dx. \tag{82}$$

Its inverse $\mathcal{F}_{L^*}^{-1} : \mathcal{S}(\mathcal{I}) \to C_{L^*}^{\infty}(M)$ is given by $(\mathcal{F}_{L^*}^{-1}h)(x) := \sum_{\xi\in\mathcal{I}} h(\xi)v_\xi(x)$ so that the conjugate Fourier inversion formula is given by

$$f(x) := \sum_{\xi\in\mathcal{I}} \hat{f}_*(\xi)v_\xi(x),\ \ f \in C_{L^*}^{\infty}(M). \tag{83}$$

The space $\mathcal{D}'_L(M) := \mathcal{L}(C_{L^*}^{\infty}(M, \mathbb{C}))$ of linear continuous functionals on $C_{L^*}^{\infty}(M)$ is called the space of L-*distribution*. By dualizing the inverse L-Fourier transform

$\mathcal{F}_L^{-1} : \mathcal{S}(\mathcal{I}) \to C_L^\infty(M)$, the L-Fourier transform extends uniquely to the mapping

$$\mathcal{F}_L : \mathcal{D}'_L(M) \to \mathcal{S}'(\mathcal{I}) \tag{84}$$

by the formula $\langle \mathcal{F}_L w, \phi \rangle := \langle w, \overline{\mathcal{F}_{L^*}^{-1}\overline{\phi}} \rangle$ with $w \in \mathcal{D}'_L(M)$, $\phi \in \mathcal{S}(\mathcal{I})$.

Moreover, the authors define Sobolev space adapted to L_M. The space $l_L^2 := \mathcal{F}_L\big(L^2(M)\big)$ is defined as the image of $L^2(M)$ under the L-Fourier transform. Then the space of l_L^2 is a Hilbert space with the linear product

$$(a, b)_{l_L^2} := \sum_{\xi \in \mathcal{I}} a(\xi)\overline{(\mathcal{F}_{L^*} \circ \mathcal{F}^{-1} b(\xi))}. \tag{85}$$

Then the space l_L^2 consists of the sequences of the Fourier coefficients of function in $L^2(M)$, in which Plancherel identity holds, for $a, b \in l_L^2$,

$$(a, b)_{l_L^2} = (\mathcal{F}_L^{-1} a, \mathcal{F}_L^{-1} b)_{L^2}. \tag{86}$$

For $f \in \mathcal{D}'_L(M) \cap \mathcal{D}'_{L^*}(M)$ and $s \in \mathbb{R}$, we say that

$$f \in \mathcal{H}_L^s(M) \text{ if and only if } \langle \xi \rangle^s \hat{f}(\xi) \in l_L^2, \tag{87}$$

provided with the norm

$$\|f\|_{\mathcal{H}_L^s} := \Big(\sum_{\xi \in \mathcal{I}} \langle \xi \rangle^{2s} \hat{f}(\xi)\overline{\hat{f}_*(\xi)}\Big)^{1/2}. \tag{88}$$

Proposition 2.4 (*[26], Proposition 6.3.*) *For every $s \in \mathbb{R}$, the Sobolev space $\mathcal{H}_L^s(M)$ is a Hilbert space with the inner product*

$$(f, g)_{\mathcal{H}_L^s(M)} := \sum_{\xi \in \mathcal{I}} \langle \xi \rangle^{2s} \hat{f}(\xi)\overline{\hat{g}_*(\xi)}, \tag{89}$$

and the Sobolev space $\mathcal{H}_L^s(M)$ and $\mathcal{H}^s(M)$ are isometrically isomorphic.

L-Quantization and L-symbol: It can be shown that, see Theorem 9.2. in [31], any continuous linear operator $A : C_L^\infty(M) \to C_L^\infty(M)$ can be expressed in terms the L-Fourier and its inverse transform as

$$Af(x) = \sum_{\xi \in \mathcal{I}} u_\xi(x)\sigma_{A,L}(x, \xi)\hat{f}(\xi) \tag{90}$$

for every $f \in C_L^\infty(M)$ and $x \in M$, where $\sigma_{A,L}(x,\xi)$ is called *the L-symbol* of A and can be computed as (see [31, Theorem 9.2])

$$\sigma_{A,L}(x,\xi) = u_\xi(x)^{-1}(Au_\xi)(x) \quad \text{for all } x \in M \text{ and } \xi \in \mathcal{I}. \tag{91}$$

In particular, if $A : C_L^\infty(M) \to C_L^\infty(M)$ be a continuous linear operator such that

$$\mathcal{F}_L(Af)(\xi) = \sigma(\xi)\mathcal{F}_L(f)(\xi), \quad \text{for } f \in C_L^\infty(M) \text{ and } \xi \in \mathcal{I}, \tag{92}$$

for some $\sigma : \mathcal{I} \to \mathbb{C}$; then A is called a *L-Fourier multiplier*.

The *adjoint operator* A^* is defined by the equation

$$\langle Au_\xi, v_\xi\rangle_{L^2} = \langle u_\xi, A^*v_\xi\rangle_{L^2}. \tag{93}$$

It follows from the Parceval identity, see Proposition 6.1. in [31] that A is an L-Fourier multiplier by $\sigma(\xi)$ if and only if A^* is an L^*-Fourier multiplier by $\overline{\sigma(\xi)}$.

Moreover, in [31], the authors provided several results in the same asymptotic spirit of Hörmander calculus for global symbols, e.g. asymptotic formulas for the symbol of the adjoint operator A^* and for the symbol of any parametrix R of A, in terms of the symbol of the operator A; for the symbol of the product AB in terms of the symbol of A and B.

3 The Dixmier Trace and the Noncommutative Residue of Fourier Multipliers on Manifolds Without Boundary

In this section we proof our main results concerning to the Dixmier trace and the noncommutative residue for invariant operators on a closed manifold M (a compact manifold without boundary).

Remark 3.1 We are interested in bounded E-multipliers on $L^2(M)$. We recall that the following inequality

$$\sup_{l\in\mathbb{N}_0} \|\sigma_{A,E}(l)\|_{op} < \infty \tag{94}$$

is a necessary and sufficient condition for the boundedness of A on $L^2(M)$. This is an immediate consequence of the Plancherel Formula.

Now, we present our main theorem of this section.

Theorem 3.1 *Let M be a compact manifold without boundary and let E in the set $\Psi_{+e}^\nu(M)$ of positive elliptic pseudo-differential operators on M. If $A : L^2(M) \to L^2(M)$ is a bounded E- invariant operator with matrix-valued symbol $\sigma_{A,E}(l)$, then*

A is Dixmier traceable if and only if

$$\tau(A) := \frac{1}{\dim(M)} \lim_{N\to\infty} \frac{1}{\log N} \sum_{l:(1+\lambda_l)^{\frac{1}{\nu}} \leq N} \mathrm{Tr}(|\sigma_{A,E}(l)|) < \infty. \tag{95}$$

Moreover, if A is positive, $\tau(A) = \mathrm{Tr}_w(A)$.

Proof We want to show that (95) is a necessary and sufficient condition for the Dixmier traceability of a E-invariant operator A with full symbol $\sigma_{A,E}(l)$. As a consequence of the relation

$$\mathcal{F}_E(Af)(l) = \sigma_{A,E}(l)(\mathcal{F}_E f)(l), \ \ l \in \mathbb{N}_0, \tag{96}$$

we obtain

$$\mathrm{Spec}_p(A) = \bigcup_{l\in\mathbb{N}_0} \mathrm{Spec}_p(\sigma_{A,E}(l)). \tag{97}$$

In fact, if $\lambda \in \mathrm{Spec}_p(A) = \ker(A - \lambda I)$, there exists $u \neq 0$, such that $Au = \lambda u$. If we take the E-Fourier transform to both sides, we have $\sigma_{A,E}(l)\widehat{u}(l) = \lambda\widehat{u}(l)$. There exists $l_0 \neq 0$, such that $\widehat{u}(l_0) \neq 0$. Hence we obtain that λ is a eigenvector of the matrix $\sigma(l_0)$. For the proof of the converse, let us assume that for some $l_0 \in \mathbb{N}_0$, the complex number $\lambda \in \mathrm{Spec}_p(\sigma(l_0))$ is an eigenvalue with eigenvector $u(l_0)$. By contradiction let us assume that λ is not a element of $\Sigma_p(A)$. It follows that A is a Fourier multiplier it have not residual or continuous spectrum, therefore λ is in the resolvent set of A, in particular $A - \lambda I$ is an injective operator. So we have that $(A-\lambda I)v = 0$ implies that $v = 0$. Let us consider the sequence of matrices $(v_l)_{l\in\mathbb{N}_0}$, defined by $v_l = 0$ (zero matrix of dimension d_l) if $l \neq l_0$, and $v_{l_0} = u(l_0)$. If u is the inverse E-Fourier transform of this sequence, we have $\widehat{u}(l) = v_l$. It is clear that for every l, $\sigma_{A,E}(l)\widehat{u}(l) = \lambda\widehat{u}(l)$. By taking the inverse E-Fourier transform, we have $Au = \lambda u$ with $u \neq 0$ which is a contradiction.

Applying (97) to the operator $\sqrt[2]{A^*A}$ we obtain the following relation for the singular values of A :

$$s(A) = \bigcup_{l\in\mathbb{N}_0} s(\sigma_{A,E}(l)). \tag{98}$$

Moreover, if $\lambda_{1,d_l}, \lambda_{2,d_l}, \cdots \lambda_{k_l,d_l}$ are the singular values of $\sigma_{A,E}(l)$ with corresponding multiplicities $m_{1,d_l}, m_{2,d_l}, \cdots m_{k_l,d_l}$, then

$$m_{1,d_l} + m_{2,d_l} + \cdots m_{k_l,d_l} = d_l \tag{99}$$

and

$$m_{1,d_l}\lambda_{1,d_l} + m_{2,d_l}\lambda_{2,d_l} + \cdots m_{k_l,d_l}\lambda_{k_l,d_l} = \mathrm{Tr}[|\sigma_{A,E}(l)|], \tag{100}$$

where $|\sigma_{A,E}(l)| = \sqrt[2]{\sigma_{A,E}(l)^*\sigma_{A,E}(l)}$. With notations above, A is Dixmier traceable (by definition) if and only if

$$\tau(A) := \lim_{N\to\infty} \frac{\sum_{l\leq N} \mathrm{Tr}(|\sigma_{A,E}(l)|)}{\log(\sum_{l\leq N} d_l)} = \lim_{N\to\infty} \frac{\sum_{l\leq N}\sum_{j=1}^{k_l} m_{j,d_l}\lambda_{j,d_l}}{\log(\sum_{l\leq N}\sum_{j=1}^{k_l} m_{j,d_l})} < \infty. \tag{101}$$

In this case $\tau(A) = \mathrm{Tr}_w(A)$. Since

$$\lim_{N\to\infty} \frac{\sum_{l\leq N} \mathrm{Tr}(|\sigma_{A,E}(l)|)}{\log(\sum_{l\leq N} d_l)} = \lim_{N\to\infty} \frac{\sum_{(1+\lambda_l)^{\frac{1}{\nu}}\leq N} \mathrm{Tr}(|\sigma_{A,E}(l)|)}{\log(\sum_{(1+\lambda_l)^{\frac{1}{\nu}}\leq N} d_l)}, \tag{102}$$

by using the Weyl Eigenvalue Counting formula for the operator E we have

$$\sum_{(1+\lambda_l)^{\frac{1}{\nu}}\leq L} d_l = C_0 L^{\varkappa} + O(L^{\varkappa-1}), \tag{103}$$

where $\varkappa = \dim(M)$. So, we obtain

$$\mathrm{Tr}_w(A) = \frac{1}{\dim(M)} \lim_{N\to\infty} \frac{1}{\log N} \sum_{(1+\lambda_l)^{\frac{1}{\nu}}\leq N} \mathrm{Tr}(|\sigma_{A,E}(l)|) \tag{104}$$

which is the desired result. □

An important fact in the formulation of our analysis is the following result proved by A. Connes in the 1980s, (see [4, pag. 307]).

Theorem 3.2 (A. Connes [4]) *Let M be a closed manifold of dimension $\varkappa$. Then every classical pseudo-differential operator A of order $-\varkappa$ lies in $\mathcal{L}^{(1,\infty)}(L^2(M))$ and*

$$\mathrm{Tr}_w(A) = \mathrm{res}(A). \tag{105}$$

Now, we apply the preceding theorem and some tools of representation theory in order to study the Dixmier trace and the noncommutative residue for the operator on a compact Lie group.

Theorem 3.3 *Let $M = G$ be a compact Lie group of dimension $\varkappa$ and $\widehat{G}$ be the unitary dual of G. If we denote by $\sigma_A(x,\xi) \equiv \sigma_{A,-\mathcal{L}_G}(x,\xi)$ the matrix valued symbol associated to A, then under the condition*

$$\|\Delta_\xi^\alpha \partial_x^\beta \sigma_A(x,\xi)\|_{op} \leq C\langle\xi\rangle^{-\varkappa-|\alpha|}, \quad x\in G, [\xi]\in\widehat{G}, \tag{106}$$

the operator A is Dixmier measurable. Moreover, if A is left-invariant and positive, its Dixmier trace is given by

$$\mathrm{Tr}_w(A) = \frac{1}{\dim(G)} \lim_{N\to\infty} \frac{1}{\log N} \sum_{\xi:\langle\xi\rangle\leq N} d_\xi \mathrm{Tr}(|\sigma_A(\xi)|). \tag{107}$$

Proof We observe that by condition (38), the operator A is a pseudo-differential operator of order $-\varkappa$. It follows from Theorem 5.2 in [33] that A is bounded on $L^2(G)$. Then, by Connes' Theorem in the form of Theorem 3.2, A is Dixmier traceable. Now, if A is a Fourier multiplier on G, from Remark 2.1 and Eq. (48), we deduce

$$\mathrm{Tr}(|\sigma_A(l)|) = d_{\xi_l} \mathrm{Tr}(|\sigma_A(\xi_l)|). \tag{108}$$

Hence (95) becomes

$$\begin{aligned} \mathrm{Tr}_w(A) &= \lim_{N\to\infty} \frac{\sum_{l\leq N} \mathrm{Tr}(|\sigma_{A,E}(l)|)}{\log[\sum_{l\leq N} d_l]} \\ &= \lim_{N\to\infty} \frac{\sum_{l\leq N} d_{\xi_l} \mathrm{Tr}(|\sigma_A(\xi_l)|)}{\log[\sum_{l\leq N} d_{\xi_l}^2]} = \lim_{N\to\infty} \frac{\sum_{\langle\xi\rangle\leq N} d_\xi \mathrm{Tr}(|\sigma_A(\xi)|)}{\log[\sum_{\langle\xi\rangle\leq N} d_{\xi_l}^2]} \end{aligned} \tag{109}$$

which proves (107). Finally, that (107) is a necessary and sufficient condition for the Dixmier traceability of A it follows from (95). By the Weyl Eigenvalue Counting Formula (see [9, pag. 539]), we have

$$\sum_{\langle\xi\rangle\leq N} d_\xi^2 = C_0 L^\varkappa + O(L^{\varkappa-1}), \tag{110}$$

where $\varkappa = \dim(G)$. Hence

$$\mathrm{Tr}_w(A) = \frac{1}{\dim(G)} \lim_{N\to\infty} \frac{1}{\log N} \sum_{\langle\xi\rangle\leq N} d_\xi \mathrm{Tr}(|\sigma_A(\xi)|) \tag{111}$$

which completes the proof. □

With an analogous analysis as in the previous result we obtain the following characterization of invariant pseudo-differential operators belonging to the Marcinkiewicz ideal $\mathcal{L}^{(p,\infty)}(L^2(M))$.

Theorem 3.4 *Let M be a compact manifold without boundary and let A be a bounded E-invariant operator on $L^2(M)$. Then $A \in \mathcal{L}^{(p,\infty)}(L^2(M))$ if only if*

$$\gamma_p(A) := \sup_{N\geq 1} N^{\dim M(\frac{1}{p}-1)} \sum_{l:(1+\lambda_l)^{\frac{1}{\nu}}\leq N} \mathrm{Tr}(|\sigma_{A,E}(l)|) < \infty, \tag{112}$$

for all $1 < p < \infty$. In this case $\gamma_p(A) \asymp \|A\|_{\mathcal{L}^{(p,\infty)}(L^2(M))}$.

Proof With the notation above we can give a short proof for this fact. Indeed, for $N > 0$ and $S_N := \sum_{l\leq N} d_l$, $A \in \mathcal{L}^{(p,\infty)}(L^2(M))$ if and only if

$$\sum_{l\leq N} \mathrm{Tr}(|\sigma_{A,E}(l)|) = O(S_N^{1-\frac{1}{p}}). \tag{113}$$

By the Weyl counting eigenvalue function, (113) becomes

$$\sum_{l:(1+\lambda_l)^{\frac{1}{\nu}}\leq N} \mathrm{Tr}(\,|\sigma_{A,E}(l)|\,) = O((C_0 N^{\varkappa})^{1-\frac{1}{p}}) = O(N^{\varkappa(1-\frac{1}{p})}), \tag{114}$$

where $\varkappa = \dim(M)$. This observation completes the proof, indeed,

$$\|A\|_{\mathcal{L}^{(p,\infty)}(H)} \asymp \sup_{N\geq 1} N^{\frac{1}{p}-1} \sum_{l:(1+\lambda_l)^{\frac{1}{\nu}}\leq N} \mathrm{Tr}(|\sigma_{A,E}(l)|) =: \gamma_p(A) < \infty. \tag{115}$$

Corollary 3.1 *If $M = G$ is a compact Lie group, and A is a Fourier multiplier,* (125) *gives that*

$$\|\sigma_A\|_{\mathcal{L}^{(p,\infty)}(\widehat{G})} := \sup_{N\geq 1} N^{(\frac{1}{p}-1)\dim G} \sum_{[\xi]:\langle\xi\rangle\leq N} d_\xi \mathrm{Tr}(|\sigma_A(\xi)|) < \infty, \tag{116}$$

is a sufficient and necessary condition in order that $A \in \mathcal{L}^{(p,\infty)}(L^2(G))$ for all $1 < p < \infty$.

Remark 3.2 The classification of E-invariant operators A on Schatten-von Neumann classes $S_p(L^2(M)) = \mathcal{L}^{(p,p)}(L^2(M))$ and on the class of r-nuclear operators $N_r(L^2(M))$ have been considered in [9–13]. A remarkable result in [11] shows that the inequality

$$\sum_{[\xi]\in\hat{G}} \|a(\xi)\|_{S_r}^r d_\xi < \infty, \tag{117}$$

where $\|\sigma_A(\xi)\|_{S_r} := \mathrm{Tr}(|\sigma_A(\xi)|^r)^{\frac{1}{r}}$, is a necessary and sufficient condition for the r-nuclearity of A. The r-nuclearity and boundedness of global pseudo-differential operators on compact Lie groups in the general setting of Besov spaces has been studied by one of the authors in [5, 6].

Our result on the noncommutative residue can be enunciated in the following way:

Theorem 3.5 *Let G be a compact Lie group of dimension $\varkappa$ and $A : C^{\infty}(G) \to D'(G)$ be a positive left invariant continuous linear operator. Then if $A \in \Psi^{-\varkappa}(G)$ is a positive classical pseudo-differential operator, then the noncommutative residue of A, is given in terms of representations on G by*

$$\mathrm{res}(A) = \frac{1}{\dim(G)} \lim_{N\to\infty} \frac{1}{\log N} \sum_{\xi:\langle\xi\rangle\leq N} d_\xi \mathrm{Tr}(|\sigma_A(\xi)|) \tag{118}$$

Proof From the Connes trace theorem in the form of Theorem 3.2 we observe that the Dixmier trace of A coincides with its noncommutative residue and (118) now follows from the Eq. (107). □

Now, we use our results on multipliers in order to provide formulae for a class of non-invariant operators (operators with symbols depending on the spatial variable $x \in G$).

Proposition 3.1 *Let G be a compact Lie group and $\varkappa = \dim G$. Let us assume that $A \in \Psi^m_{cl}(G)$ is a positive classic pseudo-differential operator with local symbol, admitting homogeneous components, of the form:*

$$\sigma^A(x,\xi) \sim \sum_{j=0}^{\infty} a_{m-j}(x)\sigma^{A_{m-j}}(\xi). \tag{119}$$

Then the noncommutative residue of A is given by

$$\mathrm{res}(A) = \frac{1}{\dim(G)} \int_G a_{-\varkappa}(x)dx \times \|\sigma_{A_{-\varkappa}}\|_{\mathcal{L}^{(1,\infty)}(\widehat{G})}. \tag{120}$$

Proof By definition of the noncommutative residue, considering the measure of Haar of a compact Lie group is normalized, and the Connes theorem we have

$$\begin{aligned}\mathrm{res}(A) &:= \frac{1}{\varkappa(2\pi)^{\varkappa}} \int_G \int_{|\xi|=1} a_{-\varkappa}(x)\sigma^{A_{-\varkappa}}(\xi)d\xi\, dx \\ &= \int_G a_{-\varkappa}(x)\mathrm{vol}_{\varkappa}(x) \times \mathrm{res}(A_{-\varkappa}) = \int_G a_{-\varkappa}(x)dx \times \mathrm{Tr}_{\omega}(A_{-\varkappa}) \\ &= \int_G a_{-\varkappa}(x)dx \times \frac{1}{\dim(G)} \lim_{N\to\infty} \frac{1}{\log N} \sum_{\xi:\langle\xi\rangle\leq N} d_\xi \mathrm{Tr}(|\sigma_{A_{-\varkappa}}(\xi)|).\end{aligned} \tag{121}$$

We finish the proof by observing that

$$\|\sigma_{A_{-\varkappa}}\|_{\mathcal{L}^{(1,\infty)}(\widehat{G})} = \frac{1}{\log N} \sum_{\xi:\langle \xi \rangle \leq N} d_\xi \mathrm{Tr}(|\sigma_{A_{-\varkappa}}(\xi)|). \tag{122}$$

We end this section with the following remark about the noncommutative residue for invariant operators. In order to distinguish the local and global symbol, let us use an upper notation to denote the local symbol of A by $\sigma^A(\xi)$, i.e, the symbol of A defined as a section of the cotangent bundle T^*G on G (see [23]), with asymptotic expansion $\sigma^A(\xi) \sim \sum \sigma^A_{m-j}(\xi)$ in positive-homogeneous components σ^A_{m-j} of degree $m-j$ then by combining of the Eqs. (118) and (12) we deduce the following relation between the global and the local symbol of A.

Corollary 3.2 *Let A be a classical operator satisfying the conditions of the previous result. So,*

$$\begin{aligned} \mathrm{res}\,(A) &:= \frac{1}{\varkappa(2\pi)^\varkappa} \int_{|\xi|=1} \sigma^A_{-\varkappa}(\xi)\, d\xi \\ &= \lim_{N\to\infty} \frac{1}{\varkappa \log N} \sum_{\xi:\langle \xi \rangle \leq N} d_\xi \mathrm{Tr}(|\sigma_A(\xi)|), \end{aligned} \tag{123}$$

where $\sigma^A_{-\varkappa}$ denotes the component of degree $-\varkappa$ in the asymptotic expansion of σ^A.

4 The Dixmier Trace of Fourier Multipliers on Compact Homogeneous Manifolds

In Theorem 3.1, on a closed manifold, we give a characterization of Dixmier traceable E-invariant operators through its global symbols. In this section, we consider the special case of multipliers on compact homogeneous manifolds and we describe such characterization in terms of the representation theory of such manifolds. We generalize the corresponding assertion on multipliers given in Theorem 3.3, but in contrast with such result we do not consider the case of Hörmander classes as in such theorem.

Now we present our main result of this section.

Theorem 4.1 *Let us assume that A is a Fourier multiplier as (50) on $M := G/K$. Then A is Dixmier traceable on $L^2(M)$ if and only if*

$$\tau(A) := \frac{1}{\dim(M)} \lim_{N\to\infty} \frac{1}{\log N} \sum_{[\pi]\in \widehat{G}_0:\langle \pi \rangle \leq N} d_\pi \mathrm{Tr}(|\sigma_A(\pi)|) < \infty. \tag{124}$$

In this case, if A is positive, $\tau(A) = \mathrm{Tr}_w(A)$.

Proof If A is a Fourier multiplier on M, from Remark 2.1 and the Eq. (48), we obtain

$$\mathrm{Tr}(|\sigma_{A,\mathcal{L}_{G/K}}(l)|) = d_{\pi_l}\mathrm{Tr}(|\sigma_A(\pi_l)|), \tag{125}$$

where $\{[\pi_l] : l \in \mathbb{N}_0\}$ is a enumeration of $\widehat{G}_0$. If we use (95) we obtain

$$\begin{aligned}\mathrm{Tr}_w(A) &= \lim_{N\to\infty} \frac{\sum_{l\le N}\mathrm{Tr}(|\sigma_{A,E}(l)|)}{\log[\sum_{l\le N} d_l]} \\ &= \lim_{N\to\infty} \frac{\sum_{l\le N} d_{\pi_l}\mathrm{Tr}(|\sigma_A(\pi_l)|)}{\log[\sum_{l\le N} d_{\pi_l}k_{\pi_l}]} = \lim_{N\to\infty} \frac{\sum_{\langle\pi_l\rangle\le N} d_\pi\mathrm{Tr}(|\sigma_A(\pi)|)}{\log[\sum_{\langle\pi\rangle\le N} d_{\pi_l}k_{\pi_l}]}\end{aligned} \tag{126}$$

which proves (107). Finally, that (107) is a necessary and sufficient condition for the Dixmier traceability of A it follows from (95). By the Weyl Eigenvalue Counting Formula (see [1, pag. 7] or [37]), we have

$$\sum_{[\pi_l]\in\widehat{G}_0:\langle\pi_l\rangle\le N} d_{\pi_l}k_{\pi_l} = O(C_0 L^{\varkappa}), \tag{127}$$

where

$$\varkappa = \dim(M), \text{ and } C_0 = \frac{1}{(2\pi)^{\varkappa}}\int_{\sigma_1(x,w)<1} dx\,dw, \tag{128}$$

and the integral is taken with respect to the measure on the cotangent bundle T^*M induced by the canonical symplectic structure (c.f. [37]). Thus, we have,

$$\mathrm{Tr}_w(A) = \frac{1}{\dim(M)}\lim_{N\to\infty}\frac{1}{\log N}\sum_{[\pi]\in\widehat{G}_0:\langle\pi\rangle\le N} d_\pi\mathrm{Tr}(|\sigma_A(\pi)|) \tag{129}$$

which completes the proof. □

5 The Dixmier Trace and the Noncommutative Residue of Pseudo-Differential Operators on Manifolds with Boundary

In this section, we prove a result about the Dixmier trace of pseudo-differential boundary value problems where we link the two different approaches: representation by local and global symbols.

Let us note that in the simplest case of a L-Fourier multiplier A, the operator A^* is a L^*-Fourier multiplier, but not necessarily a L-Fourier multiplier, and in general, the operators A and A^* satisfy different conditions on the boundary. Thus, unless the model operator L_M (i.e. L equipped with conditions on the boundary ∂M) will be considered self-adjoint, we cannot compose A and A^* on their domains. Since we are interested in the singular values of an operator A, we will assume that L is self-adjoint. In our proof, we assume the spectrum of L to be enumerated by the set $\mathcal{I} = \{\xi_l : l \in \mathbb{N}_0\}$.

Remark 5.1 A Fourier multiplier on $C_L^\infty(M)$ is an operator with symbol $\sigma_{A,L} : \mathcal{I} \to \mathbb{C}$. In this section we consider bounded Fourier multipliers on $L^2(M)$. It is easy to see (by the Plancherel formula which holds in a suitable spaces $l_L^2(\mathcal{I})$, see also [31, Theorem 14.1]), that

$$\sup_{l \in \mathbb{N}_0} |\sigma_{A,L}(\xi_l)| < \infty \tag{130}$$

is a necessary and sufficient condition for the boundedness of A on $L^2(M)$.

Theorem 5.1 *Let M be a compact manifold with boundary ∂M. If $A : L^2(M) \to L^2(M)$ is a bounded Fourier multiplier, and L is self-adjoint operator on $L^2(M)$, then A is Dixmier traceable if and only if*

$$\tau(A) = \lim_{N \to \infty} \frac{1}{\log N} \sum_{l \le N} |\sigma_{A,L}(\xi_l)| < \infty. \tag{131}$$

In this case, if A is positive, $\tau(A) = \mathrm{Tr}_\omega(A)$ and independent on the average function ω. If in addition L satisfies the Weyl Eigenvalue Counting Formula, we obtain

$$\mathrm{Tr}_w(A) = \frac{1}{\dim(M)} \lim_{N \to \infty} \frac{1}{\log N} \sum_{l : |\lambda_{\xi_l}|^{\frac{1}{m}} \le N} |\sigma_{A,L}(\xi_l)|, \tag{132}$$

where m is the order of L.

Proof We want to show that (107) is a necessary and sufficient condition for the Dixmier traceability of a Fourier multiplier A with full symbol $\sigma_{A,L}(\xi)$. It follows from (91) that if $\sigma_{A,L}(x, \xi) = \sigma_{A,L}(\xi)$ does not depend on x, then

$$Au_\xi = \sigma_{A,L}(\xi) u_\xi, \tag{133}$$

so that $\sigma_{A,L}(\xi)$ are the eigenvalues of the operator A corresponding to the eigenfunctions u_ξ. Moreover, the pointwise spectrum of A is (setting $\mathcal{I} = \{\xi_l : l \in \mathbb{N}_0\}$)

$$\mathrm{Spec}_p(A) = \{\sigma_{A,L}(\xi) \ : \ \xi \in \mathcal{I}\} = \{\sigma_{A,L}(\xi_l) \ : \ l \in \mathbb{N}\}. \tag{134}$$

In fact, if $\lambda \in \text{Spec}_p(A) = \ker(A - \lambda I)$, with eigenfunction $u \neq 0$ associated with λ. If we take the L-Fourier transform $\mathcal{F}_L$ to both sides of $Au = \lambda u$, we have $\sigma_{A,L}(\xi)\widehat{u}(\xi) = \lambda\widehat{u}(\xi)$. Since $\mathcal{F}_L$ is an isomorphism; there exists $l_0 \in \mathbb{N}$, such that $\widehat{u}(\xi_{l_0}) \neq 0$. Therefore, $\lambda = \sigma_{A,L}(\xi_{l_0})$.

For the converse, let us consider the complex number $\lambda = \sigma_{A,L}(\xi_{l_0})$ for some $l_0 \in \mathbb{N}$. By contradiction let us assume that λ is not a element of $\text{Spec}_p(A)$. Since A does not have residual or continuous spectrum, therefore λ is in the resolvent set of A, in particular $A - \lambda I$ is an injective operator. So we have that $(A - \lambda I)v = 0$ implies that $v = 0$. Let us consider the function in $v \in \mathcal{S}(\mathcal{I})$ defined by $v(\xi_l) = 0$ if $l \neq l_0$, and $v(\xi_{l_0}) = 1$. Because the L-Fourier transform is an isomorphism from $C_L^\infty(M)$ into $\mathcal{S}(\mathcal{I})$, we can consider the inverse L-Fourier transform of v, $u \in C_L^\infty(M)$. We have $\widehat{u}(\xi_l) = v(\xi_l)$. It is clear that for every l, $\sigma_{A,L}(\xi_l)\widehat{u}(\xi_l) = \lambda \cdot \widehat{u}(\xi_l)$. By taking the inverse L-Fourier transform, we have $Au = \lambda u$ with $u \neq 0$ which is a contradiction.

It follows that L is a self-adjoint operator and Theorem 10.10 in [31], that A^*A is well defined on $L^2(M)$ and a L-Fourier multiplier with symbol $\overline{\sigma_{A,L}(\xi)}\sigma_{A,L}(\xi) = |\sigma_{A,L}(\xi)|^2$. Applying (134) to the operator $\sqrt[2]{A^*A}$ we obtain the following relation for the set $s(A)$ of singular values of A :

$$s(A) = \{|\sigma_{A,L}(\xi)| : \xi \in \mathcal{I}|\}. \tag{135}$$

With notations above, we set $\mathcal{I} = \{\xi_l : l \in \mathbb{N}_0\}$ in such way $s(A)$ forms an increasing sequences, A is Dixmier traceable (by definition) if only if

$$\tau(A) := \lim_{N\to\infty} \frac{\sum_{l\leq N} |\sigma_{A,L}(\xi_l)|}{\log N} < \infty. \tag{136}$$

Since

$$\tau(A) = \lim_{N\to\infty} \frac{\sum_{l\leq N} |\sigma_{A,L}(\xi_l)|}{\log N} = \lim_{N\to\infty} \frac{\sum_{l:|\lambda_{\xi_l}|\leq N} |\sigma_{A,L}(\xi_l)|}{\log\left(\sum_{l:|\lambda_{\xi_l}|\leq N} 1\right)}, \tag{137}$$

if we assume that L satisfies the Weyl Eigenvalue Counting Formula:

$$\sum_{\xi:|\lambda_\xi|\leq N} 1 = \#\{\xi_l : |\lambda_{\xi_l}|^{\frac{1}{m}} \leq N\} = C_0 N^{\varkappa} + O(N^{\varkappa-1}) \tag{138}$$

where $\varkappa = \dim(M)$, we obtain

$$\text{Tr}_w(A) = \frac{1}{\dim(M)} \lim_{N\to\infty} \frac{1}{\log N} \sum_{l:|\lambda_{\xi_l}|\leq N} |\sigma_{A,L}(\xi_l)|, \tag{139}$$

it follows that $s(A)$ is organized as an increasing sequence such that the result is independent of the average function ω chosen, which prove the result. □

We observe that an adaption of the proof of Theorem 3.4 gives the following result:

Corollary 5.1 *Let A be a L-Fourier multiplier bounded on $L^2(M)$. Then $A \in \mathcal{L}^{(p,\infty)}(L^2(M))$ if and only if*

$$\gamma_p'(A) := \sup_{N\geq 1} N^{(\frac{1}{p}-1)} \cdot \sum_{l\leq N} |\sigma_{A,L}(\xi_l)| < \infty, \tag{140}$$

for all $1 < p < \infty$. In this case, $\|A\|_{\mathcal{L}^{(p,\infty)}(L^2(M))} \asymp \gamma_p'(A)$. Moreover, if L satisfies (27), we have

$$\gamma_p'(A) \asymp \sup_{N\geq 1} N^{\dim M(\frac{1}{p}-1)} \cdot \sum_{l\,:\,|\lambda_{\xi_l}|^{\frac{1}{m}}\leq N} |\sigma_{A,L}(\xi_l)|. \tag{141}$$

In order to make a clear exposition of the our next result we need to establish a more clear relationship between Ruzhansky–Tokmagambetov approach and the local approach of boundary value problems.

Remark 5.2 Any continuous operator $A : C^\infty(M) \to C^\infty(M)$ in Ruzhansky–Tokmagambetov's calculus ($C_L^\infty(M)$ is defined by means of L_M, cf. (75)) may be seem as an operator realization as follow: We have

$$\mathcal{D}(L_M) := \{u \in L^2(M) \mid u \ \text{ satisfies (BC)}\}. \tag{142}$$

It follows from (75) that $A = A_L$ indeed is the realization of A with respect to the space

$$\mathcal{D}(A) := C_L^\infty(M) = \{f \in L^2(M) \mid L^j f \in L^2(M), \ f \text{ satisfies (BC)}, \ j \in \mathbb{N}\}. \tag{143}$$

Theorem 5.2 *Let $A = \begin{bmatrix} P_+ \\ T \end{bmatrix}$ be a positive elliptic operator in the Boutet de Monvel algebra of order n, where P is a Fourier multiplier. We assume the existence such a operator L satisfying Ruzhansky–Tokmagambetov's calculus in Sect. 2.3. Then the operator realization P_T can be regarded as an operator in R-T's calculus (cf. Remark 5.2), the parametrix R of P_T is Dixmier's traceable and*

$$\mathrm{Tr}_\omega(R) = \mathrm{res}\,(R) = \lim_{N\to\infty} \frac{1}{\ln N} \sum_{l\leq N} |(\sigma_{P,L}(\xi_l))^{-1}|, \tag{144}$$

here $(\sigma_{P,L}(\xi_l))_{l\in\mathbb{N}_0}$ denotes the global symbol of the operator P with respect to L_M, see Remark 5.2. The expression is the same for all parametrices and independent of the choice of the boundary condition and independent on the average function of ω.

Proof We first see that the operator realization P_T belongs to Ruzhansky–Tokmagambetov's calculus, i.e. P_T is the restriction of P to the space $C_L^\infty(M)$ for some operator L as in Sect. 2.3. We know from Proposition 2.4 that

$$H^s(M) = H_L^s(M), \tag{145}$$

(i.e., the Sobolev space can be defined by the a operator L and independent of the choice of such operator). Since P is an operator satisfying the transmission property we have that P preserves the space of smooth up to boundary functions $C^\infty(\overline{M})$. Therefore, the operator P_T maps functions from $C_L^\infty(M)$ to $C_L^\infty(M)$, therefore P_T restricted to $C_L^\infty(M)$ belongs to Ruzhansky–Tokmagambetov's calculus. We know that $C_L^\infty(M)$ and $C_{L^*}^\infty(M)$ are dense subspaces on $L^2(M)$. Moreover, from (145) we have that P_T can be extended to (see also Remark 5.2)

$$\mathcal{D}(P_T) := \{u \in H^m(M) \mid Tu = 0\}. \tag{146}$$

Now, it follows from Proposition 2.2 that R maps $L^2(M)$ to $\mathcal{D}(P_T) \subset L^2(M)$, thus R lies in the Dixmier class. In this case, R can be considered as Fourier Multiplier up to trace class operators. It follows from Theorem 5.1 that R is Dixmier traceable if and only if the conditions given in (155) holds, and in this case, $\mathrm{Tr}_\omega(R) = \tau(R)$. Now, the assertion in (144) follows from Proposition 2.3. □

Now, we provide examples of certain elliptic operators on the Boutet de Monvel algebra satisfying conditions of the Ruzhansky–Tokmagambetov calculus. For this we follows Ivrii [24].

Let M be a compact manifold with smooth boundary ∂M, M of dimension $n \geq 2$, and let dx be a density on M. Let L be a second-order elliptic pseudo-differential operator on M, formally self-adjoint with respect to the inner product in $L^2(M, dx)$; let X be a vector field transversal to ∂M at every point, γ the operator of restriction to ∂M,

$$T = \gamma \quad \text{or} \quad T = \gamma X + T_1 X \tag{147}$$

a boundary operator where T_1 is a first-order classical pseudo-differential operator on ∂M. If we suppose that the vector-operator $(L, T) : C^\infty(M) \to C^\infty(M) \oplus C^\infty(\partial M)$ satisfies the Sapiro–Lopatinskii condition. Let us denote by $L_T : L^2(M) \to L^2(M)$ the clausure of L_T on $L^2(M)$. Then L_T is self-adjoint; its spectrum is discrete, with finite multiplicity and tends to $\pm\infty$, or $+\infty$, or $-\infty$.

Moreover, if H is a closed subspace in $L^2(M)$ such that the self-adjoint projector Π on H belongs to Boutet de Monvel's algebra. It is assumed that $\Pi L_T \subset L_T \Pi$ (i.e., H is an invariant subspace of L_T). Then the restriction $L_{T,H} : H \to H$ the L_T to H (here, $\mathcal{D}(L_{T,H}) = \mathcal{D}(L_T) \cap H)$ is self-adjoint; its spectrum is discrete, with finite multiplicity and tends to $\pm\infty$, or $+\infty$, or $-\infty$.

In the following, we present two general frameworks which guarantees the exists of the auxiliary operator L in the global calculus in [31], hence the Theorem 5.2 holds:

- If $H = H^m(M)$ is the Sobolev space of positive order m, since the Sobolev Embedding Theorem follows that $H^m(M)$ is a closed subspace of $L^2(M)$. The operator L_T defined by

$$L_T := I + \Delta_T : \mathcal{D}(L_T) \to L^2(M), \tag{148}$$

  here $\mathcal{D}(L_T) = \{H^m(M) \,|\, Tu = 0\}$ with T as above, defines a second-order elliptic operator. Since Sobolev Embedding Theorem we have that $H^m(M)$ is an invariant subspace of L_T. Therefore, L_T is a self-adjoint operator; its spectrum is discrete, with finite multiplicity and tends to $\pm\infty$, or $+\infty$, or $-\infty$.

  Therefore, by the spectral theorem for unbounded operators there exists an orthonormal basis for $L^2(M)$ formed with eigenvalues of L_T. We denote to the set of eigenvalues by $\{\lambda_{\xi_l}\}$, $l \in \mathbb{N}_0$ and the corresponding eigenvalues by $\{u_{\xi_l}\}_{l\in\mathbb{N}_0}$. We can assume that L_T satisfies the Weyl law (see [24, 25]):

$$N(\lambda) = \sum_{l:|\lambda_{\xi_l}|^{\frac{1}{m}} \leq \lambda} 1 = O(\lambda^n), \;\; \lambda \to \infty, \tag{149}$$

  then, if l is such that $|\lambda_{\xi_l}|^{\frac{1}{m}} \leq \lambda$,

$$\lambda^{-1}|\lambda_{\xi_l}|^{\frac{1}{m}} \leq \sum_{l:|\lambda_{\xi_l}|^{\frac{1}{m}} \leq \lambda} \lambda^{-1}|\lambda_{\xi_l}|^{\frac{1}{m}} \leq \sum_{l:|\lambda_{\xi_l}|^{\frac{1}{m}} \leq \lambda} 1 = O(\lambda^n), \;\; \lambda \to \infty, \tag{150}$$

  and as consequence we have the estimate for $\langle \xi_l \rangle = (1 + |\lambda_{\xi_l}|)^{\frac{1}{2}}$, $l \to \infty$,

$$1 < \langle \xi_l \rangle \asymp |\lambda_{\xi_l}|^{\frac{1}{m}} = O(\lambda^{n+1}), \qquad \langle \xi_l \rangle \to \infty. \tag{151}$$

  Thus, we can assume that some $-\infty < s_0 < 0$,

$$\sum_{l\in\mathbb{N}_0} \langle \xi_l \rangle^{s_0} < \infty. \tag{152}$$

  Because the WZ condition can be removed from the Ruzhansky–Tokmagambetov calculus (by merely technical reasons), we have that there exists such a operator L.
- Let M be a manifold of dimension 2. If $P : C^\infty(M) \to C^\infty(M)$ is a second order elliptic differential operator. We denote by P_T the restriction of P to $\ker T$ with domain

$$D(P_T) = \{f \in C^\infty(M) : Tf = 0\}. \tag{153}$$

Let us denote by P_T the clausure of P_T on $L^2(M)$. Hence, P_T has discrete spectrum with finite multiplicity and tends to either to ∞, or to $-\infty$ or to $\pm\infty$. Setting $L = P$, by similar arguments as before, the operator realization $L_M = P_T$ satisfies the condition in the Ruzhansky–Tokmagambetov calculus. In this case the boundary condition (BC) defines the vectorial space given by

$$f \in L^2(M) \cap (\text{BC}) \text{ if and only if } f \in D(P_T). \tag{154}$$

Because $C_L^\infty(M) \subset D(P_T)$ is a invariant subspace of L we have that L maps $C_L^\infty(M)$ itself. Therefore, the second-order elliptic differential operator $L := P$ (respectively $L_M := P_T|_{C_L^\infty}$) belongs to the Ruzhansky–Tokmagambetov calculus and satisfies the conditions in Sect. 2.3.

Corollary 5.2 *If* $A = \begin{bmatrix} P_+ \\ T \end{bmatrix}$ *is a positive elliptic operator in the Boutet de Monvel algebra of order n with T as (147). In this case, we may take* L_T *where* $L = I + \Delta$. *Then, any parametrix R of* P_T *is Dixmier's traceable and the formula given in (144) holds. In the particular case where M is a manifold of dimension 2 and P is a second-order differential operator as above, we have a formula independent of an auxiliary operator L,*

$$\mathrm{Tr}_\omega(R) = \mathrm{res}\,(R) = \lim_{N\to\infty} \frac{1}{\ln N} \sum_{l\le N} |\sigma_R(\xi_l)|, \tag{155}$$

here $(\sigma_R(\xi_l))_{l\in\mathbb{N}_0}$ *denotes the global symbol of the parametrix with respect to* $L = P_T$.

6 Examples

In this section we provide some examples in relation with our main results. First, we consider a Dixmier traceable Bessel potential associated to L. Also, we consider an example of a Dixmier traceable operator on the manifold with boundary $M = [0, 1]$. Later, we consider the case of Bessel potential on $\mathrm{SU}(2) \cong \mathbb{S}^3$ and $\mathrm{SU}(3)$ respectively. In our examples, we compute the Dixmier trace of certain operators showing closed formulae, but, we do not do numerical approximations of these expressions in all cases.

Example 6.1 Let M be a compact manifold with boundary ∂M and L as in the preceding sections. Let us consider the model operator L_M self adjoint on $L^2(M)$. If we assume that $-L$ is a positive operator of order ν, and some positive real number s_1 satisfies

$$\sum_{l\in\mathbb{N}_0} \langle\xi_l\rangle^{-s} = \infty, \quad \sum_{l\in\mathbb{N}_0} \langle\xi_l\rangle^{-s'} < \infty \tag{156}$$

for all $0 < s \leq s_1 < s' < \infty$ (this condition is natural in view of (70)). Then with the notations above the operator $A := (I-L)^{-\frac{s_1}{\nu}}$ is Dixmier traceable on $L^2(M)$. For the proof, we combine that the global symbol of A is given by

$$\sigma_{A,L}(\xi_l) = (1-\lambda_{\xi_l})^{-\frac{s_1}{\nu}} \asymp \langle \xi_l \rangle^{-s_1}. \tag{157}$$

As consequence of (156) we have that $\langle \xi_l \rangle^{-s_1} = O(l^{-1}),\ l \to \infty$, and

$$\sum_{l \leq N} \langle \xi_l \rangle^{-s_1} = O(\log N),\ N \to \infty. \tag{158}$$

Hence, by Theorem 5.1 we obtain

$$\mathrm{Tr}_w((I-L)^{-\frac{s_1}{\nu}}) \asymp \lim_{N\to\infty} \frac{1}{\log N} \sum_{l \leq N} \langle \xi_l \rangle^{-s_1} < \infty. \tag{159}$$

Example 6.2 Consider $M = [0,1]$ and the operator $L = -i\frac{d}{dx}$ on $M^\circ = (0,1)$ with the domain

$$D(L) = \{f \in W_2^1[0,1] \,|\, af(0) + bf(1) + \int_0^1 f(x)q(x)\,dx = 0\}, \tag{160}$$

where $a \neq 0$, $b \neq 0$, and $q \in C^1[0,1]$, here $W_2^1[0,1] := \{f \in L^2[0,1] \,|\, f' \in L^2[0,1]\}$. Under the assumption that

$$a + b + \int_0^1 q(x)\,dx = 1 \tag{161}$$

we have the inverse L^{-1} exists and is bounded from $L^2[0,1]$ to $D(L)$. The operator L has a discrete spectrum which can be enumerated by

$$\lambda_j = 2\pi j - i\ln(-a/b) + \alpha_j, \ \text{ for } \ j \in \mathbb{Z} \tag{162}$$

where the sequence α_j satisfies that for any $\epsilon > 0$, $\sum_{j\in\mathbb{Z}} |\alpha_j|^{1+\epsilon} < \infty$. If m_j denotes the multiplicity of the eigenvalue λ_j, then $m_j = 1$ for sufficiently large $|j|$, and the system of eigenfunctions are given by

$$u_{jk} = \frac{(ix)^k}{k!} e^{i\lambda_j x}, \ \text{ for } \ 0 \leq k \leq m_j - 1 \ \text{ and } \ j \in \mathbb{Z}. \tag{163}$$

Let $j_0 \in \mathbb{N}$ large enough so that $m_j = 1$ for $|j| \geq j_0$. The global symbol $\sigma_{L,L}(x, j)$ of L with respect to L is given by

$$\sigma_{L,L}(x, j) = \begin{cases} \frac{x^{2k}}{k!^2}\lambda_j \; x \in [0, 1], \; 0 \leq k \leq m_j - 1, \; |j| \leq j_o \\ \lambda_j \quad |j| \geq j_0; \end{cases} \tag{164}$$

for some $j_0 \in \mathbb{N}$. It follows from the functional calculus that L^{-1} has a discrete spectrum given by λ_j^{-1} with $j \in \mathbb{Z}$. Since $|\lambda_j| = O(j)$ for $j \in \mathbb{Z}$ we have $\sum_{|j|\leq N} |\lambda_j^{-1}| = O\big(\ln(2N + 1)\big)$. Therefore, is a L^{-1} is a pseudo-differential operator which lies in $\mathcal{L}^{(1,\infty)}(L^2[0, 1])$ with symbol $\sigma_{L^{-1},L}(\xi) = (\sigma_{L,L}(\xi))^{-1}$, $|\xi| \geq j_0$, and

$$\begin{aligned} &\mathrm{Tr}_\omega(L^{-1}) \\ &= \lim_{N\to\infty} \frac{1}{\ln(2N+1)} \sum_{|j|\leq N} |\lambda_j^{-1}| \\ &= \lim_{N\to\infty} \frac{1}{\ln(2N+1)} \sum_{j_0\leq|j|\leq N} |(\sigma_{L,L}(x, j))^{-1}| \\ &= \lim_{N\to\infty} \frac{1}{\ln(2N+1)} \sum_{|j|\leq N} \frac{2\pi j + i\ln(-a/b) + \overline{\alpha_j}}{(2\pi j + \mathrm{Re}(\alpha_j))^2 + (\mathrm{Im}(\alpha_j) + \ln(b/a))^2}. \end{aligned} \tag{165}$$

Example 6.3 Let us assume that A is a left-invariant operator on $\mathrm{SU}(2) \cong \mathbb{S}^3$. We recall that the unitary dual of SU(2) (see [32]) can be identified as

$$\widehat{\mathrm{SU}}(2) \equiv \{[t_l] : 2l \in \mathbb{N}, d_l := \dim t_l = (2l+1)\}. \tag{166}$$

There are explicit formulae for t_l as functions of Euler angles in terms of the so-called Legendre–Jacobi polynomials, see [32]. In this case, if A and is Dixmier traceable with symbol $\sigma([t_l]) \equiv \sigma(l)$ then,

$$\begin{aligned} &\mathrm{Tr}_w(A) \\ &= \lim_{N\to\infty} \frac{1}{\log\big(\sum_{l\leq\frac{N}{2}, l\in\frac{1}{2}\mathbb{N}_0} d_l^2\big)} \sum_{l\leq\frac{N}{2}, l\in\frac{1}{2}\mathbb{N}_0} d_{[t_l]}\mathrm{Tr}[|\sigma(l)|] \\ &= \lim_{N\to\infty} \frac{1}{\log\big(\sum_{l\leq\frac{N}{2}, l\in\frac{1}{2}\mathbb{N}_0} (2l+1)^2\big)} \sum_{l\leq\frac{N}{2}, l\in\frac{1}{2}\mathbb{N}_0} (2l+1)\mathrm{Tr}[|\sigma(l)|] \\ &= \lim_{N\to\infty} \frac{1}{\log[\frac{1}{6}(N+1)(2N^2+7N+1)]} \sum_{l\leq\frac{N}{2}, l\in\frac{1}{2}\mathbb{N}_0} (2l+1)\mathrm{Tr}[|\sigma(l)|]. \end{aligned} \tag{167}$$

For example, if $A = (I - \mathcal{L}_{\mathrm{SU}(2)})^{-\frac{\varkappa}{2}}$, is the Bessel potential of order $-\varkappa = -\dim \mathrm{SU}(2) = -3$, then A is Dixmier traceable, $\sigma(l) = (1 + l(l+1))^{-\frac{3}{2}}$ and

$$\begin{aligned}
&\mathrm{Tr}_w((I - \mathcal{L}_{\mathrm{SU}(2)})^{-\frac{3}{2}}) \\
&= \lim_{N\to\infty} \frac{1}{\log[\frac{1}{6}(N+1)(2N^2+7N+1)]} \sum_{l\le \frac{N}{2}, l\in \frac{1}{2}\mathbb{N}_0} (2l+1)^2[1+l(l+1)]^{-\frac{3}{2}} \\
&= \lim_{N\to\infty} \frac{1}{\log[\frac{1}{6}(N+1)(2N^2+7N+1)]} \sum_{n=0}^{N} (n+1)^2 \left[1 + \frac{n}{2}(\frac{n}{2}+1)\right]^{-\frac{3}{2}} \\
&= \lim_{N\to\infty} \frac{1}{\log[\frac{1}{6}(N+1)(2N^2+7N+1)]} \sum_{n=1}^{N+1} \frac{n^2}{8}[n^2+3]^{-\frac{3}{2}} \\
&\sim 0.03935\ldots \text{ (numerical evidence).}
\end{aligned} \tag{168}$$

A similar result can be obtained if we consider the operator $(I - \mathcal{L}_{sub})^{-\frac{3}{2}}$ where $\mathcal{L}_{sub} := D_1^2 + D_2^2$ is the sub-Laplacian on SO(3), here D_1 and D_2 are the derivatives in the corresponding variables to the Euler angles for SO(3) (see [32, 33]).

Example 6.4 Let A be a left-invariant pseudo-differential operator Dixmier traceable on SU(3). The Lie group SU(3) (see [17]) has dimension 8 and 3 positive square roots α, β and ρ with the property

$$\rho = \frac{1}{2}(\alpha + \beta + \rho). \tag{169}$$

We define the weights

$$\sigma = \frac{2}{2}\alpha + \frac{1}{3}\beta, \quad \tau = \frac{1}{3}\alpha + \frac{2}{3}\beta. \tag{170}$$

With the notations above the unitary dual of SU(3) can be identified with

$$\widehat{\mathrm{SU}}(3) \equiv \left\{\lambda := \lambda(a,b) = a\sigma + b\tau : a, b \in \mathbb{N}_0, d_\lambda = \frac{(a+1)(b+1)(a+b+2)}{2}\right\}. \tag{171}$$

The Dixmier trace of A, once denoted its full symbol by $\sigma(\lambda) \equiv \sigma(\lambda(a,b))$, is given by the expression

$$\begin{aligned}
&\mathrm{Tr}_w(A)\\
&= \lim_{N\to\infty} \frac{1}{\log[\sum_{0\leq a+b\leq N} d_{\lambda(a,b)}]}\times\\
&\qquad \sum_{0\leq a+b\leq N} \frac{1}{2}(a+1)(b+1)(a+b+2)\mathrm{Tr}[|\sigma(\lambda(a,b))|] \qquad (172)\\
&= \lim_{N\to\infty} \frac{1}{\log[\frac{1}{120}(N+1)(N+2)(N+3)(N+4)(2N+5)]}\times\\
&\qquad \sum_{(a,b):0\leq a+b\leq N} \frac{1}{2}(a+1)(b+1)(a+b+2)\mathrm{Tr}[|\sigma(\lambda(a,b))|].
\end{aligned}$$

Since, the eigenvalues of the Laplacian $\mathcal{L}_{\mathrm{SU}(3)}$ on $SU(3)$ are of the form

$$-c(\lambda) = -\frac{1}{9}(a^2+b^2+ab+3a+3b), \qquad (173)$$

the symbol of the operator $A = (I - \mathcal{L}_{\mathrm{SU}(3)})^{-\frac{8}{2}}$ (which is Dixmier traceable) is given by

$$\sigma(\lambda) = (1+c(\lambda))^{-4}. \qquad (174)$$

Hence,

$$\begin{aligned}
&\mathrm{Tr}_w((I-\mathcal{L}_{\mathrm{SU}(3)})^{-\frac{8}{2}})\\
&= \lim_{N\to\infty} \frac{1}{\log[\frac{1}{120}(N+1)(N+2)(N+3)(N+4)(2N+5)]}\times\\
&\qquad \sum_{(a,b):0\leq a+b\leq N} \frac{1}{36}(a+1)^2(b+1)^2(a+b+2)^2[1+(a^2+b^2+ab+3a+3b)]^{-4}.
\end{aligned} \qquad (175)$$

Remark 6.1 Similar examples to the considered above can be obtained if we take, for example, every n-torus $\mathbb{T}^n \cong \mathbb{R}^n/\mathbb{Z}^n$ or the compact Lie group $\mathrm{SO}(3) \cong \mathbb{RP}^3$, spheres $\mathbb{S}^n$ and arbitrary real and complex projective spaces.

We end this paper with the following example on the noncommutative residue of a classical operator on Lie groups.

Example 6.5 Let us assume that G is a compact Lie group of dimension $\varkappa$, and let us define $A := a(x)(-\mathcal{L}_G)^{-\frac{\varkappa}{2}}$. We consider a positive and smooth in order that A

will be a positive operator. The global symbol of A is given by

$$\sigma_A(x,\xi) := a(x)\lambda_{[\xi]}^{-\frac{\varkappa}{2}} I_{d_\xi}, \tag{176}$$

and by using (22) we have that

$$\mathrm{res}(A) = \frac{1}{\dim(G)}\int_G a(x)dx \times \lim_{N\to\infty}\frac{1}{\log N}\sum_{\xi:\langle\xi\rangle\leq N} d_\xi^2\lambda_{[\xi]}^{-\frac{\varkappa}{2}}. \tag{177}$$

In particular, for $G=\mathbb{T}^\varkappa$, where $\mathbb{T}\equiv[0,1)$, $d_\xi=1$, and $\lambda_{[\xi]}=1/4\pi^2|\xi|^2$. In this case,

$$\begin{aligned}\lim_{N\to\infty}\frac{1}{\log N}\sum_{\xi:\langle\xi\rangle\leq N} d_\xi^2\lambda_{[\xi]}^{-\frac{\varkappa}{2}} &\asymp \lim_{N\to\infty}\frac{1}{\log N}\sum_{\xi\neq 0:|\xi|\leq N}|\xi|^{-\varkappa}\\ &\asymp \lim_{N\to\infty}\frac{1}{\log N}\int_{1\leq|\xi|\leq N}|\xi|^{-\varkappa}d\xi\\ &\asymp \lim_{N\to\infty}\frac{1}{\log N}\log(N)=1.\end{aligned} \tag{178}$$

Consequently,

$$\mathrm{res}(A)\asymp\frac{1}{\varkappa}\int_{\mathbb{T}^\varkappa} a(x_1,\cdots,x_\varkappa)dx_1\cdots dx_\varkappa, \tag{179}$$

which is a well known fact (see e.g. Connes [4]).

Acknowledgments The authors would like to thank Alexander Cardona Guío "from Universidad de Los Andes" and Carolina Neira "from Universidad Nacional de Colombia" for several discussions about this topic. This project was partially supported by Universidad de Los Andes, Mathematics Department. The first author was supported by the FWO Odysseus 1 grant G.0H94.18N: Analysis and Partial Differential Equations.

References

1. Akylzhanov, R. Nursultanov, E. Ruzhansky, M.: Hardy–Littlewood, Hausdorff–Young–Paley inequalities, and $L^p - L^q$ Fourier multipliers on compact homogeneous manifolds. J. Math. Anal. Appl. **479**(2), 1519–1548 (2019). arXiv:1504.07043
2. Boutet de Monvel, L.: Comportement d'un opérateur pseudo-différentiel sur une variété à bord. I. La propriété de transmission. J. Analyse Math. **17**, 241–253 (1966)
3. Boutet de Monvel, L.: Boundary problems for pseudo-differential operators. Acta Math. **126**(1), 11–51 (1971)
4. Connes A.: Noncommutative Geometry. Academic Press, San Diego (1994)

5. Cardona, D. Besov continuity for Multipliers defined on compact Lie groups. Palest. J. Math. **5**(2), 35–44 (2016)
6. Cardona, D.: Nuclear pseudo-differential operators in Besov spaces on compact Lie groups. J. Fourier Anal. Appl. **23**(5), 1238–1262 (2017)
7. Cardona, A., Ducourtioux, C., Magnot, J.P., Paycha, S.: Weighted traces on algebras of pseudo-differential operators and geometry on loop groups. Infin. Dimens. Anal. Quantum Probab. Relat. Top. **5**(4), 503–540 (2002)
8. Del Corral, C.: Canonical Trace and Pseudo-differential Operators on Manifolds with Boudary. Ph.D. Thesis, Universidad de Los Andes, Los Andes (2016)
9. Delgado, J. Ruzhansky, M.: L^p-nuclearity, traces, and Grothendieck-Lidskii formula on compact Lie groups. J. Math. Pures Appl. **102**(1), 153–172 (2014)
10. Delgado, J. Ruzhansky, M.: Schatten classes on compact manifolds: Kernel conditions. J. Funct. Anal. **267**(3), 772–798 (2014)
11. Delgado, J. Ruzhansky, M.: Kernel and symbol criteria for Schatten classes and r-nuclearity on compact manifolds. C. R. Acad. Sci. Paris. Ser. I. **352**, 779–784 (2014)
12. Delgado, J. Ruzhansky, M.: L^p-bounds for pseudo-differential operators on compact Lie groups. arXiv:1605.07027
13. Delgado, J. Ruzhansky, M.: Fourier multipliers, symbols, and nuclearity on compact manifolds. J. Anal. Math. **135**(2), 757–800 (2018)
14. Delgado, J. Ruzhansky, M. Tokmagambetov, N.: Schatten classes, nuclearity and nonharmonic analysis on compact manifolds with boundary. J. Math. Pures Appl. (9) **107**(6), 758–783 (2017)
15. Dixmier, J.: Existence de traces non normales. C. R. Acad. Sci. Paris. Series B **262**, 1107A–1108A (1966)
16. Fedosov, B. Golse, F. Leichtnam, E. Schrohe, E.: The noncommutative Residue for Manifolds with Boundary. J. Funct. Anal. **142**(1), 1–31 (1996)
17. Fegan, H.: Introduction to compact Lie groups. In: Series in Pure Mathematics, vol. 13. World Scientific Publishing Company, River Edge (1991)
18. Gayral, V., Gracia-Bondía, J.M., Iochum, B. Schücker, T., Vàrilly, J.C.: Moyal planes are spectral triples. Comm. Math. Phys. **246**, 569–623 (2004)
19. Grubb, G.A.: Resolvent approach to traces and zeta Laurent expansions. Contemp. Math. **366**, 67–93 (2005). Corrected version in arXiv: math.AP/0311081
20. Grubb, G.: On the logarithm component in trace defect formulas. Commun. Partial Differ. Equations **30**, 1671–1716 (2005)
21. Grubb, G. Schrohe, E.: Trace Expansions and the Noncommutative Residue for Manifolds with Boundary. University of Copenhagen, Copenhagen (1999). Preprint Series no. 20
22. Grubb, G., Schrohe, E.: Traces and quasi-traces on the Boutet de Monvel algebra. Ann. Inst. Fourier **54**, 1641–1696 (2004)
23. Hörmander, L.: The Analysis of the Linear Partial Differential Operators, vol. III. Springer, Berlin (1985)
24. Ivrii, V.: Precise spectral asymptotics for elliptic operators acting in fiberings over manifolds with boundary. In: Lecture Notes in Mathematics, vol. 1100. Springer, Berlin (1984)
25. Ivrii, V.: 100 years of Weyl's law. Bull. Math. Sci. **6**(3), 379–452 (2016)
26. Nest, R., Schrohe, E.: Dixmier's Trace for Boundary Value Problems. In: Manuscripta Mathematica, vol. 96, pp. 203–218. Springer, Berlin (1998)
27. Nicola, F., Rodino, L.: Dixmier traceability for general pseudo-differential operators. In: C^*-algebras and Elliptic Theory II. Trends Mathematical, pp. 227–237. Birkhäuser, Basel (2008)
28. Paycha, S.: Regularised Integrals, Sums, and Traces. An Analytic Point of View, vol. 59. American Mathematical Socity, New York (2012)
29. Paycha, S., Scott, S.: A Laurent Expansion for Regularized Integrals of Holomorphic Symbols. In: GAFA Geometric And Functional Analysis, vol. 17(2), pp 491–536 (2007)
30. Pietsch, A.: Traces and residues of pseudo-differential operators on the torus. Int. Equ. Oper. Theory **83**(1), 1–23 (2015)
31. Ruzhansky, M., Tokmagambetov, N.: Nonharmonic Analysis of Boundary Value Problems: International Mathematics Research Notices. rnv243, 68 p. (2015)

32. Ruzhansky, M., Turunen, V.: Pseudo-Differential Operators and Symmetries: Background Analysis and Advanced Topics. Birkhäuser, Basel (2010)
33. Ruzhansky, M., Wirth, J.: L^p Fourier multipliers on compact Lie groups. Math. Z. **280**, 621–642 (2015)
34. Ruzhansky M., Turunen V., Wirth J.: Hormander class of pseudo-differential operators on compact Lie groups and global hypoellipticity. J. Fourier Anal. Appl. **20**, 476–499 (2014)
35. Schrohe, E.: Noncommutative Residues, Dixmier's Trace, and Heat Trace Expansions on Manifolds with Boundary. In: Booss-Bavnbek, B., Wojciechowski, K. (eds.) Geometric Aspects of Partial Differential Equations. Contemporary Mathematics, vol. 242, pp. 161–186. American Mathematical Society, Providence (1999)
36. Schrohe, E.: A Short Introduction to Boutet de Monvel's calculus. In: Operators Theory: Advances and Applications, vol. 125. Springer, Berlin (2001)
37. Shubin, M.A.: Pseudodifferential operators and spectral theory, 2nd edn. Springer, Berlin (2001). Translated from the 1978 Russian original by Stig I. Andersson
38. Simon, S.: Traces and Determinants of Pseudodifferential Operators. Oxford University, Oxford (2010)
39. Strichartz, R.S.: Spectral asymptotics revisited. J. Fourier Anal. Appl. **18**(3), 626—659 (2012)
40. Sukochev, F., Usachev, A.: Dixmier traces, and noncommutative analysis. J. Geom. Phys. **105**, 102–122 (2016)
41. Wodzicki, M.: Noncommutative residue I. In: Manin, Yu.I. (ed.) K-Theory, Arithmetic and Geometry. Lecture Notes in Mathematics, vol. 1289, pp. 320–399. Springer, Berlin (1987)

On the Focusing Energy-Critical 3D Quintic Inhomogeneous NLS

Yonggeun Cho and Kiyeon Lee

Abstract We consider the focusing energy-critical inhomogeneous nonlinear Schrödinger equation:

$$iu_t + \Delta u + g|u|^4 u = 0, \quad u(0) \in \dot{H}^1, \quad g \geq 0.$$

On the road map of Kenig–Merle [20] we show the global well-posedness and scattering of radial solutions under scaling, variational, and rigidity assumptions for g. We also provide sharp finite time blowup results for nonradial and radial solutions. For this we utilize the localized virial identity.

Keywords Focusing · Energy-critical NLS · Bounded inhomogeneous coefficient · Kenig-Merle argument · Global well-posedness

1 Introduction

In this paper, we consider the following Cauchy problem for an inhomogeneous quintic nonlinear Schrödinger equation in 3 dimensions:

$$\begin{cases} iu_t + \Delta u + g|u|^4 u = 0 \text{ in } \mathbb{R}^{1+3}, \\ u(0) = \varphi \in \dot{H}^1(\mathbb{R}^3). \end{cases} \tag{1}$$

Here g is the coefficient representing interaction among particles. In general, it is a bounded function of space variables on $\mathbb{R}^3$. The space $\dot{H}^1$ denotes the homogenous

Y. Cho
Department of Mathematics, and Institute of Pure and Applied Mathematics, Jeonbuk National University, Jeonju, Republic of Korea
e-mail: changocho@jbnu.ac.kr

K. Lee (✉)
Department of Mathematics, Jeonbuk National University, Jeonju, Republic of Korea
e-mail: leeky@jbnu.ac.kr

© Springer Nature Switzerland AG 2020
V. Georgiev et al. (eds.), *Advances in Harmonic Analysis and Partial Differential Equations*, Trends in Mathematics, https://doi.org/10.1007/978-3-030-58215-9_6

Sobolev space defined by

$$\dot{H}^1 = \{f \in L_x^6 : \|f\|_{\dot{H}^1} := \|\nabla f\|_{L_x^2} < +\infty\}. \tag{2}$$

Equation (1) can be a model of dilute Bose–Einstein condensate when the three-body interactions of the condensate are considered. For this see [2, 26]. It has also been considered to study the laser guiding in an axially nonuniform plasma channel. See [15, 24, 26].

We can define the energy E_g of the solution to (1) by

$$E_g(u(t)) := \frac{1}{2}\|\nabla u(t)\|_{L_x^2}^2 - \frac{1}{6}\int g|u(t)|^6 dx. \tag{3}$$

If the solution is sufficiently smooth, then its energy is conserved, that is, $E_g(u(t)) = E_g(\varphi)$ for all t in the existence time interval. This will be treated again briefly in Remark 2.3 below.

As is well known when g is a constant, a scaling invariance structure can be set up in $\dot{H}^1$ under a condition on g. To be more precise, we assume that

$$g \in (C \cap L_x^\infty)(\mathbb{R}^3) \cap C^1(\mathbb{R}^3\backslash\{0\}) \text{ and } |\nabla g(x)| \lesssim |x|^{-1} \text{ for any } x \neq 0. \tag{4}$$

Then the $\dot{H}^1$-scaled function u_λ for $\lambda > 0$ defined by $u_\lambda(t, x) = \lambda^{\frac{1}{2}} u(\lambda^2 t, \lambda x)$ is also the solution to the Eq. (1) with the coefficient $g_\lambda(x) := g(\lambda x)$ satisfying (4). Thus we may say that (1) is *essentially* energy-critical.

In [6, 7] it is shown that (1) is locally well-posed in $H^1 := \dot{H}^1 \cap L_x^2$ under the scaling condition (4). In the next section we show the local well-posedness (LWP) in $\dot{H}^1$ via contraction argument based on the Strichartz estimates. By the LWP we mean that there exists a maximal existence time interval I^* such that there exists a unique solution $u \in C(I^*; \dot{H}^1)$ and u depends continuously on the initial data. We say that (1) is globally well-posed if $I^* = \mathbb{R}$ and that u scatters in $\dot{H}^1$ if there exists linear solutions $u_\pm$ such that $u \to u_\pm$ in $\dot{H}^1$ as $t \to \pm\infty$. In this paper the solution is said to blow up if $\int_{I^*}\int_{\mathbb{R}^3} |u(t, x)|^{10} dxdt = +\infty$ since the $L_{t,x}^{10}$ norm controls the whole contraction argument. We also use the terminology of finite time blowup when I^* is bounded.

The aim of this paper is to show the global well-posedness (GWP), scattering, and finite time blowup of radial solutions to (1) for a *nonnegative* function g with the scaling condition (4). Since g is nonnegative, (1) has the focusing nature. In order to handle such focusing problem we catch up with the antecedent concentration-compactness argument of Kenig and Merle [19]. As in [19] our strategy is threefold: (a) Variational estimates based on the ground state (b) Existence and compactness of minimal energy blowup solutions (MEBS) (c) Rigidity to remove MEBS based on the localized virial argument. Here the ground state W_g means a positive radial solution to the stationary problem $\Delta W_g + g|W_g|^4 W_g = 0$. It is known that the equation may have no such solution in general (see [11]). Hence, instead of introducing W_g, we implement some subtle variational estimates by adopting the well-known function $W(x) = \left(1 + \frac{|x|^2}{3}\right)^{-\frac{1}{2}}$, which is the positive radial solution to the elliptic problem

$$\Delta W + |W|^4 W = 0 \tag{5}$$

and is unique up to scaling.

For the variational estimates we confine the lower and upper bounds of g as follows:

$$\sqrt{g_s}(3 - g_i) \leq 2, \tag{6}$$

where $g_i = \inf g$ and $g_s = \sup g$. This bound gives us a sharpness between GWP and blowup. However the condition (6) is very special and far away from optimal. We think it might be replaced by a more general one.

For the rigidity part we need the *radial symmetry* and additional *rigidity condition* for g such that

$$x \cdot \nabla g(x) \geq 0 \quad \text{for all} \quad x \neq 0. \tag{7}$$

This condition gets rid of the error term occurring when we deal with the lower bound for the second derivative of localized virial quantity $z_r(t) = \int b_r(x)|u(t)|^2\,dx$ with a smooth function b_r for $r > 0$. We will see the details in Sect. 6.2.

Now we are ready to state our main result.

Theorem 1.1 *Let g be a nonnegative, radial, and bounded function satisfying* (4), (6), *and* (7). *Suppose that* $\varphi \in \dot{H}^1_{rad} := \{f \in \dot{H}^1 : f \text{ is radial}\}$,

$$E_g(\varphi) < E_g(W), \quad \textit{and} \quad \sqrt{g_s}\|\varphi\|^2_{\dot{H}^1} < \|W\|^2_{\dot{H}^1}. \tag{8}$$

Then (1) *is globally well-posed in* $\dot{H}^1_{rad}$ *and the solution u scatters in* $\dot{H}^1_{rad}$.

The condition (8) implies the energy trapping and coercivity of energy that is $E_g(u) \sim \|u\|^2_{\dot{H}^1} \sim \|\varphi\|^2_{\dot{H}^1}$. In order to show the existence of MEBS and compactness of its flows we develop a profile decomposition for radial data. The condition (7) is crucial to removing MEBS, which is equivalent to $g'(|x|) \geq 0$. We may take g satisfying (4), (6), and (7) as follows:

$$g(s) = \frac{a(s+b)}{(s+c)} \quad \text{for} \ a \geq 0,\ 0 \leq b \leq c,\ c > 0,\ \sqrt{a}(3 - \frac{ab}{c}) \leq 2, \tag{9}$$

e.g., $a = \frac{4}{9}, b = 0, c = 1$, or

$$g(s) = \begin{cases} a \ (0 \le a < 3), & 0 \le s < 1, \\ \text{smooth and increasing}, & 1 < s < 2, \\ \frac{4}{(3-a)^2}, & s \ge 2. \end{cases} \tag{10}$$

It still remains open to resolve the rigidity part for the case when g is decreasing, oscillating, or increasing but $\sqrt{g_s}(3 - g_i) > 2$.

On the other hand to obtain a sharp blowup result we need to control the error term for the upper bound of the second derivative of localized virial quantity. To do so we assume that

$$x \cdot \nabla g(x) \le 6(k_g - \rho) g(x) \quad \text{for all} \quad x \neq 0, \tag{11}$$

where $k_g = \frac{2-\sqrt{g_s}(3-g_i)}{3-\sqrt{g_s}(3-g_i)}$ and for some $\rho \ge 0$. Then we get the following.

Theorem 1.2 *Let g be a nonnegative and bounded function satisfying* (4), (6), *and* (11).

(1) *Suppose that $\varphi \in \dot{H}^1$, $|x|\varphi \in L^2$,*

$$E_g(\varphi) < E_g(W), \quad \textit{and} \quad \sqrt{g_s}\|\varphi\|_{\dot{H}^1}^2 \ge \|W\|_{\dot{H}^1}^2. \tag{12}$$

Then the solution u to (1) *blows up in finite time.*

(2) *Suppose that g is radial, $\rho > 0$, and $\varphi \in H^1_{rad}$ with* (12). *Then the radial solution u to* (1) *blows up in finite time.*

The radial symmetry is not necessary in (1). The condition (12) gives us the fact $\int (|\nabla u|^2 - (1 - \eta) g |u|^6)\, dx < -C$ for some $0 \le \eta < 1$ and $C > 0$ and hence the finite time blowup through the localized virial identity (120) below. For this see Remark 3.14 of [19] and also [3]. In (2) the moment condition $|x|\varphi \in L^2_x$ has been replaced with the radial symmetry and L^2_x condition. This is due to the space-decay estimate of Strauss [23]. The condition $\rho > 0$ in (2) is required for a technical reason to handle error terms appearing in localized virial argument.

Other types of inhomogeneity have been considered for the focusing problem in many literatures. Among other things, we would like to mention Farah-Guzmán's result [13] in which GWP and scattering in $H^1_{rad} := \dot{H}^1_{rad} \cap L^2_x$ were studied to the problem with nonlinearity $|x|^{-b}|u|^2 u$ for some $0 < b < 1$. They used Holmer-Roudenko's inter-critical argument [17]. On the other hand, Dinh [12] studied the finite time blowup for the similar problem through inter-critical virial argument. For more related results on the inhomogeneous NLS see the references in [12, 13].

This paper is organized as follows: In Sect. 2 we establish the local theory on LWP and long-time perturbation. Sections 3, 4, 5, and 6 are devoted to proving GWP and scattering. Section 3 covers the variational estimates including the energy trapping and coercivity of energy. In Sect. 4 we develop a radial profile decomposition, which plays a crucial role in the proof of existence and compactness of MEBS in Sect. 5. In Sect. 6 we prove the rigidity part based on the compactness. We finally consider the finite time blowup in Sect. 7.

Notations

- Mixed-normed spaces: For a Banach space X and an interval I, $u \in L_I^q X$ iff $u(t) \in X$ for a.e.$t \in I$ and $\|u\|_{L_I^q X} := \|\|u(t)\|_X\|_{L_I^q} < \infty$. Especially, we denote $L_I^q L_x^r = L_t^q(I; L_x^r(\mathbb{R}^3))$, $L_{I,x}^q = L_I^q L_x^q$, $L_t^q L_x^r = L_{\mathbb{R}}^q L_x^r$,

$$S(I) = L_I^{10} L_x^{10}, \quad \text{and} \quad W(I) = L_I^{10} L_x^{\frac{30}{13}}. \tag{13}$$

- As usual different positive constants depending only on d, α are denoted by the same letter C, if not specified. $A \lesssim B$ and $A \gtrsim B$ means that $A \leq CB$ and $A \geq C^{-1}B$, respectively for some $C > 0$. $A \sim B$ means that $A \lesssim B$ and $A \gtrsim B$.

2 Local Theory

By Duhamel's formula, (1) is written as an integral equation

$$u = e^{it\Delta}\varphi + i\int_0^t e^{i(t-t')\Delta} g|u(t')|^4 u(t')dt'. \tag{14}$$

Here we define the linear propagator $e^{it\Delta}$ given by the solution to the linear problem $i\partial_t v = -\Delta v$ with initial data $v(0) = f$. It is formally given by

$$e^{it\Delta} f = \mathcal{F}^{-1}\left(e^{-it|\xi|^2}\mathcal{F}(f)\right) = (2\pi)^{-3}\int_{\mathbb{R}^3} e^{i(x\cdot\xi - t|\xi|^2)}\widehat{f}(\xi)d\xi, \tag{15}$$

where $\widehat{f} = \mathcal{F}(f)$ denotes the Fourier transform of f and $\mathcal{F}^{-1}(h)$ the inverse Fourier transform of h such that

$$\mathcal{F}(f)(\xi) = \int_{\mathbb{R}^3} e^{-ix\cdot\xi} f(x)\,dx, \quad \mathcal{F}^{-1}(h)(x) = (2\pi)^{-3}\int_{\mathbb{R}^3} e^{ix\cdot\xi} h(\xi)\,d\xi. \tag{16}$$

Now we introduce Strichartz estimate and Hardy–Sobolev inequality which are crucial for the local theory.

Lemma 2.1 ([18]) *Let (q, r) and $(\widetilde{q}, \widetilde{r})$ be pairs such that $2 \leq q, r, \widetilde{q}, \widetilde{r} \leq \infty$ and satisfy the equation $\frac{2}{q} + \frac{3}{r} = \frac{3}{2}$. Then we have*

$$\|e^{it\Delta}\varphi\|_{L_t^q L_x^r} \lesssim \|\varphi\|_{L_x^2}, \tag{17a}$$

$$\left\| \int_0^t e^{i(t-t')\Delta} F\, dt' \right\|_{L_t^q L_x^r} \lesssim \|F\|_{L_t^{\tilde{q}'} L_x^{\tilde{r}'}}. \tag{17b}$$

We call such pair *admissible*.

Lemma 2.2 ([4]) *Let* $f \in \dot{W}^{1,p}(\mathbb{R}^3) = \{f \in L^{\frac{3p}{3-p}} : \|f\|_{\dot{W}^p} := \|\nabla f\|_{L_x^p} < \infty\}$. *Then for* $1 < p < 3$ *we have*

$$\||x|^{-1} f\|_{L_x^p} \lesssim \|f\|_{\dot{W}^{1,p}}. \tag{18}$$

Let us first consider the local well-posedness.

2.1 Local Well-Posedness

We revisit [5, 20]. We have only to show the following proposition for LWP.

Proposition 2.1 *Let* g *satisfy* (4). *Assume that* $\varphi \in \dot{H}^1, 0 \in I$ *an interval, and* $\|\varphi\|_{\dot{H}^1} \le A$. *Then there exists* $\delta = \delta(A)$ *such that if* $\|e^{it\Delta}\varphi\|_{S(I)} < \delta$, *then there exists a unique solution* u *of* (1) *in* $I \times \mathbb{R}^3$ *with* $u \in C(I; \dot{H}^1(\mathbb{R}^3))$,

$$\|\nabla u\|_{W(I)} < \infty, \quad \text{and} \quad \|u\|_{S(I)} \le 2\delta. \tag{19}$$

If $\varphi_k \to \varphi$ *in* $\dot{H}^1$, *then the corresponding solutions* $u_k \to u$ *in* $C(I; \dot{H}^1)$ *for large* k.

Proof Fix $a, b > 0$, to be chosen later, let us define the function space $B_{a,b}$ and contraction mapping Φ as following:

$$B_{a,b} = \{v : \|v\|_{S(I)} \le a, \quad \|\nabla v\|_{W(I)} \le b\}, \tag{20}$$

$$\Phi(v) = e^{it\Delta}\varphi + i \int_0^t e^{i(t-t')\Delta} f(v) dt', \tag{21}$$

where $f(v) = g|v|^4 v$. By (4) and Lemmas 2.1 and 2.2 we estimate

$$\begin{aligned} \|\nabla\Phi(v)\|_{W(I)} &\le C\|\nabla\varphi\|_{L_x^2} + C\|\nabla f(v)\|_{L_I^2 L_x^{\frac{6}{5}}} \\ &\le C\left(A + \|v\|_{S(I)}^4 \|\nabla v\|_{W(I)} + \|v\|_{S(I)}^4 \||x|^{-1} v\|_{W(I)}\right) \\ &\le C(A + a^4 b). \end{aligned} \tag{22}$$

Choosing $b = 2AC$ and $Ca^4 \le \frac{1}{2}$, we have $\|\nabla\Phi(v)\|_{W(I)} \le b$. Now if $\delta = \frac{a}{2}$ and $Ca^3b \le \frac{1}{2}$, then by Sobolev embedding $\dot{W}^{1,\frac{30}{13}} \hookrightarrow L^{10}$ and the same argument as above we deduce that

$$\|\Phi(v)\|_{S(I)} \le \|e^{it\Delta}\varphi\|_{S(I)} + C\|\nabla f(v)\|_{L_I^2 L_x^{\frac{6}{5}}} \le \delta + Ca^4 b \le a. \tag{23}$$

Hence Φ is self-mapping on $B_{a,b}$.

Next we have only to replace some of v involved in the nonlinear parts above with $v - v'$ to show the contraction.

$$\begin{aligned} &\|\Phi(v) - \Phi(v')\|_{S(I)} + \|\nabla\Phi(v) - \nabla\Phi(v')\|_{W(I)} \\ &\quad \lesssim \|\nabla(|v|^4 v - |v'|^4 v')\|_{L_I^2 L_x^{\frac{6}{5}}} + \||x|^{-1}(|v|^4 v - |v'|^4 v')\|_{L_I^2 L_x^{\frac{6}{5}}} \\ &\quad \lesssim \||v|^4 |\nabla v - \nabla v'|\|_{L_I^2 L_x^{\frac{6}{5}}} + \sum_{k=0}^{3} \||v|^k |v'|^{3-k} |\nabla v'||v - v'|\|_{L_I^2 L_x^{\frac{6}{5}}} \\ &\quad \lesssim (a^4 + a^3 b)\|\nabla(v - v')\|_{W(I)}. \end{aligned} \tag{24}$$

The continuous dependency on initial data follows immediately from the above contraction argument. This completes the proof of Proposition 2.1. □

Remark 2.1 Proposition 2.1 shows the existence of maximal existence time interval I^*. Moreover, one can immediately deduce the blowup criterion: if $\|u\|_{S(I^*)} < +\infty$, then $I^* = \mathbb{R}$, and if I^* is bounded, then $\|u\|_{S(I^*)} = +\infty$. We also conclude that if $\|\varphi\|_{\dot{H}^1}$ is sufficiently small, then $I^* = \mathbb{R}$.

Remark 2.2 Suppose that $\|u\|_{S(I^*)} < +\infty$. Let us set

$$\varphi_\pm := \varphi + i\int_0^{\pm\infty} e^{-it'\Delta}[g|u|^4 u]\,dt'. \tag{25}$$

Then a standard duality argument shows that the solution u scatters to $e^{it\Delta}\varphi_\pm$ in $\dot{H}^1$.

Remark 2.3 Since the nonlinear term is essentially quintic, one can easily obtain H^2-regularity of u if $\varphi \in H^2$. The energy conservation can be readily shown in H^2. The energy conservation in $\dot{H}^1$ follows from the density and the continuous dependency on initial data of Proposition 2.1. We refer the readers to [6].

Remark 2.4 Let us define the mass by

$$m(u(t)) := \|u(t)\|_{L_x^2}^2. \tag{26}$$

Then by the density and continuity the mass of solution in Proposition 2.1 can also be shown to be conserved in the existence time interval.

2.2 Long-Time Perturbation

Proposition 2.2 *Let g be a radial function satisfying* (4). *Let $I \subset \mathbb{R}$ be a time interval containing* 0. *Let $\tilde{u}$ be a radial function defined on $I \times \mathbb{R}^3$ and satisfy*

$$\|\tilde{u}\|_{L_t^\infty \dot{H}^1} \le A, \quad \|\tilde{u}\|_{S(I)} \le M \tag{27}$$

for some constants $M, A > 0$. Assume that

$$i\partial_t \tilde{u} + \Delta \tilde{u} + f(\tilde{u}) = e \ \textit{for} \ (t,x) \in I \times \mathbb{R}^3, \tag{28}$$

where $f(\tilde{u}) = g|\tilde{u}|^4 \tilde{u}$ and that

$$\|\varphi - \tilde{u}(0)\|_{\dot{H}^1} \le A', \ \|\nabla e\|_{L_I^2 L_x^{\frac{6}{5}}} \le \varepsilon, \ \textit{and} \ \|\nabla e^{it\Delta}[\varphi - \tilde{u}(0)]\|_{W(I)} \le \varepsilon. \tag{29}$$

Then there exists $\varepsilon_0 = \varepsilon_0(M, A, A')$ and a unique solution $u \in C(I; \dot{H}^1_{rad})$ with $u(0) = \varphi$ in $I \times \mathbb{R}$, such that for $0 < \varepsilon < \varepsilon_0$ with

$$\|u\|_{S(I)} \le C(M, A, A') \ \textit{and} \ \|u(t) - \tilde{u}(t)\|_{\dot{H}^1} \le A' + C(M, A, A')\varepsilon \tag{30}$$

for all $t \in I$.

Remark 2.5 Since we treat the radial solutions, we do not need to use the inhomogeneous Strichartz estimate as described in [19] for the proof of long-time perturbation on $S(I)$. But instead we use the condition $\|\nabla e^{it\Delta}[\varphi - \tilde{u}(0)]\|_{W(I)} \le \varepsilon$, which simplifies the proof of above proposition and is utilized in the proof of the existence of minimal energy blowup solution. Of course, we can replace $W(I)$ with $S(I)$ by exploiting the inhomogeneous Strichartz estimate of [14].

Proof Without loss of generality we may assume that $I = [0, a)$ for some $0 < a \le +\infty$. Our assumption together with Hölder's inequality and Hardy–Sobolev inequality (Lemma 2.2) yields

$$\|e^{it\Delta}\nabla(\varphi - \tilde{u}(0))\|_{W(I)} \le \varepsilon, \quad \|\nabla(g|u|^4 u)\|_{L_I^2 L_x^{\frac{6}{5}}} \lesssim \|u\|_{S(I)}^4 \|\nabla u\|_{W(I)}. \tag{31}$$

Note that by the integral equation (14) for $\tilde{u}$ and Proposition 2.1, one can readily obtain $\|\nabla \tilde{u}\|_{W(I)} \le \widetilde{M}$ for some $\widetilde{M}$ depending on M, A.

Let us define $u = \tilde{u} + w$, so that the equation for w is written as

$$\begin{cases} iw_t + \Delta w = f(\tilde{u} + w) - f(\tilde{u}) + e, \\ w(0) = \varphi - \tilde{u}(0). \end{cases} \tag{32}$$

Then for arbitrary $\eta > 0$, there exists $I_j = \big[a_j, a_{j+1}\big)$ such that $\bigcup_{j=1}^{J} I_j = I$ and $\|\nabla \tilde{u}\|_{W(I_j)} \leq \eta$. Describing integral equation on I_j, we have

$$\begin{aligned} w(t) =& e^{i(t-a_j)\Delta} w(a_j) + i \int_{a_j}^{t} e^{i(t-t')\Delta}(f(\tilde{u}+w) - f(\tilde{u}))dt' \\ &- i \int_{a_j}^{t} e^{i(t-t')\Delta} e(t')dt'. \end{aligned} \tag{33}$$

By Sobolev embedding and Lemma 2.2, we get

$$\begin{aligned} \|\nabla w\|_{W(I_j)} \leq& (\|\nabla e^{i(t-a_j)\Delta} w(a_j)\|_{W(I_j)} + C\varepsilon) \\ &+ C\eta^4 \|\nabla w\|_{W(I_j)} + C\|\nabla w\|_{W(I_j)}^5. \end{aligned} \tag{34}$$

Thus, if $C\eta^4 \leq \frac{1}{3}$, we have

$$\|\nabla w\|_{W(I_j)} \leq \frac{3}{2}\gamma_j + C\|\nabla w\|_{W(I_j)}^5, \tag{35}$$

where $\gamma_j = \|\nabla e^{i(t-a_j)\Delta} w(a_j)\|_{W(I_j)} + C\varepsilon$.

From the standard continuity argument, we can find $C_0 > 0$ satisfying that

$$\|\nabla w\|_{W(I_j)} \leq 3\gamma_j \text{ and } C\|\nabla w\|_{W(I_j)}^5 \leq 3\gamma_j, \tag{36}$$

provided $\gamma_j \leq C_0$. Repeating the above argument for the equation

$$\begin{aligned} e^{i(t-a_{j+1})\Delta} w(a_{j+1}) =& e^{i(t-a_j)\Delta} w(a_j) + i \int_{a_j}^{a_{j+1}} e^{i(t-t')\Delta}(f(\tilde{u}+w) - f(\tilde{u}))dt' \\ &- i \int_{a_j}^{a_{j+1}} e^{i(t-t')\Delta} e(t')dt', \end{aligned} \tag{37}$$

we get

$$\begin{aligned} \|e^{i(t-a_{j+1})\Delta} w(a_{j+1})\|_{W(I_{j+1})} \leq& \|\nabla e^{i(t-a_j)\Delta} w(a_j)\|_{W(I_{j+1})} + C\varepsilon \\ &+ C\eta^4 \|\nabla w\|_{W(I_{j+1})} + C\|\nabla w\|_{W(I_{j+1})}^5. \end{aligned} \tag{38}$$

Taking a sufficiently small η to satisfy $\gamma_{j+1} \leq 10\gamma_j$ provided $\gamma_j \leq C_0$. This always happens if $C10^J \varepsilon_0 < C_0$. With this ε_0 we have that for any $0 < \varepsilon < \varepsilon_0$

$$\|w\|_{S(I)} + \|\nabla w\|_{W(I)} \leq 3C \sum_{j=1}^{J} \gamma_j \leq \frac{C}{3}(10^{J+1} - 1)\varepsilon. \tag{39}$$

Hence by letting $C(M, A, A') = C(10^{J+1} - 1)\varepsilon_0/3$ we get

$$\|u\|_{S(I)} \le \|w\|_{S(I)} + \|\tilde{u}\|_{S(I)} \le C(M, A, A'). \tag{40}$$

Using the Strichartz estimate and Hardy–Sobolev inequality once more, we lead

$$\|w\|_{L_I^\infty \dot{H}^1} \le A' + C\varepsilon + C\sum_{j=1}^{J} \|\nabla(f(\tilde{u} + w) - f(\tilde{u}))\|_{W(I_j)} \le A' + C(M, A, A')\varepsilon. \tag{41}$$

3 Variational Estimates

We now introduce some variational inequalities showing a sharpness between GWP and blowup. Let C_* be the best constant satisfying $\|u\|_{L_x^6} \le C_*\|u\|_{\dot{H}^1}$. The minimizer u_* such that $\|u_*\|_{L_x^6} = C_*\|u_*\|_{\dot{H}^1}$ is characterized by $u_* = e^{i\theta}\lambda^{\frac{1}{2}} W(\lambda(x - x_0))$ for some $\theta \in [-\pi, \pi]$, $\lambda > 0$, and $x_0 \in \mathbb{R}^3$. From the elliptic equation (5) it follows that $\int |\nabla W|^2 = \int |W|^6$. Hence, $\int |\nabla W|^2 = \frac{1}{C_*^3}$. For this see [1, 25].

Lemma 3.1 (Energy Trapping) *Let u be a solution of* (1) *with φ such that*

$$\sqrt{g_s}\|\varphi\|_{\dot{H}^1}^2 < \|W\|_{\dot{H}^1}^2 \quad \text{and} \quad E_g(\varphi) \le (1 - \delta_0)E_g(W) \tag{42}$$

for some $\delta_0 > 0$. Then there exits $\bar{\delta} = \bar{\delta}(\delta_0)$ such that

$$\sqrt{g_s}\|u(t)\|_{\dot{H}^1}^2 \le (1 - \bar{\delta})\|W\|_{\dot{H}^1}^2, \tag{43a}$$

$$\int |\nabla u(t)|^2 - g|u(t)|^6 dx \ge \bar{\delta}\int |\nabla u(t)|dx, \tag{43b}$$

$$\text{(Coercivity) } E_g(u(t)) \sim \|u(t)\|_{\dot{H}^1}^2 \sim \|\varphi\|_{\dot{H}^1}^2, \tag{43c}$$

for all $t \in I^*$, where I^* is the maximal existence time interval.

The above lemma follows from the continuity argument, energy conservation, and the following lemma on the initial energy trapping.

Lemma 3.2 *Let $E_g(\varphi) < (1 - \delta_0)E_g(W)$, $\sqrt{g_s}\|\varphi\|_{\dot{H}^1}^2 < \|W\|_{\dot{H}^1}^2$ for some $\delta_0 > 0$. Then there exists $\bar{\delta} = \bar{\delta}(\delta_0)$ such that*

$$\sqrt{g_s}\|\varphi\|_{\dot{H}^1}^2 \le (1 - \bar{\delta})\|W\|_{\dot{H}^1}^2, \tag{44a}$$

$$\int |\nabla\varphi|^2 - g|\varphi|^6 dx \geq \bar{\delta}\|\varphi\|_{\dot{H}^1}^2, \tag{44b}$$

$$E_g(\varphi) \geq 0. \tag{44c}$$

Proof *(of Lemma 3.2)* Let $f(y) = \frac{1}{2}y - \frac{C_*^6}{6}y^3$, $\bar{y} = \sqrt{g_s}\|\varphi\|_{\dot{H}^1}^2$, and $y_0 = \|W\|_{\dot{H}^1}^2$. Then we estimate the following:

$$\begin{aligned} f(\bar{y}) &= \frac{\sqrt{g_s}}{2}\|\varphi\|_{\dot{H}^1}^2 - \frac{(\sqrt{g_s})^3 C_*^6}{6}\|\varphi\|_{\dot{H}^1}^6 \leq \sqrt{g_s}\left(\frac{1}{2}\|\varphi\|_{\dot{H}^1}^2 - \frac{g_s}{6}\|\varphi\|_{L_x^6}^6\right) \\ &\leq \sqrt{g_s}E_g(\varphi) \leq \frac{1}{2}\sqrt{g_s}(3-g_i)(1-\delta_0)f(y_0) \leq (1-\delta_0)f(y_0). \end{aligned} \tag{45}$$

Since $0 \leq \bar{y} < y_0$ and f is strictly increasing on $[0, y_0]$, for some $\bar{\delta}$ we get

$$0 \leq f(\bar{y}), \quad \bar{y} \leq (1-\bar{\delta})y_0. \tag{46}$$

In particular,

$$\begin{aligned} \int |\nabla\varphi|^2 - g|\varphi|^6 &\geq \int |\nabla\varphi|^2 - g_s C_*^6 \left(\int |\nabla\varphi|^2\right)^3 \\ &\geq \|\varphi\|_{\dot{H}^1}^2(1-(1-\bar{\delta})^2) \geq \bar{\delta}\|\varphi\|_{\dot{H}^1}^2. \end{aligned} \tag{47}$$

This completes the proof of Lemma 3.2. □

4 Profile Decomposition

In this section we introduce a profile decomposition for radial data.

Lemma 4.1 *Let $\{v_{0,n}\} \subset \dot{H}^1_{rad}$, $\|v_{0,n}\| \leq A$, and $\|e^{it\Delta}v_{0,n}\|_{L_t^q L_x^r} \geq \delta > 0$, where δ is as in Proposition 2.1. Then up to a subsequence (still called $\{v_{0,n}\}$) for any $J \geq 1$ there exists a sequence $\{V_{0,j}\}_{1\leq j\leq J}$ in $\dot{H}^1_{rad}$ and a family of parameters $(\lambda_{j,n}, t_{j,n}) \in \mathbb{R}^+ \times \mathbb{R}$ with*

$$\frac{\lambda_{j,n}}{\lambda_{j',n}} + \frac{\lambda_{j',n}}{\lambda_{j,n}} + \frac{|t_{j,n} - t_{j',n}|}{\lambda_{j,n}^2} \xrightarrow{n\to\infty} 0 \quad j \neq j' \tag{48}$$

such that

(i) $\|V_{0,1}\|_{\dot{H}^1} \geq \alpha_0(A) > 0$,

(ii) $v_{0,n} = \sum_{j=1}^J \lambda_{j,n}^{-\frac{1}{2}} V_j^l\left(-\frac{t_{j,n}}{\lambda_{j,n}^2}, \frac{x}{\lambda_{j,n}}\right) + w_n^J, \quad V_j^l(t,x) = [e^{it\Delta}V_{0,j}](x),$

(iii) $\liminf_{n\to\infty} \|\nabla e^{it\Delta} w_n^J\|_{L_t^q L_x^r} \to 0$ *for any admissible pair* (q, r) *with* $2 < q < \infty$,

(iv) $\|v_{0,n}\|_{\dot H^1}^2 = \sum_{j=1}^J \|V_{0,n}\|_{\dot H^1}^2 + \|w_n^J\|_{\dot H^1}^2 + o(1)$ *as* $n \to \infty$,

(v) $E_g(v_{0,n}) = \sum_{j=1}^J E_g\left(\lambda_{j,n}^{-\frac12} V_j^l(-\frac{t_{j,n}}{\lambda_{j,n}^2}, \frac{\cdot}{\lambda_{j,n}})\right) + E_g(w_n^j) + o(1)$ *as* $n \to \infty$.

Remark 4.1 The space-frequency translation has been removed by radial symmetry in the above lemma. This fact enables us to get a strong convergence of remainder terms w_n^J in $L_t^q \dot W^{1,r}$ rather than in $L_t^q L^{\frac{3r}{3-r}}$. Together with long-time perturbation (Proposition 2.2), we will utilize this strong convergence for the proof of the existence and compactness of minimal energy blowup solution. For the nonradial case one has only to replace the convergence in the space $L_t^q L^{\frac{3r}{3-r}}$.

Proof One can show (i), (ii), (iii), and (iv) by exactly the same way as in [9, 10, 21]. We omit the details. Here we only consider the energy decomposition (v). Due to the kinetic energy decomposition (iv) it suffices to show

$$\lim_{n\to\infty}\left(\int g|v_{0,n}|^6\,dx - \sum_{1\le j\le J}\int g|G_{j,n}|^6\,dx - \int g|w_n^J|^6\,dx\right) = 0, \tag{49}$$

where

$$G_{j,n}(x) = \lambda_{j,n}^{-\frac12} V_j^l\left(-\frac{t_{j,n}}{\lambda_{j,n}^2}, \frac{x}{\lambda_{j,n}}\right). \tag{50}$$

For this we consider

$$\int g(x)|v_{0,n}|^6 dx - \int g(x)|v_{0,n} - G_{1,n}|^6\,dx - \int g(x)|G_{1,n}|^6 dx \xrightarrow{n\to\infty} 0. \tag{51}$$

Then by repeating the same argument w.r.t. j we conclude the proof.

Let $s_n = -\frac{t_{1,n}}{\lambda_{1,n}^2}$. Suppose that $\lim_{n\to\infty}|s_n| = \infty$. Then by the time-decay estimate such that $\|e^{it_n\Delta} f\|_{L_x^p} \to 0$ as $|t_n| \to \infty$ for any $p > 2$ and $f \in C_0^\infty$, we get

$$\begin{aligned}&\int g(x)|v_{0,n}|^6 dx - \int g(x)|v_{0,n} - G_{1,n}|^6\,dx - \int g(x)|G_{1,n}|^6 dx\\ &\lesssim \sum_{k=1}^6 \|G_{1,n}\|_{L_x^6}^k \|v_{0,n}\|_{L_x^6}^{6-k} \to 0.\end{aligned} \tag{52}$$

Up to a subsequence we now assume that $\lim_{n\to\infty} s_n = \bar{t}$. And let

$$S_{1,n}(v_{0,n}) = \lambda_{j,n}^{\frac{1}{2}} v_{0,n}(\lambda_{j,n}\cdot). \tag{53}$$

Then we have that

$$\begin{aligned} &S_{1,n}(v_{0,n}) \to e^{i\bar{t}\Delta} V_{0,1} \quad \text{weakly in } \dot{H}^1 \text{ as } n \to \infty, \\ &S_{1,n}(G_{1,n}) \to e^{i\bar{t}\Delta} V_{0,1} \quad \text{strongly in } L_x^6 \text{ as } n \to \infty. \end{aligned} \tag{54}$$

By scaling the left hand side of (51) is written as

$$\begin{aligned} &\int g(x)|v_{0,n}|^6 dx - \int g(x)|v_{0,n} - G_{1,n}|^6\, dx - \int g(x)|G_{1,n}|^6 dx \\ &= \int g(\lambda_{j,n}x)|S_{1,n}(v_{0,n})|^6 dx - \int g(\lambda_{j,n}x)|S_{1,n}(v_{0,n}) - S_{1,n}(G_{1,n})|^6 dx \\ &\quad - \int g(\lambda_{j,n}x)|S_{1,n}(G_{1,n})|^6 dx \\ &=: A_n + B_n + C_n, \end{aligned} \tag{55}$$

where

$$\begin{aligned} A_n = &\int g(\lambda_{j,n}x)|S_{1,n}(v_{0,n})|^6 dx - \int g(\lambda_{j,n}x)|S_{1,n}(v_{0,n}) - e^{i\bar{t}\Delta} V_{0,1}|^6 dx \\ &- \int g(\lambda_{j,n}x)|e^{i\bar{t}\Delta} V_{0,1}|^6\, dx, \end{aligned} \tag{56a}$$

$$\begin{aligned} B_n = &\int g(\lambda_{j,n}x)|S_{1,n}(v_{0,n}) - e^{i\bar{t}\Delta} V_{0,1}|^6 dx \\ &- \int g(\lambda_{j,n}x)|S_{1,n}(v_{0,n}) - S_{1,n}(G_{1,n})|^6 dx, \end{aligned} \tag{56b}$$

$$C_n = \int g(\lambda_{j,n}x)|e^{i\bar{t}\Delta} V_{0,1}|dx - \int g(\lambda_{j,n}x)|S_{1,n}(G_{1,n})|dx. \tag{56c}$$

Using the density by C_0^∞ and compactness of multiplication operator by $\beta \in C_0^\infty$ from $\dot{H}^1$ to L^3, one can readily show that for any $\varepsilon > 0$ there exist $\beta \in C_0^\infty$ close to $e^{i\bar{t}\Delta} V_{0,1}$ in L_x^6 and N such that if $n \geq N$, then $(S_{1,n}(v_{0,n}) - e^{i\bar{t}\Delta} V_{0,1})\beta$ close to

0 in L_x^3 and thus

$$\begin{aligned}
|A_n| &\leq \int |g(\lambda_{j,n}x)| \left| |S_{1,n}(v_{0,n})|^6 - |S_{1,n}(v_{0,n}) - e^{i\bar{t}\Delta}V_{0,1}|^6 - |e^{i\bar{t}\Delta}V_{0,1}|^6 \right| dx \\
&\leq g_s \int \left| |S_{1,n}(v_{0,n})|^6 - |S_{1,n}(v_{0,n}) - e^{i\bar{t}\Delta}V_{0,1}|^6 - |e^{i\bar{t}\Delta}V_{0,1}|^6 \right| dx \\
&= g_s \int P((S_{1,n}(v_{0,n}) - e^{i\bar{t}\Delta}V_{0,1})\beta, e^{i\bar{t}\Delta}V_{0,1} - \beta, S_{1,n}(v_{0,n}), e^{i\bar{t}\Delta}V_{0,1}, \beta)\, dx \\
&< \varepsilon,
\end{aligned} \tag{57}$$

where $P(z_1, z_2, z_3, z_4, z_5)$ is an order-5 homogeneous polynomial of z_i and their conjugates containing at least one of $z_1, z_2, \bar{z_1}, \bar{z_2}$.

On the other hand, we have

$$\begin{aligned}
|B_n| &\leq \int |g(\lambda_{j,n}x)| \left| |S_{1,n}(v_{0,n}) - S_{1,n}(G_{1,n})|^6 - |S_{1,n}(v_{0,n}) - e^{i\bar{t}\Delta}V_{0,1}|^6 \right| dx \\
&\leq g_s \int \left| |S_{1,n}(v_{0,n}) - S_{1,n}(G_{1,n})|^6 - |S_{1,n}(v_{0,n}) - e^{i\bar{t}\Delta}V_{0,1}|^6 \right| dx, \\
|C_n| &\leq \int |g(\lambda_{j,n}x)| \left| |S_{1,n}(G_{1,n})|^6 - |e^{i\bar{t}\Delta}V_{0,1}|^6 \right| dx \\
&\leq g_s \int \left| |S_{1,n}(G_{1,n})|^6 - |e^{i\bar{t}\Delta}V_{0,1}|^6 \right| dx.
\end{aligned} \tag{58}$$

The integrands of the right hand side of $|B_n|, |C_n|$ consist of algebraic terms containing at least one of $S_{1,n}(G_{1,n}) - e^{i\bar{t}\Delta}V_{0,1}$ and its conjugate. Thus Hölder's inequality and strong convergence conclude that

$$\lim_{n\to\infty}(B_n + C_n) = 0. \tag{59}$$

This completes the proof of Lemma 4.1. □

5 Minimal Energy Blowup Solution (MEBS)

In this section, we consider the existence and compactness of MEBS. Our proof is quite similar to that of [20] except for the inhomogeneous coefficient g. However, for the convenience of readers, we provide a sketch of proof.

5.1 Existence of MEBS

For each $0 < e < E_g(W)$ let

$$\mathcal{E}(e) := \left\{\varphi \in \dot{H}^1_{rad} : E_g(\varphi) < e, \sqrt{g_s}\|\varphi\|^2_{\dot{H}^1} < \|W\|^2_{\dot{H}^1}\right\} \tag{60}$$

and then

$$\mathcal{S}(e) := \sup\left\{\|v\|_{S(I^*)} : v(0) \in \mathcal{E}(e), v \text{ solution to } (1)\right\}. \tag{61}$$

Define $E_{g,c} = \sup\{e : \mathcal{S}(e) < +\infty\}$. In view of the blowup criterion and small data scattering (Remarks 2.1 and 2.2) we deduce that $0 < E_{g,c} \leq E_g(W)$. In this subsection we assume that $E_{g,c} < E_g(W)$, which will lead us to a contradiction.

By the definition of $E_{g,c}$ we deduce that

1. If $0 \leq e < E_{g,c}$, $\sqrt{g_s}\|\varphi\|^2_{\dot{H}^1} < \|W\|^2_{\dot{H}^1}$, and $E_g(\varphi) < e$, then $\|u\|_{S(I^*)} < +\infty$.
2. If $E_{g,c} \leq e < E_g(W)$, $\sqrt{g_s}\|\varphi\|^2_{\dot{H}^1} < \|W\|^2_{\dot{H}^1}$, and $E_{g,c} \leq E_g(\varphi) < e < E_g(W)$, then $\|u\|_{S(I^*)} = +\infty$.

At this point, we may expect that $E_g(W)$ is critical value between GWP and blowup.

Proposition 5.1 *Let $\varphi_c \in \dot{H}^1_{rad}$ satisfy that $E_g(\varphi_c) = E_{g,c}(< E_g(W))$ and also $\sqrt{g_s}\|\varphi_c\|^2_{\dot{H}^1} < \|W\|^2_{\dot{H}^1}$. If u_c is the corresponding solution to* (1)*, then $\|u_c\|_{S(I^*)} = +\infty$.*

The solution u_c is called the minimal energy blowup solution (MEBS). For the proof we need to introduce nonlinear profiles.

Definition 5.1 (Nonlinear Profile) Let $v_0 \in \dot{H}^1_{rad}$, $v = e^{it\Delta}v_0$, and $\{t_n\}$ a sequence with $t_n \to \bar{t} \in [-\infty, +\infty]$. We say that $u(t, x)$ is a non-linear profile associated with $(v_0, \{t_n\})$ if there exist a maximal interval I^* with $\bar{t} \in I^*$ such that u is solution of (1) on I^* and $\lim_{n\to\infty} \|u(\cdot, t_n) - v(\cdot, t_n)\|_{\dot{H}^1} = 0$.

Remark 5.1 The nonlinear profile always exists. In fact, if $\bar{t} \in (-\infty, +\infty)$, we have only to solve (1) with initial data $v(x, \bar{t})$. And if $\bar{t} = \pm\infty$, by solving the integral equation $u(t) = e^{it\Delta}v_0 + i\int_t^{\pm\infty} e^{i(t-t')\Delta}[g|u|^4u]dt'$ we get the nonlinear profile.

Remark 5.2 If u_1 and u_2 are nonlinear profiles associated with $(v_0, \{t_n\})$, then $u_1 = u_2$ by LWP. Hence the uniqueness of nonlinear profile is guaranteed.

The following lemma is useful for the proof of Propositions 5.1 and 5.2. While proving this lemma, we demonstrate how to use the profile decomposition and long-time perturbation.

Lemma 5.1 *Let* $\{z_{0,n}\} \in \dot{H}^1_{rad}$, $\sqrt{g_s}\|z_{0,n}\|^2_{\dot{H}^1} < \|W\|^2_{\dot{H}^1}$, $E_g(z_{0,n}) \to E_{g,c} < E_g(W)$, *and* $\|e^{it\Delta}z_{0,n}\|_{S(I)} \geq \delta > 0$. *Let* $\{V_{0,j}\}$ *be the linear profiles. Assume that one of*

(a) $\liminf_{n\to\infty} E_g(V_1^l(-\frac{t_{1,n}}{\lambda_{1,n}^2})) < E_{g,c}$,

(b) $\liminf_{n\to\infty} E_g(V_1^l(-\frac{t_{1,n}}{\lambda_{1,n}^2})) = E_{g,c}$
and for $s_n = -\frac{t_{1,n}}{\lambda_{1,n}^2}$, *after passing to a subsequence so that* $s_n \to \bar{s} \in [-\infty, \infty]$ *and* $E_g(V_1^l(s_n)) \to E_{g,c}$, *and if* U_1 *is the nonlinear profile associated with* $(V_{0,1}, \{s_n\})$, *then* $I = \mathbb{R}$ *and* $\|U_1\|_{S(I)} < \infty$.

If $\{z_n\}$ *is solution of* (1) *with initial data* $\{z_{0,n}\}$ *then* $\|z_n\|_{S(I)} < \infty$ *for n large.*

Proof At first we show part (b). Assume that

$$\liminf_{n\to\infty} E_g(V_1^l(-\frac{t_{1,n}}{\lambda_{1,n}^2})) = E_{g,c}. \tag{62}$$

From the energy coercivity, (43c) of Lemma 3.1, and the profile decomposition it follows that $V_{0,j} = 0$ $(j \geq 2)$ and $\|w_n^J\|_{\dot{H}^1} \to 0$. Now let us set

$$v_{0,n}(x) = \lambda_{1,n}^{\frac{1}{2}} z_{0,n}(\lambda_{1,n}x), \quad \tilde{w}_n^J(x) = \lambda_{1,n}^{\frac{1}{2}} w_n^J(\lambda_{1,n}x). \tag{63}$$

Then we have

$$v_{0,n} = V_1^l(s_n) + \tilde{w}_n^J, \quad \|\tilde{w}_n^J\|_{\dot{H}^1} \to 0, \tag{64}$$

with $\sqrt{g_s}\|v_{0,n}\|^2_{\dot{H}^1} < \|W\|^2_{\dot{H}^1}$ and $E_g(v_{0,n}) \to E_{g,c} < E_g(W)$.
Then by definition of linear and nonlinear profile we get

$$\|V_1^l(s_n) - U_1(s_n)\|_{\dot{H}^1} \to 0, \tag{65a}$$

$$v_{0,n} = U_1(s_n) + \bar{w}_n^j, \quad \|\bar{w}_n^j\|_{\dot{H}^1} \to 0, \tag{65b}$$

$$\|\nabla e^{it\Delta}[V_1^l(s_n - U_1(s_n))]\|_{W(I)} \to 0. \tag{65c}$$

Since $\|\bar{w}_n^j\|_{\dot{H}^1} \to 0$, we have $E_g(U_1) = E_{g,c} < E_g(W)$ and also from (43a) of Lemma 3.1 that

$$\sup_{t\in I}\sqrt{g_s}\|U_1(t)\|^2_{\dot{H}^1} < \|W\|^2_{\dot{H}^1}. \tag{66}$$

Therefore the case (b) follows from the long-time perturbation, Proposition 2.2.

We next assume that

$$\liminf_{n\to\infty} E_g(V_1^l(-\frac{t_{1,n}}{\lambda_{1,n}^2})) = E_{g,c}. \tag{67}$$

We will show that

$$\liminf_{n\to\infty} E_g(V_j^l(-\frac{t_{1,n}}{\lambda_{1,n}^2})) < E_{g,c} \quad (j = 2, \cdots, J). \tag{68}$$

By profile decomposition, Lemma 4.1 we have

$$\|z_{0,n}\|_{\dot{H}^1}^2 = \sum_{j=1}^{J} \|V_{0,j}\|_{\dot{H}^1}^2 + \|w_n^J\|_{\dot{H}^1}^2 + o(1) \text{ as } n \to \infty, \tag{69}$$

$$E_g(z_{0,n}) \to E_{g,c} < E_g(W). \tag{70}$$

For n large, there exists $\delta_0 > 0$ such that $E_g(z_{0,n}) \leq (1-\delta_0)E_g(W)$ and thus (43a) of Lemma 3.1 gives us that $\sqrt{g_s}\|z_{0,n}\|_{\dot{H}^1}^2 \leq (1-\bar{\delta})\|W\|_{\dot{H}^1}^2$. Then we obtain that $E_g(V_j^l(s_n)) \geq 0$ and $E_g(w_n^J) \geq 0$ for all n large, and also that $E_g(V_1^l(s_n)) \geq C\alpha_0 > 0$ for some $\alpha_0 > 0$. Thus,

$$E_g(z_{0,n}) \geq C\alpha_0 + \sum_{j=2}^{J} E_g(V_j^l(s_n)) + o(1). \tag{71}$$

Since $E_g(z_{0,n}) \to E_{g,c}$,

$$\liminf_{n\to\infty} E_g(V_j^l(s_n)) < E_{g,c} \text{ for } j \geq 2. \tag{72}$$

Let U_j be the nonlinear profile associated with $(V_{0,j}, \{s_n\})$. Our next claim is that $\|U_j\|_{S(\mathbb{R})} < +\infty$ for all $j = 1, 2, \ldots, J$. By Definition 5.1 we have

$$E_g(V_j^l(s_n)) < E_{g,c} \quad E_g(U_j) < E_{g,c}, \tag{73}$$

and

$$\sqrt{g_s}\|V_j^l(s_n)\|_{\dot{H}^1}^2 \leq \sqrt{g_s}\|z_{0,n}\|_{\dot{H}^1}^2 + o(1) \leq (1-\bar{\delta})\|W\|_{\dot{H}^1}^2 + o(1) \tag{74}$$

for n large. Hence (43a) of Lemma 3.1 shows $\sqrt{g_s}\|U_j(t)\|_{\dot{H}^1}^2 < \|W\|_{\dot{H}^1}^2$ for all $t \in I_j$. But the criticality of $E_{g,c}$ means $I_j^* = \mathbb{R}$. Hence, $\|U_j\|_{S(\mathbb{R})} < \infty$. In fact, for fixed J and large n we have

$$\sum_{j=1}^{J} \sqrt{g_s}\|V_{0,j}\|_{\dot{H}^1}^2 \leq \sqrt{g_s}\|z_{0,n}\|_{\dot{H}^1}^2 + o(1) \leq 2\|W\|_{\dot{H}^1}^2. \tag{75}$$

Then there exists j_0 such that for $j \geq j_0$, $\|V_{0,j}\|_{\dot{H}^1} \leq \tilde{\delta}$ with $\tilde{\delta} > 0$ so small that $\|e^{it\Delta}V_{0,j}\|_{S(\mathbb{R})} \leq \delta$. This shows that

$$\|U_j\|_{S(\mathbb{R})} \leq 2\delta, \tag{76a}$$

$$\|U_j\|_{L_t^\infty \dot{H}^1} + \|\nabla U_j\|_{W(\mathbb{R})} \leq C\|V_{0,j}\|_{\dot{H}^1}. \tag{76b}$$

Therefore we get $\|U_j\|_{S(\mathbb{R})} \leq C\|V_{0,j}\|_{\dot{H}^1}$ for $j \geq j_0$.

Next we define, for $\varepsilon_0 > 0$ to be chosen later,

$$H_{n,\varepsilon_0} = \sum_{j=1}^{J(\varepsilon_0)} (\lambda_{j,n})^{-\frac{1}{2}} U_j\left(\frac{t - t_{j,n}}{\lambda_{j,n}^2}, \frac{x}{\lambda_{j,n}}\right). \tag{77}$$

Then from the density of U_j by $C_0^\infty(\mathbb{R}^{1+3})$ functions and the orthogonality of $(\lambda_{j,n}, t_{j,n})$ one can readily obtain that there exists constant $M > 0$ such that $\|H_{n,\varepsilon_0}\|_{S(\mathbb{R})} \leq M$ uniformly in ε_0 for $n \geq n(\varepsilon_0)$.

Let us now define

$$R_{n,\varepsilon_0} = g|H_{n,\varepsilon_0}|^4 H_{n,\varepsilon_0} - \sum_{j=1}^{J} g|\tilde{U}_{j,n}|^4 \tilde{U}_{j,n}, \tag{78}$$

where

$$\tilde{U}_{j,n}(t,x) = \lambda_{j,n}^{-\frac{1}{2}} U_j\left(\frac{t - t_{j,n}}{\lambda_{j,n}^2}, \frac{x}{\lambda_{j,n}}\right). \tag{79}$$

Using the Lemmas 2.1, 2.2 and definition of R_{n,ε_0}, $\|\nabla R_{n,\varepsilon_0}\|_{L_t^2 L_x^{\frac{6}{5}}} \to 0$ as $n \to \infty$. Since $\|U_j\|_{S(\mathbb{R})} < \infty$, we deduce that $\|\nabla U_j\|_{W(\mathbb{R})} < \infty$. To apply Proposition 2.2 let us set $\tilde{u} = H_{n,\varepsilon_0}$, $e = R_{n,\varepsilon_0}$ and choose $J(\varepsilon_0)$ large that for n large $\|\nabla e^{it\Delta} w_n^{J(\varepsilon_0)}\|_{W(\mathbb{R})} \leq \frac{\varepsilon_0}{2}$. Then, for n large,

$$z_{0,n} = H_{n,\varepsilon_0}(0) + \tilde{w}_n^{J(\varepsilon_0)}, \tag{80}$$

where $\|\nabla e^{it\Delta}\tilde{w}_n^{J(\varepsilon_0)}\|_{W(\mathbb{R})} \leq \varepsilon_0$. We can show that $\|\nabla H_{n,\varepsilon_0}\|_{L_t^\infty \dot{H}^1} \leq A$ uniformly in ε_0 by the same way as of H_{n,ε_0}. Then for n large, $\|\tilde{w}_n^{J(\varepsilon_0)}\|_{\dot{H}^1} \leq 2\|W\|_{\dot{H}^1} =: A'$. Choose $\varepsilon < \varepsilon_0 = \varepsilon_0(M, A, A')$. Then the long-time perturbation, Proposition 2.2 leads us to the case (a). □

Now we prove the Proposition 5.1.

Proof *(of Proposition 5.1)* Let $u_{0,n} \in \dot{H}^1_{rad}$ be such that

$$\sqrt{g_s}\|u_{0,n}\|^2_{\dot{H}^1} < \|W\|^2_{\dot{H}^1}, \; E_g(u_{0,n}) \to E_{g,c}, \tag{81a}$$

$$\|e^{it\Delta}u_{0,n}\|_{S(I_n^*)} \geq \delta, \text{ and } \|u_n\|_{S(I_n^*)} = +\infty, \tag{81b}$$

where I_n^* are maximal intervals. Since $E_{g,c} < E_g(W)$, we deduce that $E_g(u_{0,n}) \leq (1-\delta_0)E_g(W)$ for large n. Using the Lemma 3.1, we can find a $\bar{\delta}$ such that

$$\sqrt{g_s}\|u_n(t)\|^2_{\dot{H}^1} \leq (1-\bar{\delta})\|W\|^2_{\dot{H}^1} \text{ for all } t \in I_n^*. \tag{82}$$

We fix $J \geq 1$ and apply the profile decomposition to $\{u_{0,n}\}$. Then we get

$$\|u_{0,n}\|^2_{\dot{H}^1} = \sum_{j=1}^{J}\|V_{0,j}\|^2_{\dot{H}^1} + \|w_n^J\|^2_{\dot{H}^1} + o(1), \tag{83}$$

$$E_g(u_{0,n}) = \sum_{j=1}^{J} E_g(V_j^l(s_n)) + E_g(w_n^J) + o(1), \tag{84}$$

where $s_n = -\frac{t_{1,n}}{\lambda_{1,n}^2}$.

For n large, $\sqrt{g_s}\|w_n^J\|^2_{\dot{H}^1} \leq (1-\frac{\bar{\delta}}{2})\|W\|^2_{\dot{H}^1}$ and $\sqrt{g_s}\|V_{0,j}\|^2_{\dot{H}^1} \leq (1-\frac{\bar{\delta}}{2})\|W\|^2_{\dot{H}^1}$. These imply that $E_g(w_n^J) \geq 0, E_g(V_j^l(s_n)) \geq 0$ if n is large. Thus we get $E_g(V_1^l(s_n)) \leq E_g(u_{0,n}) + o(1)$ and then

$$\liminf_{n\to\infty} E_g(V_1^l(s_n)) \leq E_c. \tag{85}$$

We assume that $\liminf_{n\to\infty} E_g(V_1^l(s_n)) < E_c$. Then this contradicts (a) of Lemma 5.1. Therefore,

$$\liminf_{n\to\infty} E_g(V_1^l(s_n)) = E_c. \tag{86}$$

Let U_1 be the non-linear profile associated with $(V_{0,1}, \{s_n\})$. By (84) and the facts that

$$E_g(u_{0,n}) \to E_{g,c} \text{ and } E_g(V_1^l(s_n)) \to E_{g,c}, \tag{87}$$

we see that $E_g(w_n^J) \to 0$, $E_g(V_j^l(s_n)) \to 0$ for $j \geq 2$.

But the energy coercivity ((43c) of Lemma 3.1) shows that

$$\sum_{j=2}^{J}\|V_j^l(s_n)\|^2_{\dot{H}^1} + \|w_n^j\|^2_{\dot{H}^1} \xrightarrow{n\to\infty} 0. \tag{88}$$

Since $\|V_j^l(s_n)\|_{\dot H^1} = \|V_{0,j}\|_{\dot H^1}$, $V_{0,j} = 0$ $(j > 2)$, and $\|w_n^J\|_{\dot H^1}^2 \to 0$, so that

$$u_{0,n} = \lambda_{1,n}^{-\frac{1}{2}} V_1^l\left(s_n, \frac{x}{\lambda_{1,n}}\right) + w_n^J. \tag{89}$$

Rescaling as $v_{0,n}(x) = \lambda_{1,n}^{\frac{1}{2}} u_{0,n}(\lambda_{1,n}x)$, we have $v_{0,n} = V_0^l(s_n) + \tilde{w}_n^J$ with $\|\tilde{w}_n^J\|_{\dot H^1} \to 0$. By definition of nonlinear profile, we get

$$E_g(U_1(s_n)) = E_g(V_1^l(s_n)) + o(1) = E_{g,c} + o(1), \tag{90}$$

$$\begin{aligned} \sqrt{g_s}\|U_1(s_n)\|_{\dot H^1}^2 &= \sqrt{g_s}\|V_1^l(s_n)\|_{\dot H^1}^2 + o(1) = \sqrt{g_s}\|V_{0,1}\|_{\dot H^1}^2 + o(1) \\ &= \sqrt{g_s}\|u_{0,n}\|_{\dot H^1}^2 + o(1) < \|W\|_{\dot H^1}^2 \quad \text{as} \quad n \to \infty. \end{aligned} \tag{91}$$

For a fixed $\bar{s} \in I_1^*$, which I_1^* is maximal interval of U_1, the energy conservation yields

$$E_g(U_1(\bar{s})) = E_g(U_1(s_n)) \to E_{g,c} \tag{92}$$

and hence

$$E_g(U_1(\bar{s})) = E_{g,c}. \tag{93}$$

From the energy trapping it follows that $\sqrt{g_s}\|U_1(\bar{s})\|_{\dot H^1}^2 < \|W\|_{\dot H^1}^2$. If $\|U_1\|_{S(I_1^*)} < \infty$, it is a contradiction to (b) of Lemma 5.1. Therefore we deduce

$$\|U_1\|_{S(I_1^*)} = \infty. \tag{94}$$

By setting $U_1 = u_c$ we conclude the proof of Proposition 5.1. □

5.2 *Compactness of the MEBS Flows*

Proposition 5.2 *For any u_c as in Proposition 5.1, with $\|u_c\|_{S(I^*)} = +\infty$, there exist $\lambda(t) \in \mathbb{R}^+$, $t \in I_+^*$ with $I_+^* := I^* \cap [0, \infty)$ such that*

$$\mathcal{U} = \left\{ v(x,t) := \lambda(t)^{-\frac{1}{2}} u_c\left(t, \frac{x}{\lambda(t)}\right) : t \in I_+^* \right\} \tag{95}$$

has compact closure in $\dot H^1_{rad}$.

Proof If $\mathcal{U}$ does not have a compact closure in $\dot{H}^1_{rad}$, then there exists $\eta_0 > 0$ and $\{t_n\}$ with $t_n > 0$ such that for all $\lambda_0 \in \mathbb{R}^+$,

$$\left\| (\lambda_0)^{-\frac{1}{2}} u\left(t_n, \frac{x}{\lambda_0}\right) - u(t_{n'}, x) \right\|_{\dot{H}^1} \geq \eta_0 \quad (n \neq n'), \tag{96}$$

where $u = u_c$. Passing through a subsequence (still called $\{t_n\}$), $\{t_n\}$ is assumed to have a limit $\bar{t} \in \overline{I_+^*} = [0, T_+]$. Then $\bar{t} = T_+$ by the continuity of flow. We now may assume that $\|e^{it\Delta}u(t_n)\|_{S(0,\infty)} \geq \delta$, where δ is as in Proposition 2.1. Since

$$E_g(u(t)) = E_g(\varphi) = E_{g,c} < E_g(W), \tag{97a}$$

$$\sqrt{g_s}\|\varphi\|^2_{\dot{H}^1} < \|W\|^2_{\dot{H}^1}, \tag{97b}$$

we get

$$\sqrt{g_s}\|u(t)\|^2_{\dot{H}^1} \leq (1 - \bar{\delta})\|W\|^2_{\dot{H}^1} \tag{98}$$

for $t \in I_+^*$.

Now we apply the profile decomposition to $v_{0,n} = u(t_n)$. In view of the proof of Proposition 5.1, we conclude that

$$\liminf_{n\to\infty} E_g(V_1^l(s_n)) \leq E_{g,c}, \tag{99}$$

where $s_n = -\frac{t_{1,n}}{\lambda_{1,n}^2}$. Since $\|e^{it\Delta}v_{0,n}\|_{S(0,\infty)} \geq \delta$, if $\liminf_{n\to\infty} E_g(V_1^l(s_n)) < E_{g,c}$, then $\|v_n\|_{S(0,\infty)} < \infty$. This contradicts the fact that u is MEBS. Thus

$$\liminf_{n\to\infty} E_g(V_1^l(s_n)) = E_{g,c}, \tag{100}$$

which implies that $V_{0,j} = 0$ for $j \geq 2$ and $\|w_n^J\|_{\dot{H}^1} \to 0$. Therefore we deduce that

$$u(t_n) = (\lambda_{1,n})^{-\frac{1}{2}} V_1^l\left(s_n, \frac{x}{\lambda_{1,n}}\right) + w_n^J. \tag{101}$$

If $\{s_n\}$ is unbounded, then from the profile decomposition (101) one can show that there exists subsequences (still called $\{s_n\}$, $\{t_n\}$) such that $\|e^{it\Delta}u(t_n)\|_{S(0,\infty)} \leq \delta$ or $\|e^{it\Delta}u(t_n)\|_{S(-\infty,0)} < \delta$ for large n. From the LWP it follows that $\|u\|_{S(t_n,+\infty)} \leq 2\delta$ or $\|u\|_{S(-\infty,t_n)} \leq 2\delta$, respectively. This contradicts the fact that u is MEBS. Therefore $\{s_n\}$ should be bounded.

By passing to a subsequence we may assume that $s_n \to t_0 \in \mathbb{R}$. Since $\|w_n^J\|_{\dot{H}^1} \to 0$, for arbitrary $\lambda_0 > 0$

$$\left\|(\lambda_0)^{-\frac{1}{2}}(\lambda_{1,n})^{-\frac{1}{2}} V_1^l\left(s_n, \frac{x}{\lambda_{1,n}\lambda_0}\right) - (\lambda_{1,n'})^{-\frac{1}{2}} V_1^l\left(s_{n'}, \frac{x}{\lambda_{1,n'}}\right)\right\|_{\dot{H}^1} \geq \frac{\eta_0}{2} \text{ if } n \neq n'. \tag{102}$$

By the change of variables $x \mapsto \lambda_{1,n'} y$ we get

$$\left\|\left(\frac{\lambda_{1,n'}}{\lambda_0\lambda_{1,n}}\right)^{-\frac{1}{2}} V_1^l\left(s_n, \frac{\lambda_{1,n'} y}{\lambda_{1,n}\lambda_0}\right) - V_1^l(s_{n'}, y)\right\|_{\dot{H}^1} \geq \frac{\eta_0}{2} \text{ if } n \neq n'. \tag{103}$$

Since $s_n \to t_0$, by choosing $\lambda_0 = \frac{\lambda_{1,n'}}{\lambda_{1,n}}$, we reach another contradiction. This completes the proof of Proposition 5.2. □

6 A Rigidity Theorem

In this section, we will remove the MEBS u_c by a rigidity theorem under the condition $E_{g,c} < E_g(W)$.

Proposition 6.1 (Rigidity) *Suppose g is nonnegative, bounded radial function satisfying the conditions* (4), (6), *and* (7). *Let $\varphi \in \dot{H}^1_{rad}$ satisfy that $E_g(\varphi) < E_g(W)$ and $\sqrt{g_s}\|\varphi\|^2_{\dot{H}^1} < \|W\|^2_{\dot{H}^1}$. Let u be the corresponding solution to* (1) *with φ and let $I^* = (-T_-, T_+)$ be the maximal existence time interval. Assume there exists $\lambda(t) > 0$ such that*

$$\mathcal{U} := \left\{v(t,x) = (\lambda(t))^{-\frac{1}{2}} u\left(t, \frac{x}{\lambda(t)}\right) : t \in [0, T_+)\right\} \tag{104}$$

has compact closure in $\dot{H}^1_{rad}$. Then $T_+ = +\infty$ and $\varphi = 0$.

Remark 6.1 If there exists a sequence $\{t_i\} \subset [0, T_+)$ such that $\lambda(t_i) \to 0$ as $i \to \infty$, then from the compactness of $\mathcal{U}$ it follows that there exists a subsequence $\{t_{i_j}\}$ such that $u(t_{i_j}, \cdot) = \lambda(t_{i_j})^{\frac{1}{2}} v(t_{i_j}, \lambda(t_{i_j})\cdot) \to 0$ as $j \to \infty$. This implies that $T_+ = +\infty$ and $\varphi = 0$. Therefore, we may assume that $\lambda(t) \geq A_0$ for some constant $A_0 > 0$.

Remark 6.2 Assume that $\lambda(t) \geq A_0$. Then the scaling invariance and compactness of $\mathcal{U}$ bring us to the fact that for any $\varepsilon > 0$ there exists $r = r(\varepsilon)$ such that

$$\int_{|x|>r} \left(|\nabla u|^2 + |u|^6 + \frac{|u|^2}{|x|^2}\right) dx = \int_{|y|>r\lambda(t)} \left(|\nabla v|^2 + |v|^6 + \frac{|v|^2}{|x|^2}\right) dx$$
$$\leq \int_{|y|\geq A_0 r} (\cdots) \leq \varepsilon \tag{105}$$

for all $t \in I^*$.

The proof is divided into two cases: $T_+ < +\infty$ and $T_+ = +\infty$.

6.1 Case: $T_+ < \infty$

Suppose that $T_+ < +\infty$ and there exists a sequence $\{t_i\}$ such that $t_i \to T_+$ and $\lambda(t_i) \to \lambda_0 > 0$ as $i \to \infty$. Then by the compactness of $\mathcal{U}$, LWP, and long-time perturbation one can deduce that $\|u\|_{S(T_+-\delta,T_++\delta)} < +\infty$ for some $\delta > 0$. This is a contradiction to the maximality of T_+ and hence implies that if $T_+ < +\infty$, then $\lambda(t) \to +\infty$ as $t \to T^+$. For details see p.667 of [20].

Now let $a(x) \in C_0^\infty(\mathbb{R}^3)$ and $a_r(x)$ be as follows:

$$a(x) := \begin{cases} 1, & |x| \leq 1, \\ 0, & |x| \geq 10, \end{cases} \qquad a_r(x) := a(\frac{x}{r}). \tag{106}$$

And we define that

$$y_r(t) = \int a_r |u(t)|^2 dx \tag{107}$$

for $t \in [0, T_+)$. Then the density by H^2 data yields

$$y_r'(t) = 2\mathrm{Im}\left(\int \bar{u}\nabla u \nabla a_r dx - \int a_r g |u|^6 dx\right). \tag{108}$$

Since g is bounded, by Hardy–Sobolev inequality and energy trapping (Lemma 3.1) we have

$$
\begin{aligned}
|y_r'(t)| &\le C\left|\int \bar{u}\frac{1}{|x|}\nabla u \nabla a_r dx\right| + C\left|\int a_r |u|^6 dx\right| \\
&\le C\|\nabla u\|_{L_x^2}\left\|\frac{u}{|x|}\right\|_{L_x^2} + C\|u\|_{L_x^6}^6 \\
&\le C(\|W\|_{\dot{H}^1}^2 + \|W\|_{\dot{H}^1}^6).
\end{aligned}
\tag{109}
$$

Next we will show that

$$
\lim_{t\to T_+}\int_{|x|\le r}|u(t,x)|^2 dx = 0 \ \text{ for all } \ r > 0. \tag{110}
$$

Since $u(t,x) = (\lambda(t))^{\frac{1}{2}} v(t, \lambda(t)x)$, we get

$$
\begin{aligned}
\int_{|x|<r}|u|^2 dx &= \int_{|y|<r\lambda(t)} (\lambda(t))^{-2}|v(t,y)|^2 dy \\
&= (\lambda(t))^{-2}\int_{|y|<\varepsilon r\lambda(t)} |v(t,y)|^2 dy \\
&\quad + (\lambda(t))^{-2}\int_{\varepsilon r\lambda(t)<|y|<r\lambda(t)} |v(t,y)|^2 dy \\
&=: A + B,
\end{aligned}
\tag{111}
$$

where ε will be determined later. Fixing r, by Hölder's inequality, we obtain

$$
A \le C\varepsilon^2 r^2 \|v(t)\|_{L_x^6}^2 \le C\varepsilon^2 r^2\|W\|_{\dot{H}^1}^2 < \frac{\varepsilon}{2} \tag{112}
$$

for any $\varepsilon < \frac{1}{2}(Cr^2\|W\|_{\dot{H}^1}^2)^{-1}$. On the other hand, since $\lambda(t) \to +\infty$ as $t \to T_+$, B is estimated as

$$
B \le r^2\|v(t)\|_{L^6(|y|\ge\varepsilon r\lambda(t))}^2 < \frac{\varepsilon}{2} \ \ (t \ \text{ close to } \ T_+) \tag{113}
$$

by Remark 6.2. Therefore we get (110).

Since $|y_r'(t)| \le C$ from (109), we have $y_r(0) \le y_r(t) + Ct$ for all $t \in [0, T_+)$ and thus

$$
y_r(0) \le \lim_{t\to T_+} y_r(t) + CT_+ = CT_+. \tag{114}
$$

Taking the limit $r \to \infty$, we get that $\varphi \in L^2$. For any $\varepsilon > 0$, choose α small enough that

$$\int_{T_+-\alpha}^{T_+} |y_r'| dx \leq C\alpha < \frac{\varepsilon}{2}. \tag{115}$$

Then the conservation of mass (Remark 2.4) and (110) yield

$$\begin{aligned}\|\varphi\|_{L_x^2}^2 = \|u(T_+ - \alpha)\|_{L_x^2}^2 &\leq \|u(T_+ - \alpha)\|_{L_x^2(|x|<r)}^2 + \frac{\varepsilon}{2} \leq y_r(T_+ - \alpha) + \frac{\varepsilon}{2} \\ &\leq \lim_{t \to T_+} \int_t^{T_+-\alpha} y_r'(s) ds + \frac{\varepsilon}{2} < \varepsilon\end{aligned} \tag{116}$$

for a large r. Since ε is arbitrary, we have that $\varphi = 0$, which contradicts $T_+ < \infty$.

6.2 Case: $T_+ = \infty$

Choose $\delta > 0$ such that $E_g(\varphi) \leq (1 - \delta)E_g(W)$. Then from Lemma 3.1 (energy trapping) and Remark 6.2 we deduce that there exists $r_0 > 0$ such that for $r > r_0$ and $t \in [0, \infty)$,

$$\int_{|x|<r} (|\nabla u|^2 - g|u|^6)\, dx \geq C_\delta \|\varphi\|_{\dot{H}^1}^2 > 0. \tag{117}$$

Let $b(x) \in C_0^\infty(\mathbb{R}^3)$ and $b_r(x)$ be as follows:

$$b(x) := \begin{cases} |x|^2 & (|x| \leq 1) \\ 0 & (|x| \geq 10) \end{cases}, \qquad b_r(x) := r^2 b(\frac{x}{r}). \tag{118}$$

Set $z_r(t) = \int b_r |u(t)|^2 dx$. Then from the density by H^2 data and continuous dependency of solutions it follows that

$$\frac{d}{dt} z_r = 2\mathrm{Im} \int \nabla b_r \cdot \nabla u \bar{u}\, dx \tag{119}$$

and

$$\begin{aligned}\frac{d^2}{dt^2}z_r &= 2\mathrm{Im}\int [-\Delta b_r u_t \bar{u} - (\nabla b_r \cdot \nabla \bar{u})\, u_t + (\nabla b_r \cdot \nabla u)\, \bar{u_t}]\, dx \\ &= 4\mathrm{Re}\int (\nabla^2 b_r \cdot \nabla \bar{u})\nabla u dx - \frac{4}{3}\int (\Delta b_r) g |u|^6 dx \\ &\quad + \frac{2}{3}\int (\nabla b_r \cdot \nabla g)\, |u|^6 dx - \int (\Delta^2 b_r)|u|^2 dx.\end{aligned} \tag{120}$$

In case of constant g we refer the readers to [22] and also [16]. Here (120) has been obtained without regard to the radial symmetry.

By (119) we deduce that

$$|z_r'(t)| \le Cr^2 \int |\nabla u|^2 dx \le C_{\delta_0} r^2 \|W\|_{\dot{H}^1}^2. \tag{121}$$

Since $x \cdot \nabla g \ge 0$, by Remark 6.2, (120) can be estimated as follows:

$$\begin{aligned} z_r''(t) &\ge 8\left[\int\limits_{|x|\le r} |\nabla u|^2 - g|u|^6\right] + \frac{4}{3}\int\limits_{|x|\le r} (x \cdot \nabla g)|u|^6 dx \\ &\quad - C\left[\int\limits_{r\le |x|\le 10r} |\nabla u|^2 + g|u|^6 + \frac{|u|^2}{|x|^2} dx\right] \\ &\ge C_{\delta_0}\|\varphi\|_{\dot{H}^1}^2 \end{aligned} \tag{122}$$

for r sufficiently large. Therefore, by integrating (122) over $[0, t]$, we get

$$|z_r'(t) - z_r'(0)| \le 2C_{\delta_0} r^2 \|W\|_{\dot{H}^1}^2, \quad z_r'(t) - z_r'(0) \ge \left(\tilde{C}_{\delta_0}\|\varphi\|_{\dot{H}^1}^2\right) t, \tag{123}$$

which are not compatible for arbitrarily large t. This completes the proof of Proposition 6.1 and hence Theorem 1.1 holds.

7 Proof of the Blow-up Result

In this section, we show the finite time blowup via localized virial identity (120). To this end, we introduce a variational estimate which is fundamental part of the proof for Theorem 1.2.

Lemma 7.1 *Let* $E_g(\varphi) < (1-\delta_0)E_g(W)$ *and* $\sqrt{g_s}\|\varphi\|^2_{\dot{H}^1} \geq \|W\|^2_{\dot{H}^1}$ *for some* $\delta_0 > 0$*. Then there exists* $\bar{\delta} = \bar{\delta}(\delta_0)$ *such that*

$$\sqrt{g_s}\|\varphi\|^2_{\dot{H}^1} \geq (1+\bar{\delta})\|W\|^2_{\dot{H}^1}, \tag{124a}$$

$$\int (|\nabla\varphi|^2 - (1-\eta)g|\varphi|^6)\,dx \leq -\frac{(2-3\eta)\bar{\delta}}{\sqrt{g_s}}\|W\|^2_{\dot{H}^1}, \tag{124b}$$

for $0 \leq \eta \leq k_g$ *where* $k_g = \frac{2-\sqrt{g_s}(3-g_i)}{3-\sqrt{g_s}(3-g_i)}$.

Proof Let us invoke from the proof of Lemma 3.2 that

$$f(\bar{y}) \leq (1-\delta_0)f(y_0). \tag{125}$$

where $f(y) = \frac{1}{2}y - \frac{C_*^6}{6}y^3$, $\bar{y} = \sqrt{g_s}\|\varphi\|^2_{\dot{H}^1}$, and $y_0 = \|W\|^2_{\dot{H}^1}$. Since $y_0 \leq \bar{y}$ and f is strictly decreasing on $[y_0, +\infty)$, we find out that $\bar{y} \geq (1+\bar{\delta})y_0$ for $\bar{\delta}$ as in Lemma 3.2. Hence we reach the following:

$$\begin{aligned}
&\int |\nabla\varphi|^2 - (1-\eta)g|\varphi|^6 dx \\
&= 6(1-\eta)E_g(\varphi) - (2-3\eta)\int |\nabla\varphi|^2 \\
&\leq 6(1-\eta)E_g(W) - \frac{2-3\eta}{\sqrt{g_s}}\|W\|^2_{\dot{H}^1} - \frac{(2-3\eta)\bar{\delta}}{\sqrt{g_s}}\|W\|^2_{\dot{H}^1} \\
&\leq \frac{\sqrt{g_s}(3-g_i) - 2 - \eta(\sqrt{g_s}(3-g_i)-3)}{\sqrt{g_s}}\|W\|^2_{\dot{H}^1} - \frac{(2-3\eta)\bar{\delta}}{\sqrt{g_s}}\|W\|^2_{\dot{H}^1} \\
&\leq -\frac{(2-3\eta)\bar{\delta}}{\sqrt{g_s}}\|W\|^2_{\dot{H}^1}.
\end{aligned} \tag{126}$$

Therefore we get the desired result. □

The following holds immediately from the Lemma 7.1 together with the energy conservation and continuity argument.

Corollary 7.1 *Let* u *be a solution of* (1) *with* φ *such that*

$$E_g(\varphi) \leq (1-\delta_0)E_g(W) \;\; \text{and} \;\; \sqrt{g_s}\|\varphi\|^2_{\dot{H}^1} \geq \|W\|^2_{\dot{H}^1} \tag{127}$$

for some $\delta_0 > 0$. *Then there exits* $\bar{\delta} = \bar{\delta}(\delta_0)$ *such that*

$$\sqrt{g_s}\|u(t)\|_{\dot{H}^1}^2 \geq (1+\bar{\delta})\|W\|_{\dot{H}^1}^2, \tag{128a}$$

$$\int |\nabla u(t)|^2 - (1-\eta)g|u(t)|^6 dx \leq -\frac{(2-3\eta)\bar{\delta}}{\sqrt{g_s}}\|W\|_{\dot{H}^1}^2 \tag{128b}$$

for all $t \in I^*$ *and* $0 \leq \eta \leq k_g$, *where* I^* *is the maximal existence time interval and* $k_g = \frac{2-\sqrt{g_s}(3-g_i)}{3-\sqrt{g_s}(3-g_i)}$.

Proof *(of Theorem 1.2)* We first show the part (1). Let us invoke the localized virial identities (119) and (120). By integrating and taking limit $r \to \infty$ on both sides of (119) and (120), Fatou's lemma yields

$$\int |x|^2|u(t)|^2\,dx \leq 8\int_0^t\int_0^s\int \left(|\nabla u(t')|^2 - (1-k_g)g|u(t')|^6\right)dx dt' ds$$
$$+ 2t\mathrm{Im}\int (\nabla\varphi\cdot x)\varphi\,dx + \int |x|^2|\varphi|^2\,dx. \tag{129}$$

Then from Corollary 7.1 it follows that

$$\int |x|^2|u(t)|^2\,dx \leq -C_g\bar{\delta}t^2 + 2t\mathrm{Im}\int (\nabla\varphi\cdot x)\varphi\,dx + \int |x|^2|\varphi|^2\,dx \tag{130}$$

for some constant C_g. The last inequality gives us that the maximal interval is bounded.

For the part (2), we need another b_r. Let us introduce the function $\beta \in C^4([0,\infty))$ such that $\beta(s) = s$ for $0 \leq s \leq 1$, smooth for $1 < s < 10$, and 0 for $s \geq 10$ and further that $0 \leq \beta \leq 1$ and $\beta'(s) \leq 1$ for all $s \geq 0$. For the construction of such function see Appendix B of [3].

Now let $\beta_r(s) = r\beta(\frac{s}{r})$ and $b_r(|x|) = \int_0^{|x|}\beta_r(s)\,ds$. Then, by (120) and radial symmetry of u, we get

$$\frac{d^2}{dt^2}z_r = 4\int \beta_r'(|x|)|\nabla u|^2\,dx - 4\int \frac{1}{3}\nabla\cdot\left(\frac{x}{|x|}\beta_r(|x|)\right)g|u|^6\,dx$$
$$+ 4\int \frac{\beta_r(|x|)}{|x|}\frac{x\cdot\nabla g}{6}|u|^6\,dx - \int \Delta\nabla\cdot\left(\frac{x}{|x|}\beta_r(|x|)\right)|u|^2\,dx. \tag{131}$$

Since $\beta_r'(s) \le 1$ and $x \cdot \nabla g \le 6(k_g - \rho)g$, we then have

$$\begin{aligned}
&\frac{d^2}{dt^2} z_r(t) \\
&\le 4\int \left(|\nabla u(t)|^2 - (1 - k_g + \rho) g|u(t)|^6\right) dx \\
&\quad + 4\int \left[1 - \frac{1}{3}\nabla \cdot \left(\frac{x}{|x|}\beta_r(|x|)\right)\right] g|u|^6 - \int \Delta\nabla \cdot \left(\frac{x}{|x|}\beta_r(|x|)\right) |u|^2 dx \\
&\le 4\int \left(|\nabla u(t)|^2 - (1 - k_g + \rho) g|u(t)|^6\right) dx + C\left[\int_{|x|\ge r} g|u|^6 + \frac{|u|^2}{|x|^2} dx\right] \\
&\le 4\int \left(|\nabla u(t)|^2 - (1 - k_g + \rho) g|u(t)|^6\right) dx \\
&\quad + C g_s \|u\|^4_{L^\infty_x(|x|\ge r)} \|u\|^2_{L^2_x} + C r^{-2} \|u\|^2_{L^2_x}.
\end{aligned} \tag{132}$$

To control the second term we use the decay estimate of radial function f [8, 23]:

$$\||x|^{\frac{1}{2}} f\|_{L^\infty_x} \le C_0 \|f\|^{\frac{1}{2}}_{L^2_x} \|\nabla f\|^{\frac{1}{2}}_{L^2_x}. \tag{133}$$

The mass conservation (Remark 2.4) gives us that

$$\begin{aligned}
\frac{d^2}{dt^2} z_r(t) \le& 4(1+\varepsilon(r)) \int \left(|\nabla u(t)|^2 - \frac{(1 - k_g + \rho)}{1+\varepsilon(r)} g|u(t)|^6\right) dx \\
&+ C r^{-2} \|\varphi\|^2_{L^2_x}
\end{aligned} \tag{134}$$

where $\varepsilon(r) = \frac{1}{4} C C_0^4 g_s \|\varphi\|^4_{L^2_x} r^{-2}$. Hence if we choose r large enough, then since $\rho > 0$, by Corollary 7.1 we deduce that

$$\frac{d^2}{dt^2} z_r(t) \le -\frac{C_g \bar{\delta}}{2}. \tag{135}$$

By the same argument as of part (1) we obtain the desired result.

Acknowledgments The authors would like to thank the referee for careful reading. This work was supported by NRF-2018R1D1A3B07047782 (Republic of Korea).

References

1. Aubin, T.: Équations différentielles non linéaires et probléme de Yamabe concernant, la courbure scalaire. J. Math. Pures Appl. IX. Sér. **55**, 269–296 (1976)
2. Belmonte-Beitia, J., Pérez-García, V.M., Vekslerchik, V., Torres, P.J.: Lie symmetries and solitons in nonlinear systems with spatially inhomogeneous nonlinearities. Phys. Rev. Lett. **98**, 064102 (2007)
3. Boulenger, T., Lenzmann, E.: Blowup for biharmonic NLS. Ann. Sci. Éc. Norm. Supér. **50**, 503—544 (2017)
4. Cazenave, T.: Semilinear Schrödinger Equations. In: Courant Lecture Notes in Mathematics, vol. 10. New York University/Courant Institute of Mathematical Sciences/American Mathematicla Society, New York/Providence (2003)
5. Cazenave, T., Weissler, F.B. The Cauchy problem for the critical nonlinear Schrödinger equation in H^s. Nonlinear Anal. Theory Methods Appl. **14**, 807–836 (1990)
6. Cho, Y.: Well-posedness and scattering of inhomogeneous cubic-quintic NLS. J. Math. Phys. **60**(8), 081513, 13 (2019)
7. Cho, Y., Lee, K.: Small data scattering of the inhomogeneous cubic-quintic NLS in 2 dimentions. Nonlinear Anal. **188**, 142–157 (2019)
8. Cho, Y., Ozawa, T.: Sobolev inequalities with symmetry. Commun. Contem. Math. **11**, 355–365 (2009)
9. Cho, Y., Hwang, G., Kwon, S., Lee, S.: Profile decompositions and blowup phenomena of mass critical fractional Schrödinger equations. Nonlinear Anal. Theory Methods Appl. **86**, 12–29 (2013)
10. Cho, Y., Hwang, G., Ozawa, T.: On the focusing energy-critical fractional nonlinear Schrödinger equations. Adv. Differ. Equ. **23**, 161–192 (2018)
11. Ding, W.-Y., Ni, W.-M.: On the elliptic equation $\Delta u + Ku^{(n+2)/(n-2)} = 0$ and related topic. Duke Math. J. **52**, 485–506 (1985)
12. Dinh, V.D.: Blowup of H^1 solutions for a class of the focusing inhomogeneous nonlinear Schrödinger equation. Nonlinear Anal. Theory Methods Appl. **174**, 169—188 (2018)
13. Farah, L.G., Guzmán, C.M.: Scattering for the radial 3D cubic focusing inhomogeneous nonlinear Schrödinger equation. J. Differ. Equations **262**, 4175—4231 (2017)
14. Foschi, D.: Inhomogeneous Strichartz estimates. J. Hyperbolic Differ. Equ. **2**, 1–24 (2005)
15. Gill, T.S.: Optical guiding of laser beam in nonuniform plasma. Pramana J. Phys. **55**, 842–845 (2000)
16. Glassey, R.T.: On the blowing up of solutions to the Cauchy problem for nonlinear Schrödinger equations. J. Math. Phys. **18**, 1794–1797 (1977)
17. Holmer, J., Roudenko, S.: A sharp condition for scattering of the radial 3D cubic nonlinear Schrödinger equation. Comm. Math. Phys. **282**, 435—467 (2008)
18. Keel, M., Tao, T.: Endpoint Strichartz estimates. Am. J. Math. **120**, 955—980 (1998)
19. Kenig, C.: Global Well-posedness, Scattering and Blow Up for the Energy-Critical, Focusing, Nonlinear Schrödinger and Wave Equations. http://math.uchicago.edu/~cek/Kenigrev1.pdf
20. Kenig, C., Merle, F.: Global well-posedness, scattering and blow up for the energy critical focusing, nonlinear Schrödinger equation in the radial case. Invent. Math. **166**, 645–675 (2006)
21. Keraani, S.: On the defect of compactness for the Strichartz estimates of the Schrödinger equations. J. Differ. Equ. **175**, 353–392 (2001)
22. Merle, F.: Determination of Blow-up Solutions with Minimal Mass for Nonlinear Schrödinger Equations with Critical Power. Duke Math. J. **69**, 427—454 (1993)
23. Strauss, W.A.: Existence of solitary waves in higher dimensions. Commun. Math. Phys. **55**, 149–162 (1977)
24. Sulem, C., Sulem, P.L.: The Nonlinear Schrödinger Equation, Self-focusing and Wave Collapse. Springer, Berlin (1993)

25. Talenti, G.: Best constant in Sobolev inequality. Ann. Mat. Pura Appl. **110**, 353–372 (1976)
26. Tang, X.-Y., Shukla, P.K.: Solution of the one-dimensional spatially inhomogeneous cubic-quintic nonlinear Schrödinger equation with an external potential. Phys. Rev. A **76**, 013612-1–013612-10 (2007)

Lifespan of Solutions to Nonlinear Schrödinger Equations with General Homogeneous Nonlinearity of the Critical Order

Hayato Miyazaki and Motohiro Sobajima

Abstract This paper is concerned with the upper bound of the lifespan of solutions to nonlinear Schrödinger equations with general homogeneous nonlinearity of the critical order. In Masaki and the first author (Differ Integr Equ 32(3–4):121–138, 2019), they obtain the upper bound of the lifespan of solutions to our equation via a test function method introduced by Zhang (Duke Math J 97(3):515–539, 1999; C R Acad Sci Paris Sér I Math 333(2):109–114, 2001). Their nonlinearity contains a non-oscillating term $|u|^{1+2/d}$ which causes difficultly for constructing an even small data global solution. The non-oscillating term corresponds to the L^1-scaling critical. In this paper, it turns out that the upper bound can be refined by employing a unified test function by Ikeda and the second author (Nonlinear Anal 182:57–74, 2019).

Keywords upper bound of lifespan · small data blow-up · nonlinear Schrödinger equations.

Mathematics Subject Classification 2010: 35Q55, 35B44

1 Introduction

In this paper, we consider the lifespan of solutions to nonlinear Schrödinger equations

$$i\partial_t u + \Delta u = F(u), \ (t,x) \in \mathbb{R}^{1+d} \tag{1}$$

H. Miyazaki (✉)
Teacher Training Courses, Faculty of Education, Kagawa University, Takamatsu, Kagawa, Japan
e-mail: miyazaki.hayato@kagawa-u.ac.jp

M. Sobajima
Department of Mathematics, Faculty of Science and Technology, Tokyo University of Science, Noda-shi, Chiba, Japan

© Springer Nature Switzerland AG 2020

V. Georgiev et al. (eds.), *Advances in Harmonic Analysis and Partial Differential Equations*, Trends in Mathematics, https://doi.org/10.1007/978-3-030-58215-9_7

where $u = u(t, x)$ is a complex-valued unknown function, and the nonlinearity F is homogeneous of degree $1 + 2/d$, that is, F satisfies the condition

$$F(\lambda z) = \lambda^{1+\frac{2}{d}} F(z) \tag{2}$$

for any $z \in \mathbb{C}$ and $\lambda > 0$. It is well-known that the exponent $1+2/d$ is critical in view of large time behavior of solutions to (1). Indeed, considering $F(u) = \eta|u|^{p-1}u$ with $p > 1$ and $\eta \in \mathbb{R} \setminus \{0\}$, the solution asymptotically behaves like a free solution as $t \to \infty$, when $p > 1+2/d$. On the other hand, if $p = 1+2/d$, then the solution behaves like a free solution with a logarithmic phase correction

$$(2it)^{-\frac{d}{2}} e^{i\frac{|x|^2}{4t}} \widehat{u_+}\left(\frac{x}{2t}\right) \exp\left(-\eta i \left|\widehat{u_+}\left(\frac{x}{2t}\right)\right|^{\frac{2}{d}} \log t\right) \tag{3}$$

for a suitable function u_+ as $t \to \infty$, where $\widehat{u_+}$ denotes the Fourier transform of u_+ with respect to x (cf. [2, 3, 12, 15]). Furthermore, in the critical case, the behavior of the solutions depends on the shape of the nonlinearity. More precisely, under (2) and some summability assumption on $\{g_n\}_n$, Masaki and the first author [7] introduce a decomposition of the nonlinearity

$$F(u) = g_0|u|^{1+\frac{2}{d}} + g_1|u|^{\frac{2}{d}}u + \sum_{n\neq 0,1} g_n|u|^{1+\frac{2}{d}-n}u^n \tag{4}$$

with the coefficients

$$g_n = \frac{1}{2\pi}\int_0^{2\pi} F(e^{i\theta})e^{-in\theta}\,d\theta. \tag{5}$$

By using this decomposition, they prove that if $g_0 = 0$ and $g_1 \in \mathbb{R}$, then (1) admits a solution which asymptotically behaves like (3) with $\eta = g_1$ (see also Masaki, the first author and Uriya [9], and references therein).

Later on, in [8], they try to deal with the large time behavior of the solution to (1) under the case $g_0 \neq 0$ in which the nonlinearity contains the non-oscillating term $|u|^{1+2/d}$. However, according to the previous works such as [4, 6, 13, 14], we do not always have even small data global solutions to (1) in this case. Hence it seems difficult to find certain large time behavior of the solution. For this reason, instead of specifying the large time behavior, they denies the existence of the solution which behaves like a free solution with or without a logarithmic phase correction like (3) with $\eta = g_1$. They further obtain the upper bound of the lifespan of the solution with a general initial data corresponding to a finite time blowup result for small data (see Theorem 2.1). In this paper, we aim to refine the upper bound of the lifespan given by Masaki and the first author [8]. Here, let us briefly review previous analysis of blowup phenomena and the lifespan of solutions to the nonlinear Schrödinger equation with non-oscillating nonlinearity

$$i\partial_t u + \Delta u = \rho|u|^p, \quad \rho \in \mathbb{C} \setminus \{0\}, \quad p > 1. \tag{6}$$

Ikeda and Wakasugi [6] discovered the blowup phenomena of solutions to (6) when $p \leqslant 1 + 2/d$. In [4], Ikeda and Inui obtained the upper bound of the lifespan of solutions to (6) under slowly decaying initial data if $p < 1+2/d$. Further, Fujiwara and Ozawa [1] give an estimate of the lifespan of solutions to (6) in d-dimensional Torus. Also, Oh et al. [11] investigate stability of the finite time blowup solution constructed in [4] under a stochastic setting. In the L^1-scaling critical case $p = 1 + 2/d$, Ikeda and the second author [5] give an upper bound of the lifespan of solutions to (6) which can be regarded as a refinement of [6] and [1]. They introduce a unified test function, in order to treat the lifespan estimate of blowup solutions to semilinear evolution equations in L^1-scaling critical setting such as (6). Their test function method is an improvement of that given by Mitidieri and Pokhozhaev [10].

2 Main Results

Before stating the main results, we recall the technique for the decomposition of the nonlinearity like (4) developed by Masaki and the first author [7]. We identify a homogeneous nonlinearity F and 2π-periodic function g as follows: A homogeneous nonlinearity F is written as

$$F(u) = |u|^{1+\frac{2}{d}} F\left(\frac{u}{|u|}\right). \tag{7}$$

Let us here introduce a 2π-periodic function $g(\theta) = g_F(\theta)$ by $g_F(\theta) = F(e^{i\theta})$. Conversely, for a given 2π-periodic function g, we can construct a homogeneous nonlinearity $F = F_g : \mathbb{C} \to \mathbb{C}$ by $F_g(u) = |u|^{1+\frac{2}{d}} g\,(\arg u)$ if $u \neq 0$ and $F_g(u) = 0$ if $u = 0$. Since $g(\theta)$ is 2π-periodic function, it holds that

$$g(\theta) = \sum_{n\in\mathbb{Z}} g_n e^{in\theta}, \quad g_n := \frac{1}{2\pi} \int_0^{2\pi} g(\theta) e^{-in\theta}\, d\theta \tag{8}$$

under at least $\{g_n\}_n \in \ell^1(\mathbb{Z})$. Hence, the above expansion gives us (4).

In this paper, we refine the upper bound of the lifespan in [8] by employing the technique in [5]. To state the main results, let us give the definition of a weak solution and the lifespan of the solution.

Definition 2.1 (Weak Solution) Suppose that $F(z)$ is locally uniformly bounded. We say that a function $u(t, x) \in \mathcal{S}'((-\infty, T) \times \mathbb{R}^d)$ is a weak solution to (1) with initial data $u(0, x) = u_0(x) \in L^1_{\mathrm{loc}}(\mathbb{R}^d)$ on $[0, T)$, $T > 0$, if $u \in L^{(d+2)/d}_{\mathrm{loc}}((0, T) \times$

$\mathbb{R}^d$) and the identity

$$\begin{aligned}&\int_{(0,T)\times\mathbb{R}^d} u(t,x)(-i\partial_t\psi(t,x)+\Delta\psi(t,x))dxdt\\&=i\int_{\mathbb{R}^d} u_0(x)\psi(0,x)dx+\int_{(0,T)\times\mathbb{R}^d} F(u(t,x))\psi(t,x)dxdt\end{aligned} \tag{9}$$

holds for any test function $\psi \in C_0^\infty((-\infty, T)\times\mathbb{R}^d)$.

Definition 2.2 (Lifespan of Solutions) For a given data $u_0 \in L^1_{\text{loc}}(\mathbb{R}^d)$, we define the maximal existence time by

$$T_{\max} = T_{\max}(u_0) := \sup\left\{ T>0\ ;\ \begin{array}{c}\text{There exists a weak solution } u(t)\\ \text{to (1) with } u(0)=u_0 \text{ on } [0,T)\end{array}\right\}. \tag{10}$$

By the test function method introduced by Zhang [16, 17], Masaki and the first author [8] give an upper bound of the lifespan under $g_0 = 1$. Remark that without loss of generality, we may let $g_0 = 1$ by change of variable.

Theorem 2.1 ([8]) *Let $d \geqslant 1$ and $\varepsilon > 0$. Suppose that $\{g_n\}_n \in \ell^1(\mathbb{Z})$ satisfies $g_0 = 1$ and $\mu := g_0 - \sum_{n\neq 0}|g_n| > 0$. If $f \in L^1_{\text{loc}}(\mathbb{R}^d)$ satisfies*

$$-\operatorname{Im} f(x) \geqslant \begin{cases}|x|^{-k} & |x|>R_0,\\ 0 & |x|\leqslant R_0\end{cases} \tag{11}$$

for some $k \leqslant d$ and $R_0 > 0$, then there exist $C = C(k, R_0, \mu) > 0$ and $\varepsilon_0 > 0$ such that

$$T_{\max}(\varepsilon f) \leqslant \begin{cases} C\varepsilon^{-\frac{2}{d-k}} & k<d,\\ \exp(C/\varepsilon) & k=d\end{cases} \tag{12}$$

holds for any $\varepsilon \in (0, \varepsilon_0)$.

In the case $k = d$, as a main result, we give a refinement of the upper bound of the lifespan.

Theorem 2.2 (Refinement of the Lifespan) *Let $d \geqslant 1$ and $\varepsilon > 0$. Suppose that $\{g_n\}_n \in \ell^1(\mathbb{Z})$ satisfies $g_0 = 1$ and $\mu := g_0 - \sum_{n\neq 0}|g_n| > 0$. If $f \in L^1_{\text{loc}}(\mathbb{R}^d)$ satisfies*

$$-\operatorname{Im} f(x) \geqslant \begin{cases}|x|^{-d}(\log|x|)^{-\alpha} & |x|>R_0,\\ 0 & |x|\leqslant R_0\end{cases} \tag{13}$$

for some $\alpha \in \mathbb{R}$ *and* $R_0 > 0$, *then there exist* $C = C(\alpha, R_0, \mu) > 0$ *and* $\varepsilon_0 > 0$ *such that*

$$T_{\max}(\varepsilon f) \leqslant \begin{cases} \exp(C\varepsilon^{-\frac{2}{d}}) & \text{if } \alpha > 1, \\ \exp\left(C\left(\varepsilon \log \varepsilon^{-1}\right)^{-\frac{2}{d}}\right) & \text{if } \alpha = 1, \\ \exp(C\varepsilon^{-\frac{2}{d+2(1-\alpha)}}) & \text{if } \alpha < 1 \end{cases} \tag{14}$$

holds for any $\varepsilon \in (0, \varepsilon_0)$.

Remark 2.1 When $\alpha = 0$, the estimate (14) is a refinement of (12). In fact, we obtain the additional blowup rate $\frac{d}{d+2}$, compared with (12).

Remark 2.2 In addition to the assumption of the theorem, let us suppose that $F(e^{i\theta})$ is Lipschitz continuous and $f \in L^2(\mathbb{R}^d)$. Then, as mentioned in [8], a standard contraction argument yields a unique solution $u(t) \in C_t(I; L^2_x(\mathbb{R}^d)) \cap L^{\frac{2(d+2)}{d}}_{t,\mathrm{loc}}(I; L^{\frac{2(d+2)}{d}}_x(\mathbb{R}^d))$ in the sense of the integral equation corresponding to (1)

$$u(t) = U(t)\varepsilon f - i\int_0^t U(t-s)F(u(s))ds \tag{15}$$

in $L^2(\mathbb{R}^d)$ for any $t \in I$, where $I = (-\tau, T)$ for some $\tau, T > 0$ and $U(t) = e^{it\Delta}$ is the free Schrödinger group. The solution is a weak solution on $[0, T)$ by a suitable extension of u in $(-\infty, -\tau/2) \times \mathbb{R}^d$. Thus, denoting $I_{\max}$ by a maximal existence interval of the solution, $\widetilde{T} = \sup I_{\max}$ coincides with $T_{\max}(\varepsilon f)$ in Theorem 2.2 and $u(t)$ blows up at $t = \widetilde{T}$ in such a sense that $\lim_{t\to\widetilde{T}-0} \|u(t)\|_{L^2} = \infty$.

3 Proof of the Main Result

We follow the test function method argument as in [5]. Set $\eta \in C^\infty([0, \infty))$ satisfying

$$\eta(s) = \begin{cases} 1 & \text{if } s \in [0, 1/2] \\ \text{decreasing} & \text{if } s \in (1/2, 1) \\ 0 & \text{if } s \in [1, \infty), \end{cases} \tag{16}$$

$$\eta^*(s) = \begin{cases} 0 & \text{if } s \in [0, 1/2) \\ \eta(s) & \text{if } s \in [1/2, \infty). \end{cases} \tag{17}$$

We further define the cut-off functions

$$\psi_R(t,x) = \left[\eta\left(\frac{|x|^2+t}{R}\right)\right]^{2p_0'}, \quad \psi_R^*(t,x) = \left[\eta^*\left(\frac{|x|^2+t}{R}\right)\right]^{2p_0'} \tag{18}$$

for any $R > 0$ and all $(t,x) \in [0,\infty) \times \mathbb{R}^N$, where $p_0 = 1 + \frac{2}{d}$. Let us here show the following estimate for the initial data:

Lemma 3.1 *Let $f \in L^1_{\text{loc}}(\mathbb{R}^d)$ satisfy* (13) *for some $\alpha \in \mathbb{R}$ and $R_0 > 0$. Then there exist constants $C = C(d, R_0, \alpha) > 0$ and $R_1 = R_1(d, R_0, \alpha) > 0$ such that*

$$-\int_{\mathbb{R}^d} \operatorname{Im} f(x)\psi_R(0,x)dx \geqslant \begin{cases} C & \textit{if } \alpha > 1, \\ C\log\log R & \textit{if } \alpha = 1, \\ C(\log R)^{1-\alpha} & \textit{if } \alpha < 1 \end{cases} \tag{19}$$

holds as long as $R_1 < R$.

Proof of Lemma 3.1 We may assume $2R_0^2 < R$. A direct computation shows

$$\begin{aligned} &-\int_{\mathbb{R}^d} \operatorname{Im} f(x)\psi_R(0,x)dx \\ &\geqslant \int_{\{|x|>R_0\}} |x|^{-d}(\log|x|)^{-\alpha}\left[\eta\left(\frac{|x|^2}{R}\right)\right]^{2p_0'} dx \\ &\geqslant \int_{\{R_0<|x|<\sqrt{R/2}\}} |x|^{-d}(\log|x|)^{-\alpha}\left[\eta\left(\frac{|x|^2}{R}\right)\right]^{2p_0'} dx \\ &= C\int_{R_0}^{\sqrt{R/2}} r^{-1}(\log r)^{-\alpha}\, dr \end{aligned} \tag{20}$$

for any $R > 2R_0^2$. Further we see that

$$\int_{R_0}^{\sqrt{R/2}} r^{-1}(\log r)^{-\alpha}\, dr \geqslant \begin{cases} C & \text{if } \alpha > 1, \\ C\log\log R & \text{if } \alpha = 1, \\ C(\log R)^{1-\alpha} & \text{if } \alpha < 1 \end{cases} \tag{21}$$

when R is large enough. Indeed, when $\alpha = 1$, ones easily has

$$\int_{R_0}^{\sqrt{R/2}} r^{-1}(\log r)^{-\alpha}\, dr = \log(\log R - \log 2) - \log 2 - \log(\log R_0). \tag{22}$$

Then it holds that

$$\log R - \log 2 \geqslant \frac{1}{2}\log R, \quad \log\log R - 2\log 2 - \log R_0 \geqslant \frac{1}{2}\log\log R \tag{23}$$

as long as R is sufficiently large. Thus, we have the estimate (21). The other cases are similar way, taking into account

$$\begin{aligned} &\frac{1}{2}(\log R - \log 2) \geqslant 2\log R_0 \quad \text{if } \alpha < 1, \\ &\frac{1}{2}\log R - \log 2 \geqslant 2\log R_0 \quad \text{if } \alpha > 1 \end{aligned} \tag{24}$$

as long as R is sufficiently large. In conclusion, taking $R_1 > 0$ satisfying $R_1 > 2R_0^2$ and (21), we have (19) if $R > R_1$. □

Let us go back to the proof of Theorem 2.2. Put $T_\varepsilon = T_{\max}(\varepsilon f)$. let $u(t,x)$ be a weak solution on $[0, T_\varepsilon)$ with initial condition $u(0) = \varepsilon f$. For each $\varepsilon > 0$, it is deduced that $T_\varepsilon \leqslant R_1$ or $T_\varepsilon > R_1$, where R_1 is given by Lemma 3.1. Since the former case is trivial, we may suppose that $T_\varepsilon > R_1$. Taking the real part of the both side and $\psi = \psi_R$ in (9), from density argument and the integration by part, we deduce that weak solutions to (1) satisfies

$$\begin{aligned} &\mathrm{Re}\iint_{(0,T_\varepsilon)\times\mathbb{R}^d} u(t,x)\{-i\partial_t(\psi_R(t,x)) + \Delta(\psi_R(t,x))\}dtdx \\ =& -\varepsilon\int_{\mathbb{R}^d} \mathrm{Im}\, f(x)\psi_R(0,x)dx \\ &+ \mathrm{Re}\sum_{n\in\mathbb{Z}} g_n \iint_{(0,T_\varepsilon)\times\mathbb{R}^d} F_n(u(t,x))\psi_R(t,x)dtdx \end{aligned} \tag{25}$$

for any $R \in [R_0, T_\varepsilon)$, where $F_n(u) = |u|^{p_0-n}u^n$. Let us here introduce

$$I_n(R) = \iint_{(0,T_\varepsilon)\times\mathbb{R}^d} F_n(u(t,x))\psi_R(t,x)\,dxdt. \tag{26}$$

Then since $|I_n(R)| \leqslant I_0(R)$, we have

$$\begin{aligned} &\mathrm{Re}\sum_{n\in\mathbb{Z}} g_n \iint_{(0,T_\varepsilon)\times\mathbb{R}^d} F_n(u(t,x))\psi_R(t,x)dtdx \\ &\geqslant I_0(R) - \sum_{n\neq 0}|g_n| I_0(R) = (1-\mu)I_0(R). \end{aligned} \tag{27}$$

Hence, we see from the above and (25) that

$$\begin{aligned} & -\varepsilon \int_{\mathbb{R}^d} \operatorname{Im} f(x)\psi_R(0,x)dx + (1-\mu)I_0(R) \\ & \leqslant \iint_{(0,T_\varepsilon)\times\mathbb{R}^d} |u(t,x)| \left(|\partial_t(\psi_R(t,x))| + |\Delta(\psi_R(t,x))|\right) dtdx \\ & \leqslant \iint_{(0,T_\varepsilon)\times\mathbb{R}^d} |u(t,x)| \left(\frac{C_1}{R} + C_2\frac{|x|^2}{R^2}\right) \left(\psi_R^*(t,x)\right)^{\frac{1}{p_0}} dtdx. \end{aligned} \tag{28}$$

Taking $A = C_1 + C_2/R_0$, combining the Hölder inequality with the definition of p_0, it is obtained that

$$\begin{aligned} & -\varepsilon \int_{\mathbb{R}^d} \operatorname{Im} f(x)\psi_R(0,x)dx + (1-\mu)I_0(R) \\ & \leqslant A \left(\iint_{(0,T_\varepsilon)\times\mathbb{R}^d} |u(t,x)|^{p_0}\psi_R^*(t,x)\,dxdt\right)^{\frac{1}{p_0}} \left(\int_{(0,R)\times B(0,\sqrt{R})} \frac{1}{R^{p_0'}} dtdx\right)^{\frac{1}{p_0'}} \\ & \leqslant C \left(\iint_{(0,T_\varepsilon)\times\mathbb{R}^d} |u(t,x)|^{p_0}\psi_R^*(t,x)\,dxdt\right)^{\frac{1}{p_0}}, \end{aligned} \tag{29}$$

which implies

$$\begin{aligned} & -\varepsilon \int_{\mathbb{R}^d} \operatorname{Im} f(x)\psi_R(0,x)dx + (1-\mu)I_0(R) \\ & \leqslant C \left(\iint_{(0,T_\varepsilon)\times\mathbb{R}^d} |u(t,x)|^{p_0}\psi_R^*(t,x)\,dxdt\right)^{\frac{1}{p_0}}. \end{aligned} \tag{30}$$

Following the idea in [5, Lemma 3.10], we here define two functions by

$$y(r) = \iint_{(0,T_\varepsilon)\times\mathbb{R}^d} |u(t,x)|^{p_0}\psi_r^*(t,x)\,dxdt, \quad Y(R) = \int_0^R y(r)\frac{dr}{r}. \tag{31}$$

Note that $y \in C([0,T_\varepsilon))$ and $Y \in C^1([0,T_\varepsilon))$ satisfying

$$RY'(R) = \iint_{(0,T_\varepsilon)\times\mathbb{R}^d} |u(t,x)|^{p_0}\psi_R^*(t,x)\,dxdt. \tag{32}$$

A direct calculation shows

$$
\begin{aligned}
Y(R) &= \int_0^R \left(\iint_{(0,T_\varepsilon)\times\mathbb{R}^d} |u(t,x)|^{p_0} \psi_r^*(t,x)\,dxdt \right) \frac{dr}{r} \\
&= \iint_{(0,T_\varepsilon)\times\mathbb{R}^d} |u(t,x)|^{p_0} \left(\int_0^R \left[\eta^* \left(\frac{|x|^2+t}{r} \right) \right]^{2p_0'} \frac{dr}{r} \right) dxdt \qquad (33) \\
&= \iint_{(0,T_\varepsilon)\times\mathbb{R}^d} |u(t,x)|^{p_0} \left(\int_{(|x|^2+t)/R}^\infty \left[\eta^*(s) \right]^{2p_0'} \frac{ds}{s} \right) dxdt.
\end{aligned}
$$

Noting that

$$
\int_\sigma^\infty \left[\eta^*(s) \right]^{2p_0'} \frac{ds}{s} \leqslant [\eta(\sigma)]^{2p_0'} \log 2 \qquad (34)
$$

for any $\sigma \geqslant 0$, one sees that

$$
\begin{aligned}
Y(R) &\leqslant \log 2 \iint_{(0,T_\varepsilon)\times\mathbb{R}^d} |u(t,x)|^{p_0} \left[\eta\left(\frac{|x|^2+t}{R} \right) \right]^{2p_0'} dxdt \\
&= \log 2 \iint_{(0,T_\varepsilon)\times\mathbb{R}^d} |u(t,x)|^{p_0} \psi_R(t,x)\,dxdt \qquad (35) \\
&= \log 2 \, I_0(R).
\end{aligned}
$$

Therefore, plugging these estimates and (19) into (30), it holds that

$$
CRY'(R) \geqslant \begin{cases} \left(C\varepsilon + \frac{(1-\mu)}{\log 2} Y(R) \right)^{p_0} & \text{if } \alpha > 1, \\ \left(C\varepsilon \log\log R + \frac{(1-\mu)}{\log 2} Y(R) \right)^{p_0} & \text{if } \alpha = 1, \\ \left(C\varepsilon (\log R)^{1-\alpha} + \frac{(1-\mu)}{\log 2} Y(R) \right)^{p_0} & \text{if } \alpha < 1 \end{cases} \qquad (36)
$$

for any $R \in (R_1, T_\varepsilon)$. Hence we reach to

$$
-\frac{C}{R} \geqslant \begin{cases} \frac{d}{dR} \left(\varepsilon + Y(R) \right)^{1-p_0} & \text{if } \alpha > 1, \\ \frac{d}{dR} \left(\varepsilon \log\log R + Y(R) \right)^{1-p_0} & \text{if } \alpha = 1, \\ \frac{d}{dR} \left(\varepsilon (\log R)^{1-\alpha} + Y(R) \right)^{1-p_0} & \text{if } \alpha < 1. \end{cases} \qquad (37)
$$

Let us handle the case $\alpha = 1$. Integrating the both side in $(R^{1/2}, R)$, it holds that

$$
(\varepsilon \log\log R + Y(R))^{1-p_0} - \left(\varepsilon \log\log R^{1/2} + Y(R^{1/2}) \right)^{1-p_0} \leqslant -C \log R^{1/2}, \qquad (38)
$$

which yields

$$C \log R^{1/2} \leqslant \left(\varepsilon \log\log R^{1/2} + Y(R^{1/2})\right)^{1-p_0} \leqslant \left(\varepsilon \log\log R^{1/2}\right)^{1-p_0}. \tag{39}$$

Putting $S = \log R^{1/2}$, we can rewrite the above inequality as

$$\varepsilon^{p_0-1} S (\log S)^{p_0-1} \leqslant C^{-1}. \tag{40}$$

Set $\varepsilon_* = \varepsilon^{p_0-1}$ and $S_* = c_* \varepsilon_*^{-1} (\log \varepsilon_*^{-1})^{1-p_0}$ for any $\varepsilon \in (0, 1)$, where c_* will be chosen later. Then it follows that

$$\begin{aligned} &\varepsilon^{p_0-1} S_* (\log S_*)^{p_0-1} \\ &= \varepsilon_* \Big(c_* \varepsilon_*^{-1} (\log \varepsilon_*^{-1})^{1-p_0}\Big)\Big(\log c_* + \log \varepsilon_*^{-1} - (p_0 - 1)\log(\log \varepsilon_*^{-1})\Big)^{p_0-1} \\ &= c_* \left(1 + \frac{\log c_* - (p_0-1)\log(\log \varepsilon_*^{-1})}{\log \varepsilon_*^{-1}}\right)^{p_0-1} \geqslant c_* \delta_* \end{aligned} \tag{41}$$

for some $\delta_* > 0$. Therefore taking $c_* = (\delta_* C)^{-1}$, we have

$$\varepsilon^{p_0-1} S (\log S)^{p_0-1} \leqslant C^{-1} \leqslant \varepsilon^{p_0-1} S_* (\log S_*)^{p_0-1}. \tag{42}$$

This provides that $S \leqslant S_*$ and therefore

$$R \leqslant \exp(2S_*) = \exp\left(2c_* \varepsilon_*^{-1} (\log \varepsilon_*^{-1})^{1-p_0}\right) \tag{43}$$

holds. Taking $R \to T_\varepsilon$, we conclude that

$$T_\varepsilon \leqslant \exp\left(C \left(\varepsilon \log \varepsilon^{-1}\right)^{-\frac{2}{d}}\right). \tag{44}$$

One also chooses $\varepsilon_0 \in (0, 1)$ satisfying $\exp\left(C \left(\varepsilon_0 \log \varepsilon_0^{-1}\right)^{-\frac{2}{d}}\right) = R_1$. Then, (44) is true for all $\varepsilon \in (0, \varepsilon_0)$. The other cases are easier. Indeed, we establish

$$\begin{aligned} &\log R \leqslant C \varepsilon^{\frac{1-p_0}{1-(1-\alpha)(1-p_0)}} \quad \text{if } \alpha < 1, \\ &\log R \leqslant C \varepsilon^{1-p_0} \quad \text{if } \alpha > 1, \end{aligned} \tag{45}$$

which implies the desired estimate (14). The proof of Theorem 2.2 is completed.

Acknowledgments The first author was supported by JSPS KAKENHI grant number 19K14580. The second author was supported by JSPS KAKENHI grant number 18K13445.

References

1. Fujiwara, K., Ozawa, T.: Lifespan of strong solutions to the periodic nonlinear Schrödinger equation without gauge invariance. J. Evol. Equ. **17**(3), 1023–1030 (2017)
2. Ginibre, J., Ozawa, T.: Long range scattering for nonlinear Schrödinger and Hartree equations in space dimension $n \geq 2$. Commun. Math. Phys. **151**(3), 619–645 (1993)
3. Hayashi, N., Naumkin, P.I.: Asymptotics for large time of solutions to the nonlinear Schrödinger and Hartree equations. Am. J. Math. **120**(2), 369–389 (1998)
4. Ikeda, M., Inui, T.: Small data blow-up of L^2 or H^1-solution for the semilinear Schrödinger equation without gauge invariance. J. Evol. Equ. **15**(3), 571–581 (2015)
5. Ikeda, M., Sobajima, M.: Sharp upper bound for lifespan of solutions to some critical semilinear parabolic, dispersive and hyperbolic equations via a test function method. Nonlinear Anal. **182**, 57–74 (2019)
6. Ikeda, M., Wakasugi, Y.: Small-data blow-up of L^2-solution for the nonlinear Schrödinger equation without gauge invariance. Differ. Integr. Equ. **26**(11–12), 1275–1285 (2013)
7. Masaki, S., Miyazaki, H.: Long range scattering for nonlinear Schrödinger equations with critical homogeneous nonlinearity. SIAM J. Math. Anal. **50**(3), 3251–3270 (2018)
8. Masaki, S., Miyazaki, H.: Nonexistence of scattering and modified scattering states for some nonlinear Schrödinger equation with critical homogeneous nonlinearity. Differ. Integr. Equ. **32**(3–4), 121–138 (2019)
9. Masaki, S., Miyazaki, H., Uriya, K.: Long-range scattering for nonlinear Schrödinger equations with critical homogeneous nonlinearity in three space dimensions. Trans. Am. Math. Soc. **371**(11), 7925–7947 (2019)
10. Mitidieri, E., Pokhozhaev, S.I.: A priori estimates and the absence of solutions of nonlinear partial differential equations and inequalities. Trans. Mat. Inst. Steklova **234**, 1–384 (2001)
11. Oh, T., Okamoto, M., Pocovnicu, O.: On the probabilistic well-posedness of the nonlinear Schrödinger equations with non-algebraic nonlinearities. Discrete Contin. Dyn. Syst. **39**(6), 3479–3520 (2019)
12. Ozawa, T.: Long range scattering for nonlinear Schrödinger equations in one space dimension. Commun. Math. Phys. **139**(3), 479–493 (1991)
13. Shimomura, A.: Nonexistence of asymptotically free solutions for quadratic nonlinear Schrödinger equations in two space dimensions. Differ. Integr. Equ. **18**(3), 325–335 (2005)
14. Shimomura, A., Tsutsumi, Y.: Nonexistence of scattering states for some quadratic nonlinear Schrödinger equations in two space dimensions. Differ. Integr. Equ. **19**(9), 1047–1060 (2006)
15. Tsutsumi, Y., Yajima, K.: The asymptotic behavior of nonlinear Schrödinger equations. Bull. Am. Math. Soc. **11**(1), 186–188 (1984)
16. Zhang, Q.S.: Blow-up results for nonlinear parabolic equations on manifolds. Duke Math. J. **97**(3), 515–539 (1999)
17. Zhang, Q.S.: A blow-up result for a nonlinear wave equation with damping: the critical case. C. R. Acad. Sci. Paris Sér. I Math. **333**(2), 109–114 (2001)

Spectral Theory for Magnetic Schrödinger Operators in Exterior Domains with Exploding and Oscillating Long-Range Potentials

Kiyoshi Mochizuki

Abstract In this paper we present an approach to growth estimates of generalized eigenfunctions for exterior magnetic Schrödinger operators with exploding and oscillating long-range potentials. Also we apply them to show the principle of limiting absorption.

Keywords Magnetic Schrödinger operators · Long range potentials · Limiting absorption principle

1 Introduction

Growth properties of the generalized eigenfunctions of the Schrödinger operators have been studied by many authors: e.g., Kato [13] for short-range potentials, Simon [21], Ikebe-Uchiyama [8] for long-range potentials, Jäger-Rejto [11], Uchiyama [22] for exploding potentials, and Mochizuki-Uchiyama [19], Jäger-Rejto [12], Mochizuki [16] for oscillating long-range potentials. These results are applied to show the principle of limiting absorption: e.g., Eidus [6], Mochizuki [15] for short-range potentials, Ikebe-Saito [7] for long range potentials, Lavine [14], Ben-Artzi [1], [2], Jäger-Rejto [11], Yamada [23] for exploding potentials, and Mochizuki-Uchiyama [20] for oscillating long-range potentials.

In these works growth properties are proved by formulating a differential inequality for a functional of solutions of the eigenvalue problems. The principle of limiting absorption is based on this result applied to another functional identity for the inhomogeneous equations. The latter identity is constructed in considering approximate phase, leading us to radiation condition, of solutions. Note that Jäger-Rejto [11] (see also [12]) successfully used the second type of identity to the study of growth properties. Suggested by this work a unified approach is established in

K. Mochizuki (✉)
Emeritus, Department of Mathematics, Chuo University, Bunkyo-ku, Tokyo, Japan
e-mail: mochizuk@math.chuo-u.ac.jp

© Springer Nature Switzerland AG 2020

V. Georgiev et al. (eds.), *Advances in Harmonic Analysis and Partial Differential Equations*, Trends in Mathematics, https://doi.org/10.1007/978-3-030-58215-9_8

Mochizuki [16], to oscillating long-range potentials. This paper will generalize this approach to wide-ranging two-body problems including exploding and oscillating long-range potentials.

We consider the Schrödinger operator

$$L = -\Delta_b + c(x) = -\sum_{j=1}^{n}\{\partial_j + ib_j(x)\}^2 + c(x) \tag{1}$$

in an exterior domain in $\mathbf{R}^n$ with Dirichlet or Robin boundary condition. Here $b(x) = (b_1(x), \cdots, b_n(x))$ is a magnetic potential. The electric potential $c(x)$ is decomposed as

$$c(x) = c_0(r) + c_1(x) + c_2(x) + c_3(x) \qquad (r = |x|), \tag{2}$$

where $c_0(r)$ is exploding, $c_1(x)$ is oscillating long-range, $c_2(x)$ is (non-oscillating) long-range, and $c_3(x)$ is short-range in a generalized sense. Also, the magnetic field $\nabla \times b(x) = (\partial_j b_k(x) - \partial_k b_j(x))_{j>k}$ is assumed to be short-range. They all are sufficiently smooth.

As a whole $c(x)$ is assumed to satisfy

$$c(x) \geq -Cr^{\alpha} \ \ (0 \leq \alpha \leq 2) \ \text{ near infinity.} \tag{3}$$

Thus, our theory includes the case of critical exponent $\alpha = 2$ to determine L as a selfadjoint operator. Moreover, we do not require the main part $\tilde{c}(x) = c(x) - c_3(x)$ being repulsive. Instead, it is assumed to satisfy

$$\frac{\partial_r \tilde{c}(x)}{\tilde{c}(x)} \geq -\frac{2\beta}{r} \ \ (0 \leq \beta < 1) \ \text{ near infinity.} \tag{4}$$

Most results listed above exclude the critical case $\alpha = 2$ and also non-repulsive case. To explain more about the modification given, let us consider some examples.

For the potential (2) with

$$c_0(r) = -\{1 + \psi_0 \sin(\gamma \log r)\}r^{\alpha} \ \text{ and } \ c_1(x) = 0, \tag{5}$$

where $0 \leq \alpha \leq 2$, $\psi_0 > 0$ and $\gamma \neq 0$ real, we have the following: In case $\alpha > 0$, assume that

$$\psi_0 < \frac{2+\alpha}{\sqrt{(2+\alpha)^2 + \gamma^2}}. \tag{6}$$

Then the whole line $\mathbf{R}$ consists of the absolutely continuous spectrum of L. On the other hand, if $\alpha = 0$, the essential spectrum of L consists of the half line

$[-1-\psi_0, \infty)$ among which

$$\left(-1+\frac{\psi_0}{2}\sqrt{4+\gamma^2}, \infty\right) \tag{7}$$

is contained in the absolutely continuous spectrum of L for each $\psi_0 > 0$.

Similar results are also obtained for the potential (2) with

$$c_0(r)+c_1(x) = -\left\{1+\psi(x)\frac{\sin(\gamma r)}{r}\right\}r^{\alpha}, \tag{8}$$

where $0 \le \alpha \le 2$, $\gamma \ne 0$ and $\psi(x)$ is a real function such that

$$\psi(x) = O(1),\ \nabla^{\ell}\psi(x) = O(r^{-1})\ (\ell = 1, 2). \tag{9}$$

In this case, if $\alpha > 0$, assuming

$$\psi_0 = \limsup_{r\to\infty} |\psi(x)| < \frac{2+\alpha}{|\gamma|} \ \text{ and } \ \alpha < 2 \tag{10}$$

(the equality $\alpha = 2$ is admitted if $\psi = \psi(r)$ (radial) or ψ_0 is sufficiently small), we see that the whole line **R** consists of the absolutely continuous spectrum. On the other hand, if $\alpha = 0$, the essential spectrum fills the half line $[-1, \infty)$ among which both

$$\left(-1+\frac{1}{4}\gamma^2+\frac{1}{2}\psi_0|\gamma|, \infty\right) \text{ and } \left(-1, -1+\frac{1}{4}\gamma^2-\frac{1}{2}\psi_0|\gamma|\right) \tag{11}$$

become absolutely continuous for each $\psi_0 > 0$.

In the above cases of oscillating long-range potentials (when $\alpha = 0$), (7) and (11) do not cover all the essential spectrum, and there remains unsolved some intervals. But our new results improve the previous ones. The absolute continuity is covered in [16] to

$$\left(-1+\frac{\psi_0}{2}\sqrt{4+\gamma^2/(\min\{2\delta, 1\})^2}, \infty\right) \tag{12}$$

for (7), and

$$\left(-1, -1+\frac{1}{4}\gamma^2-\frac{1}{2}\psi_0|\gamma|/\min\{2\delta, 1\}\right) \cup \left(-1+\frac{1}{4}\gamma^2+\frac{1}{2}\psi_0|\gamma|/\min\{2\delta, 1\}, \infty\right) \tag{13}$$

for (11), where $\delta > 0$ appears e.g., in the exponent $O(r^{-1-\delta})$ of the short range part. These intervals shrink to empty sets as $\delta \to 0$. On the other hand, our new intervals are independent of $\delta > 0$. Moreover, we can adopt the exponent of the form $O(r^{-1}[\log r]^{-1-\delta})$ for short-range perturbations.

The main reason of the predominance stated above is in the new formulation of the radiation condition, and the use of the functional equation given in consideration to this condition, cf., Mochizuki [17] where is already formulated this new functional equation, but is not applied to the critical case $\alpha = 2$, though a more general operator

$$L = -\sum_{j,\ell=1}^{n} \{\partial_j + ib_j(x)\}a_{j\ell}(x)\{\partial_\ell + ib_\ell(x)\} + c(x), \tag{14}$$

is treaded there.

Note here that a complete different commutator method have been developed by Itakura [9] to repulsive potentials including $c_0(r) = -r^\alpha$ with $\alpha \leq 2$, where the Rellich theorem (absence of generalized eigenfunctions) is proven in the Agmon-Hörmander space. The space is optimal in the sense that a generalized eigenfunctions are actually constructed in any larger spaces with weight like $O(r^{-1-\delta})$ (cf., also Ito-Skibsted [10]). To apply his methods to our exterior problems, however, seems not easy.

Since we require $c_0(r)$ being radial, the exploding potential derived from Stark effect is not covered in this paper (also in [9]). On the other hand, the commutator method developed in [10] is adaptable to the Stark Hamiltonian in $\mathbf{R}^n$. Spectral and scattering for this Hamiltonian has been studied e. g., by Ben-Artzi [3].

As is seen in (11), Our results for oscillating long-range potentials ($\alpha = 0$) remains some unsolved part. The absolute continuity of the whole essential spectrum is obtained by Ben-Artzi-Devinatz [4] and Devinatz-Rejto [5] for radial oscillating potentials of the form

$$c(r) = \psi_0 \sin(\gamma r^\nu)/r^\ell + c_3(r) \qquad (0 < \nu \leq 1, \ell > 0) \tag{15}$$

in the whole space $\mathbf{R}^N$. The problem is then reduced to a direct sum of ordinary differential operators of Schrödinger type. The treatment of the case $\ell > \nu$ is excluded in this paper.

The remainder of this paper will be organized as follows. In the next Sect. 2 we summarize the assumptions required on the potentials, and state the main results (three theorems) without proof. The approximate phase of [20] and [16] is generalized there and the radiation conditions are newly defined by using it. The fundamental functional equation of solutions will be formulated in Sect. 3. The proof of Theorems 2.1–2.3 are respectively given in Sects. 4–6. Finally, in Sect. 7 we reconsider the examples given above to check how our assumptions are satisfied by them.

2 Assumptions and Results

Let $\Omega \subset \mathbf{R}^n$ ($n \geq 2$) be an exterior domain with smooth boundary $\partial\Omega$. We consider in Ω the boundary value problem

$$Lu - \zeta u = f(x) \ \text{ in } \Omega, \ \ \mathscr{B}u = 0 \ \text{ on } \partial\Omega, \tag{16}$$

where $L = -\Delta_b + c(x)$ with the magnetic Laplacian

$$\Delta_b = \nabla_b \cdot \nabla_b = -\sum_{j=1}^{n} \{\partial_j + ib_j(x)\}^2, \tag{17}$$

$\zeta \in \mathbf{C}$, $\partial_j = \partial/\partial x_j$ and $i = \sqrt{-1}$, and $\mathscr{B}u|_{\partial\Omega} = 0$ is the Dirichlet or Robin boundary condition.

$$\mathscr{B}u = u \ \text{ or } \ = \tilde{x} \cdot \nabla_b u + d(x)u \qquad (\tilde{x} = x/|x|). \tag{18}$$

We do not exclude the case when $\partial\Omega$ is empty and $\Omega = \mathbf{R}^n$. In case $\partial\Omega$ is not empty, let $R_0 > 0$ be chosen so that $\partial\Omega \subset \{x;\, |x| < R_0\}$.

In the following we put $S_t = \{x \in \mathbf{R}^n;\, |x| = t\}$, $\Omega_t = \{x \in \Omega;\, |x| \leq t\}$ and $\Omega_t' = \{x \in \Omega;\, |x| > t\}$ for $t > 0$. Moreover, we use the notation $f' = df(r)/dr$, $\partial_r g = \partial g(x)/\partial r = \tilde{x} \cdot \nabla g(x)$.

The coefficients are all real and sufficiently smooth. To define the behavior of coefficients near infinity we introduce a decreasing weight function $\mu = \mu(r)$ ($r > 0$) verifying

$$(\mu.1) \qquad \mu(r) = o(r^{-1}), \ \text{decreasing and} \ \int_0^\infty \mu(r)dr < \infty.$$

Without loss of generality we can assume $\mu(r) \geq r^{-2}$ near infinity.

Assumptions $c(x)$ is divided into $c(x) = c_0(r) + c_1(x) + c_2(x) + c_3(x)$, where $r = |x|$, and each $b_j(x)$ and $c_j(x)$ satisfy the following conditions: There exist an interval $I \subset \mathbf{R}$, $R_1 \geq R_0$, $0 < \tilde{\beta} < 1$ and $a \geq 0$, such that we have in $(x, \lambda) \in \Omega_{R_1}' \times I$,

$(A0)$ There exist $C_j > 0$ ($j = 0, 1, 2$) and $0 \leq \alpha \leq 2$ such that

$$C_0 \leq \lambda - c_0(r) \leq C_1(1 + r)^\alpha \ \text{ and}$$

$$\frac{|c_0'(r)|^2}{(\lambda - c_0(r))^2} + \frac{|c_0''(r)|}{\lambda - c_0(r)} = O(\mu)\sqrt{\lambda - c_0(r)}.$$

$(A1)$

$$\frac{|c_1(x)|^2}{(\lambda - c_0(r))^{3/2}} \leq \frac{d_0}{r} \text{ for some } d_0 > 0,$$

if $c(x)$ is non-radial we require more $d_0 < \sqrt{32(1-\tilde{\beta})/3}$,

$$\frac{|\partial_r c_1(x)|}{\lambda - c_0(r)} = O(r^{-1}) \text{ and } \frac{|(\partial_r^2 + a)c_1(x)|}{\lambda - c_0(r)} = O(\mu).$$

$(A2)$

$$\frac{r^{-1}|c_2(x)| + |\partial_r c_2(x)|}{\lambda - c_0(r)} = O(\mu), \quad \frac{|\partial_r^2 c_2(x)|}{\lambda - c_0(r)} = O(\mu)\sqrt{\lambda - c_0(r)}.$$

$(A3)$

$$|\nabla \times b(x)| + |c_3(x)| = O(\mu)\sqrt{\lambda - c_0(r)}.$$

$(A4)$

$$\frac{-c_0'(r)}{\lambda - c_0(r)} + \frac{-\partial_r c_1(x)}{\lambda - c_0(r) - a/4} \geq -\frac{2\tilde{\beta}}{r}.$$

$(A5)$

$$\frac{|c_2(x)c_0'(r)|}{(\lambda - c_0(r))^2} = O(\mu).$$

$(A6)$

$$\frac{|\tilde{\nabla} c_1(x)| + |\tilde{\nabla} c_2(x)|}{\lambda - c_0(r)} = O(\mu), \text{ where } \tilde{\nabla} = \nabla - \tilde{x}\partial_r, \text{ and}$$

$$\frac{|\tilde{\nabla}\partial_r c_1(x)| + |\tilde{\nabla}\partial_r c_2(x)|}{\lambda - c_0(r)} = O(\mu)\sqrt{\lambda - c_0(r)}.$$

Remark 2.1 Combining $(A0)$ and $(A1)$, we see that

$$\partial_r\left[\frac{c_1(x)}{\lambda - c_o(r)}\right] = O(r^{-1}) \qquad \text{and} \qquad \partial_r^2\left[\frac{c_1(x)}{\lambda - c_o(r)}\right] = O(\mu), \tag{19}$$

which imply $\dfrac{c_1(x)}{\lambda - c_o(r)} = O(r^{-1})$ at infinity if $a > 0$ (cf., Lemma 8.1 of [20]).

Under these conditions, since $c(x) \geq -C|x|^\alpha$ ($\alpha \leq 2$), operator L determines a selfadjoint operator in $L^2(\Omega)$ with domain

$$\mathcal{D}(L) = \{u \in H^2_{\mathrm{loc}}(\overline{\Omega}) \cap L^2(\Omega); -\Delta_b u + cu \in L^2(\Omega), \mathcal{B}u|_{\partial\Omega} = 0\}. \tag{20}$$

The selfadjointness of L shows that if $\zeta \in \mathbf{C}\backslash\mathbf{R}$, (16) has a unique L^2-solution for given any $f \in L^2(\Omega)$. To extend the unique existence theorem to real ζ contained in the continuous spectrum of L, it is necessary to consider the operator $-\Delta_b + c(x)$ in a wider class of functions in $H^2_{\mathrm{loc}}(\overline{\Omega})$. The domain will be characterized by a new type of radiation condition at infinity.

Let $I \subset \mathbf{R}$ be an interval such that each $\lambda \in \overline{I}$ verifies the above Assumptions with common C_j ($j = 1\,3$), $\tilde{\beta}$ and d_0. We put $\Gamma_\pm = \{\zeta = \lambda \pm i\epsilon; \lambda \in I, 0 < \epsilon \leq \epsilon_0\}$ for sufficiently small $\epsilon_0 > 0$ and choose $\zeta \in \Gamma_\pm$ in Eq. (16). For solution u we define the vector-valued function

$$\theta = \theta(x, \zeta) = \nabla_b u + \tilde{x} K(x, \zeta) u, \quad \tilde{x} = \frac{x}{|x|}. \tag{21}$$

Then (16) is rewritten as

$$-\nabla_b \cdot \theta + K\tilde{x} \cdot \theta + q_K u = f; \tag{22}$$

with

$$q_K(x, \zeta) = c(x) - \zeta + \partial_r K + \frac{n-1}{r} K - K^2. \tag{23}$$

The function $K(x, \zeta)$ is chosen so that $q_K(x, \zeta)$ behaves like short-range. For this aim, we are able to generalize the argument of Mochizuki-Uchiyama [18] and Mochizuki [15] to define

$$K(x, \zeta) = -i\sqrt{k(x, \zeta)} + \frac{n-1}{2r} + \frac{\partial_r k(x, \zeta)}{4k(x, \zeta)}, \tag{24}$$

where

$$k(x, \zeta) = \zeta - c_0(r) - \eta(r, \zeta) c_1(x) - c_2(x), \tag{25}$$

with

$$\eta(r, \zeta) = \frac{\zeta - c_0(r)}{\zeta - c_0(r) - a}. \tag{26}$$

In [18] a primitive form of (24) is given by solving a Riccati type equation. Note that it essentially coincides with the approximate phase introduced by Jäger-Rejto [11] (also [12]) in terms of their exploding potentials.

Note that when $\zeta \in \Gamma_\pm$ goes to $\lambda \in I$ each limit of the square root $\sqrt{k(x,\zeta)}$ should be distinguished as

$$\sqrt{k(x,\lambda \pm i0)} = \pm\sqrt{k(x,\lambda)}, \quad \text{where } \sqrt{k(x,\lambda)} \geq 0. \tag{27}$$

Theorem 2.1 *Under the above Assumption and* $\lambda \in I$*, let* $u = u(x,\lambda) \in H^2_{\mathrm{loc}}$ *be a solution of the homogeneous equation*

$$-\Delta_b u + c(x)u - \lambda u = 0 \ \ in\ \Omega, \ \ \mathcal{B}u|_{\partial\Omega} = 0. \tag{28}$$

If the support of u *is not compact, then we have*

$$\liminf_{t\to\infty} \int_{S_t} \frac{1}{\sqrt{k(x,\lambda)}} |\tilde{x} \cdot \theta(x,\lambda \pm i0)|^2 dS > 0. \tag{29}$$

To enter into the principle of limiting absorption, we introduce another weight function $\mu_0 = \mu_0(r) > 0$ verifying

$(\mu_0.1)$ $\qquad \mu(r) \leq \mu_0(r) = o(r^{-1})$, decreasing and $\int_0^\infty \mu_0(r)dr < \infty$.

$(\mu_0.2)$ $\qquad \varphi_0'(r) \leq \varphi'(r)$ and $\dfrac{\varphi_0'(r)}{\varphi_0(r)} \leq \dfrac{1}{r} + \mathrm{Re}\dfrac{\partial_r k(x,\zeta)}{2k(x,\zeta)}$ for $r > R_0$,

where

$$\varphi(r) = \left(\int_r^\infty \mu(s)ds\right)^{-1} \text{ and } \varphi_0(r) = \left(\int_r^\infty \mu_0(s)ds\right)^{-1}, \tag{30}$$

and

$(\mu_0.3)$ $\qquad \varphi_0'(r)\sqrt{\lambda - c_0(r)}$ is bounded in $r > R_0$.

Remark 2.2

(1) If $\mu = (1+r)^{-1-\delta}$ $(0 < \delta \leq 1)$ or $= (1+r)^{-1}[\log(e+r)]^{-1-\delta_0}]$ $(0 < \delta_0 \leq 1)$ for $r > R_0$ in Assumptions, then the function

$$\mu_0(r) = (1+r)^{-1}[\log(e+r)]^{-1-\delta_0} \tag{31}$$

verifies the above $(\mu_0.1) \sim (\mu_0.3)$ since we have $\varphi_0'(r) = \delta_0^2(1+r)^{-1}[\log(e+r)]^{-1+\delta_0}$.

(2) If $\mu(r) = (1+r)^{-1-\delta}$ and $\alpha < 2$ in Assumptions, then we can choose $\mu_0(r) = (1+r)^{-1-\delta_0}$ with $\delta_0 \leq \min\{\delta, 1-\alpha/2\}$.

Definition 2.1 The solution of (16) in $H^2_{\rm loc}(\overline{\Omega})$ is said to satisfy the radiation conditions if

$$\int_\Omega \mu_0(r)\sqrt{k_0}|u(x,\zeta)|^2dx < \infty, \quad \int_{\Omega'_{R_1}} \frac{\varphi_0'(r)}{\sqrt{k_0}}|\tilde{x}\cdot\theta(x,\zeta)|^2dx < \infty, \tag{32}$$

where $k_0 = k_0(r) = \lambda^* - c_0(r)$ with fixed any $\lambda^* \in \overline{I}$. A solution of (16) which also satisfies the radiation condition is called a radiative solution.

As is required by (25), the weight $k_0(r)$ is equivalent to $|k(x,\zeta)|$ for each $\zeta \in \Gamma_\pm$. Note that we can choose $k_0(r) = 1$ when $c_0(r)$ is bounded.

For nonnegative function $\xi = \xi(x)$ we define the weighted L^2-space $L^2_\xi(\Omega)$ with norm

$$\|f\|^2_\xi = \int_\Omega \xi(x)|f(x)|^2dx. \tag{33}$$

The case $\xi(x) \equiv 1$ is the norm of $L^2(\Omega)$ and we simply write it as $\|f\|$.

Now, let $\zeta \in \Gamma_\pm$. Then the resolvent $R(\zeta) = (L-\zeta)^{-1}$ forms a bounded operator in $L^2(\Omega)$ which depends continuously on ζ, and the absolute continuity of L in the interval I is proved through the following theorems.

Theorem 2.2 *For each $f \in L^2(\Omega)$ and $\zeta \in \Gamma_\pm$ the solution $u = R(\zeta)f$ satisfies the radiation conditions.*

Theorem 2.3 *Let $\zeta \in \Gamma_\pm$ and $f \in L^2_{(\mu_0\sqrt{k_0})^{-1}}(\Omega)$. Then there exists $C_4 = C_4(\Gamma_\pm) > 0$ such that*

$$\sup_{\zeta\in\Gamma_\pm} \|R(\zeta)f\|_{\mu_0\sqrt{k_0}} \le C_4\|f\|_{(\mu_0\sqrt{k_0})^{-1}}, \tag{34}$$

and as an operator from $L^2_{(\mu_0\sqrt{k_0})^{-1}}(\Omega)$ to $L^2_{\mu_0\sqrt{k_0}}(\Omega)$, $R(\zeta)$ is extended continuously to $\overline{\Gamma}_\pm$. Moreover, for $f \in L^2_{(\mu_0\sqrt{k_0})^{-1}}(\Omega)$, $u = R(\lambda \pm i0)f$ becomes a unique (outgoing (+) or incoming (−)) radiative solution of (16) with $\zeta = \lambda$.

Corollary 2.1 *Let $\{E(\lambda); \lambda \in \boldsymbol{R}\}$ denote the spectral resolution of L. Let $I = [\lambda_0, \lambda_1] \subset \boldsymbol{R}$ be any interval as given in Assumptions and put $E(I) = E(\lambda_1) - E(\lambda_0)$. Then the part of L in $E(I)L^2(\Omega)$ is absolutely continuous, and we have for any $f, g \in L^2_{(\mu_0\sqrt{k_0})^{-1}}(\Omega)$*

$$(E(I)f, g) = \frac{1}{2\pi i}\int_I (R(\lambda + i0)f - R(\lambda - i0)f, g)d\lambda, \tag{35}$$

where the integrand means the duality between $L^2_{\mu\sqrt{k_0}}(\Omega)$ and $L^2_{(\mu\sqrt{k_0})^{-1}}(\Omega)$

$$(R(\lambda \pm i0)f, g) = \int_\Omega [R(\lambda \pm i0)f](x)\overline{g(x)}dx. \tag{36}$$

3 A Quadratic Identity and Basic Estimates of $k(x, \zeta)$

The proof of these theorems are based on a quadratic identity for the function $\theta(x, \zeta)$. By use of this function the first equation of (16) is rewritten as (22). More generally, in the following we put

$$u_\sigma = e^\sigma u, \quad f_\sigma = e^\sigma f \ \text{ and } \ \theta_\sigma = \nabla_b u_\sigma + \tilde{x} K u_\sigma, \tag{37}$$

where $\sigma = \sigma(r)$ is a nonnegative function of $r > 0$. Then u_σ satisfies

$$-\Delta_b u_\sigma + 2\sigma' \tilde{x} \cdot \nabla_b u_\sigma + \left\{ c(x) - \zeta + \sigma'' - \sigma'^2 + \sigma' \frac{n-1}{r} \right\} u_\sigma = f_\sigma, \tag{38}$$

or corresponding to (21) we have

$$-\nabla_b \cdot \theta_\sigma + (K + 2\sigma') \tilde{x} \cdot \theta_\sigma + q_{K,\sigma} u_\sigma = f_\sigma; \tag{39}$$

with

$$q_{K,\sigma} = q_K + \sigma'' - \sigma'^2 + \frac{n-1}{r} \sigma' - 2K\sigma'. \tag{40}$$

Note that $u_0 = u$ and $\theta_0 = \theta$.

For a smooth weight function $\Phi = \Phi(x) > 0$, let us consider the real part of equation (39) multiplied by $\Phi(\tilde{x} \cdot \overline{\theta_\sigma})$:

$$\mathrm{Re}[\Phi\{-\nabla_b \cdot \theta_\sigma + (K + 2\sigma')\tilde{x} \cdot \theta_\sigma + q_{K,\sigma} u_\sigma\}(\tilde{x} \cdot \overline{\theta_\sigma})] = \mathrm{Re}\{\Phi f_\sigma (\tilde{x} \cdot \overline{\theta_\sigma})\}. \tag{41}$$

Here

$$-\Phi \nabla_b \cdot \theta_\sigma (\tilde{x} \cdot \overline{\theta_\sigma}) = -\nabla \cdot [\Phi \theta_\sigma (\tilde{x} \cdot \overline{\theta_\sigma})] + \nabla \Phi \cdot \theta (\tilde{x} \cdot \overline{\theta_\sigma}) + \Phi \theta_\sigma \cdot \overline{\nabla_b (\tilde{x} \cdot \theta_\sigma)}. \tag{42}$$

In the third term of the right we use the identity

$$\begin{aligned}
(\partial_\ell + ib_\ell)(\tilde{x} \cdot \theta_\sigma) &= \frac{1}{r} \sum_j (\delta_{j\ell} - \tilde{x}_j \tilde{x}_\ell) \theta_{\sigma,j} + i\tilde{x} \cdot B'_\ell u_\sigma + \partial_\ell K u_\sigma \\
&\quad + (\tilde{x} \cdot \nabla_b + K)(\partial_\ell + ib_\ell) u_\sigma \\
&= \frac{1}{r} \sum_j (\delta_{j\ell} - \tilde{x}_j \tilde{x}_\ell) \theta_{\sigma,j} + i\tilde{x} \cdot B'_\ell u_\sigma + \partial_\ell K u_\sigma \\
&\quad + (\tilde{x} \cdot \nabla_b + K)\theta_{\sigma,\ell} - \tilde{x}_\ell (\tilde{x} \cdot \nabla K) u_\sigma - \tilde{x}_\ell K \tilde{x} \cdot \theta_\sigma,
\end{aligned} \tag{43}$$

where $B'_\ell = (\nabla \times b)_\ell = (\partial_\ell b_m - \partial_m b_\ell)_{m=1,\cdots,n}$. Then since

$$\begin{aligned}
&\mathrm{Re}\Phi \sum_\ell \theta_{\sigma,\ell}\overline{(\tilde{x}\cdot\nabla_b + K)\theta_{\sigma,\ell}} \\
&\qquad = \frac{1}{2}\left\{\nabla\cdot[\Phi\tilde{x}|\theta_\sigma|^2] - \tilde{x}\cdot\nabla\Phi|\theta_\sigma|^2 - \Phi\frac{n-1}{r}|\theta_\sigma|^2\right\} + \mathrm{Re}\Phi\overline{K}|\theta_\sigma|^2,
\end{aligned} \tag{44}$$

it follows that

$$\begin{aligned}
&-\mathrm{Re}\Phi\nabla_b\cdot\theta_\sigma(\tilde{x}\cdot\overline{\theta_\sigma}) \\
&\qquad = \mathrm{Re}\Big\{-\nabla\cdot[\Phi\theta_\sigma(\tilde{x}\cdot\overline{\theta_\sigma})] + \nabla\Phi\cdot\theta(\tilde{x}\cdot\overline{\theta_\sigma}) \\
&\qquad\quad + \frac{\Phi}{r}\left(|\theta_\sigma|^2 - |\tilde{x}\cdot\theta_\sigma|^2\right) + \Phi\sum_\ell \theta_{\sigma,\ell}\overline{(i\tilde{x}\cdot B'_\ell + \tilde{\nabla}_\ell K)u_\sigma} \\
&\qquad\quad + \frac{1}{2}\nabla\cdot[\Phi\tilde{x}|\theta_\sigma|^2] - \frac{1}{2}\partial_r\Phi|\theta_\sigma|^2 \\
&\qquad\quad + \Phi\left(\overline{K} - \frac{n-1}{2r}\right)|\theta_\sigma|^2 - \Phi K|\tilde{x}\cdot\theta_\sigma|^2\Big\}.
\end{aligned} \tag{45}$$

Substitute this to Eq. (41) and note that $\mathrm{Re}K = \mathrm{Re}\overline{K}$. Then integrating over $\Omega_{R,t}$ derives the following

Proposition 3.1 *For any solution $u \in H^2_{\mathrm{loc}}(\overline{\Omega})$ of (16) we have*

$$\begin{aligned}
&\mathrm{Re}\int_{\Omega_{R,t}} \Phi f_\sigma(\tilde{x}\cdot\overline{\theta_\sigma})dx \\
&\quad = -\left[\int_{S_t} - \int_{S_R}\right]\Phi\left(|\tilde{x}\cdot\theta_\sigma|^2 - \frac{1}{2}|\theta_\sigma|^2\right)dS \\
&\qquad + \mathrm{Re}\int_{\Omega_{R,t}} \Phi\Bigg[\frac{\nabla\Phi}{\Phi}\cdot\theta(\tilde{x}\cdot\overline{\theta_\sigma}) + \frac{1}{r}\left(|\theta_\sigma|^2 - |\tilde{x}\cdot\theta_\sigma|^2\right) - \frac{\partial_r\Phi}{2\Phi}|\theta_\sigma|^2 \\
&\qquad + \left(K - \frac{n-1}{2r}\right)|\theta_\sigma|^2 + 2\sigma'|\tilde{x}\cdot\theta_\sigma|^2 + \mathcal{B}(u_\sigma,\theta_\sigma) \\
&\qquad + (q_{K,\sigma} - q_K)u_\sigma(\tilde{x}\cdot\overline{\theta_\sigma})\Bigg]dx
\end{aligned} \tag{46}$$

where

$$\mathcal{B}(u_\sigma,\theta_\sigma) = \sum_\ell \theta_{\sigma\ell}\overline{(i\tilde{x}\cdot B'_\ell + \tilde{\nabla}_\ell K)u_\sigma} + q_K u_\sigma(\tilde{x}\cdot\overline{\theta_\sigma}). \tag{47}$$

We put

$$k_1(x,\zeta) = \zeta - c_0(r) - \eta(r,\zeta)c_1(x); \quad \eta(r,\zeta) = \frac{\zeta - c_0(r)}{\zeta - c_0(r) - a/4}. \tag{48}$$

By definition we have

$$k(x,\zeta) = k_1(x,\zeta) - c_2(x) = \zeta - c_0(r) - \eta(r,\zeta)c_1(x) - c_2(x). \tag{49}$$

Note that

$$\eta(r,\zeta) = O(1), \ \eta'(r,\zeta) = O(\sqrt{\mu}) \text{ and } \eta''(r,\zeta) = O(\mu) \tag{50}$$

as $r \to \infty$. In fact,

$$\eta'(r,\zeta) = \frac{(a/4)c_0'(r)}{(\zeta - c_0(r) - a/4)^2} \tag{51}$$

and

$$\eta''(r,\zeta) = \frac{a/4}{\zeta - c_0 - a/4}\left\{\frac{c_0''(r)}{(\zeta - c_0(r) - a/4)} + \frac{2c_0'(r)^2}{(\zeta - c_0(r) - a/4)^2}\right\}. \tag{52}$$

Proposition 3.2 *The following estimates hold for* $(x,\lambda) \in \Omega'_{R_1} \times \Gamma_\pm$ *if* $R_1 \geq R_0$ *is chosen sufficiently large and* $\epsilon_0 > 0$ *sufficiently small.*

$(K0)$ $k(x,\zeta)$ *and* $k_1(x,\zeta)$ *are equivalent to* $\zeta - c_0(r)$*:*

$$\frac{k(x,\zeta)}{\zeta - c_0(r)} = 1 + O(\sqrt{r\mu}), \quad \frac{k_1(x,\zeta)}{\zeta - c_0(r)} = 1 + O(\sqrt{r\mu}).$$

$(K1)$

$$\frac{\partial_r k(x,\zeta)}{k(x,\zeta)} - \frac{\partial_r k_1(x,\zeta)}{k_1(x,\zeta)} = O(\mu).$$

$(K2)$

$$\mathrm{Re}\frac{\partial_r k(x,\zeta)}{2k(x,\zeta)} \geq -\frac{\beta}{r} \ for \ some \ \tilde{\beta} < \beta < 1.$$

$(K3)$

$$\frac{\partial_r k(x,\zeta)}{k(x,\zeta)^{5/4}} = O(\sqrt{\mu}).$$

$(K4)$

$$\frac{\tilde{\nabla} k(x,\zeta)}{k(x,\zeta)} = O(\mu), \quad \frac{\tilde{\nabla}\partial_r k(x,\zeta)}{k(x,\zeta)^{3/2}} = O(\mu).$$

$(K5)$

$$\frac{q_K(x,\zeta)}{k(x,\zeta)^{1/2}} = O(\mu).$$

$(K6)$ *If $c_1(x)$ is non-radial,*

$$\frac{|c_1(x)|^2}{k(x,\lambda)^{3/2}} \le \frac{d_0}{r} \quad \text{with } 0 < d_0 < \sqrt{\frac{32(1-\beta)}{3}}.$$

Proof Assertion $(K0)$ is obvious from $(A1)$ and $(A2)$ since we have

$$k(x,\zeta) = (\zeta - c_0(r))\left\{1 - \frac{c_1(x)}{\zeta - c_0(r) - a/4} - \frac{c_2(x)}{\zeta - c_0(r)}\right\}. \tag{53}$$

Note that

$$\partial_r k = \frac{\eta'}{\eta} k_1 + \eta(-c_0' - \partial_r c_1) - \partial_r c_2. \tag{54}$$

Then

$$\begin{aligned}\frac{\partial_r k}{k} - \frac{\partial_r k_1}{k_1} &= \frac{\eta'}{\eta}\left(\frac{k_1}{k} - 1\right) + \eta(-c_0' - \partial_r c_1)\left(\frac{1}{k} - \frac{1}{k_1}\right) - \frac{\partial_r c_2}{k} \\ &= \frac{(a/4)\eta c_0' c_2}{(\zeta - c_0)(\zeta - c_0 - a/4)k} + \frac{\eta(-c_0' - \partial_r c_1)c_2}{k_1 k} + \frac{-\partial_r c_2}{k},\end{aligned} \tag{55}$$

and $(K0)$, $(A1)$, $(A2)$ and $(A5)$ lead to $(K1)$.

We shall show $(K2)$ for $k_1(x,\zeta)$. The same inequality for $k(x,\zeta)$ comes from $(K1)$. Note that

$$\begin{aligned}\frac{\partial_r k_1}{k_1} &= \frac{-c_0'}{\zeta - c_0} + \frac{c_0'}{\zeta - c_0 - a/4} + \frac{-c_0' - \partial_r c_1}{\zeta - c_0 - c_1 - a/4} \\ &= \frac{-c_0'}{\zeta - c_0} + \frac{-\partial_r c_1}{\zeta - c_0 - a/4} + \frac{c_0' + \partial_r c_1}{\zeta - c_0 - a/4}\frac{\eta c_1}{k_1}.\end{aligned} \tag{56}$$

Then since

$$\frac{c_0' + \partial_r c_1}{\zeta - c_0 - a/4}\frac{\eta c_1}{k_1} = O(\sqrt{\mu})O(r^{-1/2}) = o(r^{-1}), \tag{57}$$

it follows that

$$\begin{aligned} \mathrm{Re}\frac{\partial_r k_1}{k_1} &= \left\{\frac{-c_o'}{\lambda - c_0} + \frac{-\partial_r c_1}{\lambda - c_0 - a/4}\right\}\left(\frac{\lambda - c_0}{|\zeta - c_0|}\right)^2 + o(r^{-1}) \\ &\quad + \frac{-\partial_r c_1}{\lambda - c_0 - a/4}\left(\frac{-\epsilon^2}{|\zeta - c_0|^2} + \frac{\epsilon^2}{|\zeta - c_0 - a/4|^2}\right). \end{aligned} \tag{58}$$

Thus,

$$\mathrm{Re}\frac{\partial_r k_1}{k_1} + \frac{2\beta}{r} \geq \left\{\frac{-c_0'}{\lambda - c_0} + \frac{-\partial_r c_1}{\lambda - c_0 - a/4} + \frac{2\tilde{\beta}}{r}\right\} + \frac{2(\beta - \tilde{\beta} - C\epsilon^2)}{r} \geq 0. \tag{59}$$

Assertions $(K3)$ and $(K4)$ are direct from $(K0)$, $(A0) \sim (A2)$ and $(A6)$. Next, $(K5)$ follows from the identity

$$\begin{aligned} q_K &= (1 - \eta)c_1 + c_3 + \frac{(n-1)(n-3)}{4r^2} - 5\left(\frac{\partial_r k}{4k}\right)^2 + \frac{\partial_r^2 k}{4k} \\ &= O(\mu)\sqrt{\zeta - c_0} + O(r^{-2}) - 5\left(\frac{\partial_r k}{4k}\right)^2 + (1 - \eta)c_1 + \frac{\partial_r^2 k}{4k} \end{aligned} \tag{60}$$

since we have

$$\left(\frac{\partial_r k}{k}\right)^2 = \left(\frac{-c_0' - \eta' c_1 - \eta\partial_r c_1 - \partial_r c_2}{k}\right)^2 = O(\mu)\sqrt{\zeta - c_0}, \tag{61}$$

and

$$\begin{aligned} (1 - \eta)c_1 + \frac{\partial_r^2 k}{4k} &= \frac{-(a/4)c_1}{\zeta - c_0 - a/4} + \frac{-c_0'' - \partial_r^2(\eta c_1) - \partial_r^2 c_2}{4k} \\ &= -\eta\frac{\partial_r^2 c_1 + a c_1}{4k} - \frac{a c_1(\eta c_1 + c_2)}{4(\zeta - c_0 - a/4)k} \\ &\quad - \frac{c_0'' - 2\eta'\partial_r c_1 - \eta'' c_1 - \partial_r^2 c_2}{4k} \\ &= O(\mu)\sqrt{\zeta - c_0}. \end{aligned} \tag{62}$$

Here, we have used the estimate in Remark 1: $\frac{c_1}{k} = O(r^{-1})$ when $a > 0$.

The final assertion $(K6)$ is obvious from the first estimate of $(A1)$. □

4 Proof of Theorem 2.1

Let $u = u(x, \lambda)$ be a solution of the homogeneous equation (28) in Proposition 1, and choose $\Psi = \dfrac{\psi(x)}{\sqrt{k(x,\lambda)}}$ there. Then it is rewritten as

Proposition 4.1 *For each solution u of* (28)*, the following identity holds.*

$$\begin{aligned}
&\Big[\int_{S_t} - \int_{S_R}\Big]\frac{\psi}{\sqrt{k}}\Big(|\tilde{x}\cdot\theta_\sigma|^2 - \frac{1}{2}|\theta_\sigma|^2\Big)dS \\
&\quad = \mathrm{Re}\int_{\Omega_{R,t}} \frac{\psi}{\sqrt{k}}\Big[\Big(\frac{1}{r} - \frac{\partial_r\psi}{2\psi} + \frac{\partial_r k}{2k}\Big)|\theta_\sigma|^2 \\
&\qquad + \Big(-\frac{1}{r} + \frac{\partial_r\psi}{\psi} - \frac{\partial_r k}{2k}\Big)|\tilde{x}\cdot\theta_\sigma|^2 + \Big(\frac{\tilde{\nabla}\psi}{\psi} - \frac{\tilde{\nabla}k}{2k}\Big)\cdot\theta_\sigma(\tilde{x}\cdot\overline{\theta_\sigma}) \\
&\qquad + 2\sigma'|\tilde{x}\cdot\theta_\sigma|^2 + \mathcal{B}(u_\sigma,\theta_\sigma) + (q_{K,\sigma} - q_K)u_\sigma(\tilde{x}\cdot\overline{\theta_\sigma})\Big]dx,
\end{aligned} \tag{63}$$

where $\theta_\sigma = \theta_\sigma(x, \lambda \pm i0)$ *with* $K(x, \lambda \pm i0) = \mp i\sqrt{k} + \dfrac{n-1}{2r} + \dfrac{\partial_r k}{4k}$.

Other than this proposition we need the following:

Lemma 4.1 *Let* $r > R_1$*. Then for each solution u of* (28)*, we have*

$$\mathrm{Im}\Big[\int_{S_r}(\tilde{x}\cdot\nabla_b u_\sigma)\overline{u_\sigma}\,dS\Big] = 0. \tag{64}$$

Proof We multiply by $\overline{u}$ on both sides of (28) and integrate by parts over Ω_r. Then the imaginary part gives the above identity in case $\sigma = 0$. The general case also holds since we have

$$(\tilde{x}\cdot\nabla_b u_\sigma)\overline{u_\sigma} = e^{2\sigma}\tilde{x}\cdot(\nabla_b u + \tilde{x}\sigma' u)\overline{u} \tag{65}$$

and $\sigma = \sigma(r)$ is real. □

Lemma 4.2 *Let* $\psi(x)$ *be a smooth positive function of* $x \in \Omega'_{R_1}$*. Then for each solution u of* (28) *we have*

$$\int_{S_r}\psi\sqrt{k}|u_\sigma|^2 dS \le C_\psi\int_{S_r}\psi\frac{1}{\sqrt{k}}|\tilde{x}\cdot\theta_\sigma|^2 dS \;\; in\; r > R_1, \tag{66}$$

where $C_\psi = C_\psi(r) = \dfrac{\max_{\tilde{x}\in S_1}\psi(r\tilde{x})}{\min_{\tilde{x}\in S_1}\psi(r\tilde{x})}$.

Proof If $\psi = 1$, the inequality follows from Lemma 4.1 since we have

$$\frac{1}{\sqrt{k}}|\tilde{x}\cdot\theta_\sigma|^2 = \frac{1}{\sqrt{k}}|\tilde{x}\cdot\theta_\sigma \pm i\sqrt{k}u_\sigma)|^2 + \frac{1}{\sqrt{k}}|\sqrt{k}u_\sigma|^2 \pm 2\mathrm{Im}(\tilde{x}\cdot\nabla_b u_\sigma)\overline{u_\sigma}. \tag{67}$$

For general $\psi(x)$, the inequality we use

$$\begin{aligned}\int_{S_r}\psi\sqrt{k}|u_\sigma|^2 dS &\le \max_{\tilde{x}\in S_1}\psi(r\tilde{x})\int_{S_r}\sqrt{k}|u_\sigma|^2 dS\\ &\le \max_{\tilde{x}\in S_1}\psi(r\tilde{x})\int_{S_r}\frac{1}{\sqrt{k}}|\tilde{x}\cdot\theta_\sigma|^2 dS\\ &\le \frac{\max_{\tilde{x}\in S_1}\psi(r\tilde{x})}{\min_{\tilde{x}\in S_1}\psi(r\tilde{x})}\int_{S_1}\frac{\psi}{\sqrt{k}}|\theta_\sigma|^2 dS\end{aligned} \tag{68}$$

and the he statement follows. □

By use of this lemma we have the following

Lemma 4.3 *Let $\psi = 1$ or $\psi = k_1(x,\lambda)$ and $r > R_1$. Then there exists $C_5 > 0$ such that for each solution u of* (28) *we have*

$$\int_{S_r}\frac{\psi}{\sqrt{k}}\left\{\left|\left(\frac{\tilde{\nabla}\psi}{\psi} - \frac{\tilde{\nabla}k}{2k}\right)\cdot\theta_\sigma(\tilde{x}\cdot\overline{\theta_\sigma})\right| + |\mathcal{B}(u_\sigma,\theta_\sigma)|\right\}dS \le C_5\mu\int_{S_r}\frac{\psi}{\sqrt{k}}|\theta_\sigma|^2 dS, \tag{69}$$

Proof Note that $(K3)$ and $(K5)$ imply

$$\left|\frac{\tilde{\nabla}k}{2k}\cdot\theta_\sigma(\tilde{x}\cdot\overline{\theta_\sigma})\right| + \left|\frac{\tilde{\nabla}k_1}{2k_1}\cdot\theta_\sigma(\tilde{x}\cdot\overline{\theta_\sigma})\right| \le O(\mu)|\theta_\sigma|^2. \tag{70}$$

On the other hand, since $B'_\ell = \mathbf{O}(\mu)\sqrt{k},\ q_K = O(\mu)\sqrt{k}$ and

$$\tilde{\nabla}K = \mp i\frac{\tilde{\nabla}k}{2\sqrt{k}} + \frac{\tilde{\nabla}\partial_r k}{4k} - \frac{\partial_r k\tilde{\nabla}k}{4k^2} = \mathbf{O}(\mu)\sqrt{k}, \tag{71}$$

noting (47), we have $|\mathcal{B}(u_\sigma,\theta_\sigma)| \le O(\mu)|\sqrt{k}u_\sigma||\theta_\sigma|$. The use of Lemma 4.2 then gives

$$\begin{aligned}\int_{S_r}\frac{\psi}{\sqrt{k}}|\mathcal{B}(u_\sigma,\theta_\sigma)|dS &\le O(\mu)\left[\int_{S_r}\psi\sqrt{k}|u_\sigma|^2 dS\int_{S_r}\frac{\psi}{\sqrt{k}}|\theta_\sigma|^2 dS\right]^{1/2}\\ &\le O(\mu)\sqrt{C_\psi}\int_{S_t}\frac{\psi}{\sqrt{k}}|\theta_\sigma|^2 dS\end{aligned} \tag{72}$$

and the statement follows. □

Now, we define

$$F_0(t) = \int_{S_t} \frac{1}{\sqrt{k}} \left\{ |\tilde{x} \cdot \theta|^2 - \frac{1}{2}|\theta|^2 \right\} dS, \tag{73a}$$

$$F_1(t) = \int_{S_t} \frac{k_1}{\sqrt{k}} \left\{ |\tilde{x} \cdot \theta|^2 - \frac{1}{2}|\theta|^2 \right\} dS \tag{73b}$$

$$F_{\sigma,\tau}(t) = \int_{S_t} \frac{r^p k_1}{\sqrt{k}} \left\{ |\tilde{x} \cdot \theta_\sigma|^2 - \frac{1}{2}|\theta_\sigma|^2 + \frac{1}{2}(\sigma'^2 - \tau)|u_\sigma|^2 \right\} dS, \tag{73c}$$

where we choose

$$1 + \beta < p < 2, \tag{73d}$$

and $\tau = \tau(r)$ is also a positive weight function. With these auxiliary functions the proof of Theorem 2.1 will be divided in three parts:

- *Part 1* The case required that there exists a sequence $r_n \to \infty$ such that $F_0(r_n) > 0$.
- *Part 2* The case required that there exists $\tilde{R} \geq R_1$ such that $F_0(t) \leq 0$ for $t \geq \tilde{R}$ and does not have compact support, on the other hand, there exists $r_n \to \infty$ such that $F_1(r_n) > 0$.
- *Part 3* The case required the complementary conditions that there exists $\tilde{R} \geq R_1$ such that $F_1(r) \leq 0$ for $t \geq \tilde{R}$ and u does not have compact support.

Consideration for Part 1

We choose $\sigma = 0$ and $\psi = 1$ in Proposition 3. Then by definition $F_0(t)$, $t > R$, satisfies

$$\frac{d}{dt} F_0(t) = \mathrm{Re} \int_{S_t} \frac{1}{\sqrt{k}} \left[\left(\frac{1}{r} + \frac{\partial_r k}{2k} \right) (|\theta|^2 - |\tilde{x} \cdot \theta|^2) - \frac{\tilde{\nabla} k}{2k} \cdot \theta (\tilde{x} \cdot \overline{\theta}) + \mathcal{B}(u, \theta) \right] dS. \tag{74}$$

It follows from $(K2)$ that

$$\frac{1}{r} + \frac{\partial_r k}{2k} \geq \frac{1 - \beta}{r} \quad \text{for } r \geq R_1. \tag{75}$$

Thus, with the help of Lemma 4.3 with $\sigma = 0$ and $\psi = 1$, we can summarize

$$\begin{aligned}\frac{d}{dt}F_0(t) &\geq \int_{S_t} \frac{1}{\sqrt{k}}\left[\left(\frac{1}{r}+\frac{\partial_r k}{2k}\right)(|\theta|^2 - |\tilde{x}\cdot\theta|^2) - C_5\mu|\theta|^2\right]dS \\ &= \int_{S_t} \frac{1}{\sqrt{k}}\left[\left(\frac{1}{r}+\frac{\partial_r k}{2k} - 2C_5\mu\right)(|\theta|^2 - |\tilde{x}\cdot\theta|^2)\right. \\ &\quad \left. - 2C_5\mu\left(|\tilde{x}\cdot\theta|^2 - \frac{1}{2}|\theta|^2\right)\right]dS \geq -2C_5\mu(t)F_0(t)\end{aligned} \tag{76}$$

for $t \geq R_2$, if $R_2 \geq R_1$ is chosen sufficiently large to satisfy

$$\frac{1-\beta}{r} - 2C_5\mu(r) \geq 0 \quad \text{for } r \geq R_2. \tag{77}$$

By assumption there exists $r_n \geq R_2$ such that $F_0(r_n) > 0$ and hence, we conclude

$$F_0(t) \geq e^{-2C_5\int_{r_n}^{\infty}\mu(s)ds}F_0(r_n) > 0 \quad \text{for } t \geq r_n, \tag{78}$$

which proves Theorem 2.1 since we have

$$\int_{S_t}\frac{1}{\sqrt{k}}|\tilde{x}\cdot\theta|^2 dS \geq 2F_0(t). \tag{79}$$

Consideration for Part 2
First note that $F_0(t) \leq 0$ in $t \geq \tilde{R}$ implies

$$\int_{S_t}\frac{k_1}{\sqrt{k}}\{|\theta|^2 - |\tilde{x}\cdot\theta|^2\}dS \geq C_{k_1}^{-1}\int_{S_t}\frac{k_1}{\sqrt{k}}|\tilde{x}\cdot\theta|^2 dS, \tag{80}$$

where

$$\begin{aligned}C_{k_1}^{-1}(r) &= 1 - \frac{\max_{\tilde{x}\in S_1} k_1(r\tilde{x},\lambda) - \min_{\tilde{x}\in S_1} k_1(r\tilde{x},\lambda)}{\max_{\tilde{x}\in S_1} k_1(r\tilde{x},\lambda)} \\ &\geq 1 - \frac{2d_0 r^{-1/2}}{k_1(x,\lambda)^{1/4}} \geq 1 - C_6 r^{-1/2}\end{aligned} \tag{81}$$

for some $C_6 > 0$. In the first inequality we have used $(K6)$. Thus, it follows that

$$C_6 t^{-1/2}\int_{S_t}\frac{k_1}{\sqrt{k}}|\tilde{x}\cdot\theta|^2 dS \geq 2F_1(t) \quad \text{in } t \geq \tilde{R}. \tag{82}$$

Now, we return to the identity of Proposition 3 with $\sigma = 0$ and $\psi = k_1$. Differentiating this, we have

$$\begin{aligned}\frac{d}{dt}F_1(t) &= \int_{S_t} \frac{k_1}{\sqrt{k}}\left[\left(\frac{1}{r} - \frac{\partial_r k}{2k} + \frac{\partial_r k_1}{2k_1}\right)\{|\theta|^2 - |\tilde{x}\cdot\theta|^2\}\right. \\ &\qquad\qquad\qquad\left. + \frac{\partial_r k_1}{2k_1}|\tilde{x}\cdot\theta|^2 + \mathcal{B}(u,\theta)\right]dS \\ &\geq \int_{S_t} \frac{k_1}{\sqrt{k}}\left[C_{k_1}^{-1}\left(\frac{1}{r} - O(\mu)\right)|\tilde{x}\cdot\theta|^2 + \left(\frac{\partial_r k_1}{2k_1} - C_5\mu\right)|\tilde{x}\cdot\theta|^2\right]dS,\end{aligned} \tag{83}$$

where (80) is used in the last inequality. Since

$$C_{k_1}^{-1}\left(\frac{1}{r} - 2O(\mu)\right) \geq (1 - C_6 r^{-1/2})\left(\frac{1}{r} - O(\mu)\right) \geq \frac{1}{r} - O(r^{-3/2}) \tag{84}$$

and $\frac{\partial_r k_1}{2k_1} \geq -\frac{\beta}{r} - O(\mu)$ $(\beta < 1)$ by $(K1)$ and $(K2)$, it follows that

$$\frac{d}{dt}F_1(t) \geq \int_{S_t} \frac{k_1}{\sqrt{k}}\left(\frac{1-\beta}{r} - O(r^{-3/2}) - O(\mu)\right)|\tilde{x}\cdot\theta|^2 dS \geq \frac{\epsilon}{t}\int_{S_t}\frac{k_1}{\sqrt{k}}|\tilde{x}\cdot\theta|^2 dS \tag{85}$$

for some $0 < \epsilon < 1-\beta$. Choose $r_n \geq \tilde{R}$. Then this shows that $F_1(t) \geq F_1(r_n) > 0$ in $t \geq r_n$, and also in considering (82), we obtain

$$\frac{d}{dt}F_1(t) \geq 2\epsilon C_6^{-1} t^{-1/2} F_1(t) \quad \text{in } t \geq r_n. \tag{86}$$

This and (82) yield the inequality

$$\begin{aligned}\int_{S_t}\frac{1}{\sqrt{k}}|\tilde{x}\cdot\theta|^2 dS &\geq \frac{1}{\max_{\tilde{x}\in S_1} k_1(t\tilde{x},\lambda)}\int_{S_t}\frac{k_1}{\sqrt{k}}|\tilde{x}\cdot\theta|^2 dS \\ &\geq \frac{2C_6^{-1}t^{1/2}}{\max_{\tilde{x}\in S_1} k_1(t\tilde{x},\lambda)}F_1(t) \\ &\geq C_7^{-1}t^{1/2-\alpha}e^{4\epsilon C_6^{-1}(t^{1/2}-r_n^{1/2})}F_1(r_n)\end{aligned} \tag{87}$$

for some $C_7 \geq C_1C_6$ in $t \geq r_n$ if $r_n \geq \tilde{R}$ is chosen sufficiently large. The proof is thus complete.

Consideration for Part 3

We choose $\psi = r^p\sqrt{k_1}$ in Proposition 3 and differentiate both sides by t:

$$\begin{aligned}
&\frac{d}{dt}\int_{S_t}\frac{r^pk_1}{\sqrt{k}}\left\{|\tilde{x}\cdot\theta_\sigma|^2-\frac{1}{2}|\theta_\sigma|^2\right\}dS\\
&\quad=\mathrm{Re}\int_{S_t}\frac{r^pk_1}{\sqrt{k}}\left[\left(\frac{1}{r}-\frac{p}{2r}-\frac{\partial_rk_1}{2k_1}+\frac{\partial_rk}{2k}\right)|\theta_\sigma|^2\right.\\
&\qquad+\left(-\frac{1}{r}+\frac{p}{r}+\frac{\partial_rk_1}{k_1}-\frac{\partial_rk}{2k}\right)|\tilde{x}\cdot\theta_\sigma|^2\\
&\qquad\left.+\frac{\tilde{\nabla}k}{2k}\cdot\theta_\sigma(\tilde{x}\cdot\overline{\theta_\sigma})+\mathcal{B}(u_\sigma,\theta_\sigma)\right]dS\\
&\qquad+\mathrm{Re}\int_{S_t}\frac{r^pk_1}{\sqrt{k}}\left[2\sigma'|\tilde{x}\cdot\theta_\sigma|^2+(q_{K,\sigma}-q_K)u_\sigma(\tilde{x}\cdot\overline{\theta_\sigma})\right]dS\\
&\quad\equiv J_1+J_2
\end{aligned}\tag{88}$$

Lemma 4.4 *There exists $R_3 \geq R_1$ such that for $r \geq R_3$*

$$J_1 \geq \int_{S_t}\frac{r^pk_1}{\sqrt{k}}\left\{\frac{p-1-\beta}{r}|\tilde{x}\cdot\theta_\sigma|^2+\left(\frac{2-p}{2r}-o(r^{-1})\right)|\theta_\sigma|^2\right\}dS. \tag{89}$$

Proof By means of $(K1)$ and $(K2)$

$$\frac{2-p}{2r}-\frac{\partial_rk_1}{2k_1}+\frac{\partial_rk}{2k}\geq\frac{2-p}{2r}-C_8\mu, \tag{90}$$

$$\frac{p-1}{r}+\frac{\partial_rk_1}{k_1}-\frac{\partial_rk}{2k}\geq\frac{p-1-\beta}{r}-2C_8\mu \tag{91}$$

for some $C_8 > 0$. Thus, taking also account of Lemma 4.3, we have for $t \geq R_1$

$$J_1 \geq \int_{S_t}\frac{r^pk_1}{\sqrt{k}}\left[\frac{2-p}{2r}|\theta_\sigma|^2+\frac{p-1-\beta}{r}|\tilde{x}\cdot\theta_\sigma|^2-(3C_8+C_5)\mu\}|\theta_\sigma|^2\right]dS, \tag{92}$$

and the desired inequality follows. □

Next we proceed to J_2. Note that

$$q_{K,\sigma}-q_K=\sigma''-\sigma'^2-2\sigma'\left(\mp i\sqrt{k}+\frac{\partial_rk}{4k}\right). \tag{93}$$

Then

$$
\begin{aligned}
2\sigma'|\tilde{x}\cdot\theta_\sigma|^2 &+ \mathrm{Re}\{(q_{K,\sigma}-q_k)u_\sigma(\tilde{x}\cdot\overline{\theta_\sigma})\}\\
&= \mathrm{Re}\Bigg\{2\sigma'(\tilde{x}\cdot\theta_\sigma \pm i\sqrt{k}u_\sigma)(\tilde{x}\cdot\overline{\theta_\sigma})\\
&\quad + \left(\sigma''-\sigma'^2-2\sigma'\frac{\partial_r k}{4k}\right)u_\sigma(\tilde{x}\cdot\overline{\theta_\sigma})\Bigg\}\\
&= \mathrm{Re}\Bigg\{2\sigma'|\tilde{x}\cdot\theta_\sigma \pm i\sqrt{k}u_\sigma|^2 \pm 2\sigma'(\tilde{x}\cdot\theta_\sigma \pm i\sqrt{k}u_\sigma)i\sqrt{k}\overline{u_\sigma}\\
&\quad + \left(\sigma''-2\sigma'\frac{\partial_r k}{4k}\right)u_\sigma(\tilde{x}\cdot\overline{\theta_\sigma})\\
&\quad - \frac{1}{2}\nabla\cdot(\tilde{x}\sigma'^2|u_\sigma|^2) + \left(\sigma''\sigma'-\sigma'^2\frac{\partial_r k}{4k}\right)|u_\sigma|^2\Bigg\}\\
&= 2\sigma'|\tilde{x}\cdot\theta_\sigma + i\sqrt{k}u_\sigma|^2 \mp 2\sigma'\sqrt{k}\mathrm{Im}\{\tilde{x}\cdot\nabla_b u_\sigma\overline{u_\sigma}\}\\
&\quad + \mathrm{Re}\left\{\left(\sigma''-2\sigma'\frac{\partial_r k}{4k}\right)(\tilde{x}\cdot\theta_\sigma \pm i\sqrt{k}u_\sigma)\overline{u_\sigma}\right\}\\
&\quad - \frac{1}{2}\nabla\cdot(\tilde{x}\sigma'^2|u_\sigma|^2) + \left(\sigma''\sigma'-\sigma'^2\frac{\partial_r k}{4k}\right)|u_\sigma|^2.
\end{aligned}
\tag{94}
$$

We combine this and the identity

$$
0 = \frac{1}{2}\nabla\cdot(\tilde{x}\tau|u_\sigma|^2) - \mathrm{Re}\{\tau(\tilde{x}\cdot\theta_\sigma \pm i\sqrt{k}u_\sigma)\overline{u_\sigma}\} + \tau\frac{\partial_r k}{4k}|u_\sigma|^2 - \frac{\tau'}{2}|u_\sigma|^2. \tag{95}
$$

Then

$$
\begin{aligned}
2\sigma'|\tilde{x}\cdot\theta_\sigma|^2 &+ \mathrm{Re}\{(q_{K,\sigma}-q_k)u_\sigma(\tilde{x}\cdot\overline{\theta_\sigma})\}\\
&= 2\sigma'|\tilde{x}\cdot\theta_\sigma \pm i\sqrt{k}u_\sigma|^2 \mp 2\sigma'\sqrt{k}\mathrm{Im}\{\tilde{x}\cdot\nabla_b u_\sigma\overline{u_\sigma}\}\\
&\quad + \mathrm{Re}\left\{\left(\sigma''-2\sigma'\frac{\partial_r k}{4k}-\tau\right)(\tilde{x}\cdot\theta_\sigma \pm i\sqrt{k}u_\sigma)\overline{u_\sigma}\right\}\\
&\quad - \frac{1}{2}\nabla\cdot\left\{\tilde{x}(\sigma'^2-\tau)|u_\sigma|^2\right\}\\
&\quad + \left(\sigma''\sigma'-\sigma'^2\frac{\partial_r k}{4k}+\tau\frac{\partial_r k}{4k}-\frac{\tau'}{2}\right)|u_\sigma|^2.
\end{aligned}
\tag{96}
$$

Hence, it follows that

$$\begin{aligned} J_2 = & -\frac{1}{2}\frac{d}{dt}\int_{S_t}\frac{r^p k_1}{\sqrt{k}}(\sigma'^2-\tau)|u_\sigma|^2 dS + \int_{S_t}\frac{r^p k_1}{\sqrt{k}}\Big[\\ & (\sigma'^2-\tau)\left(\frac{p}{2r}+\frac{\partial_r k_1}{2k_1}-\frac{\partial_r k}{2k}\right)|u_\sigma|^2 + 2\sigma'|\tilde{x}\cdot\theta_\sigma \pm i\sqrt{k}u_\sigma|^2 \\ & \mp 2\sigma'\sqrt{k}\mathrm{Im}\{\tilde{x}\cdot\nabla_b u_\sigma \overline{u_\sigma}\} \\ & + \mathrm{Re}\left\{\left(\sigma''-2\sigma'\frac{\partial_r k}{4k}-\tau\right)(\tilde{x}\cdot\theta_\sigma \pm i\sqrt{k}u_\sigma)\overline{u_\sigma}\right\} \\ & + \left(\sigma''\sigma'-\sigma'^2\frac{\partial_r k}{4k}+\tau\frac{\partial_r k}{4k}-\frac{\tau'}{2}\right)|u_\sigma|^2\Big] dS. \end{aligned} \tag{97}$$

Noting Lemma 4.1 and definition (48) of $k_1(x,\lambda)$, we have

$$\mp\int_{S_t}\frac{r^p k_1}{\sqrt{k}}2\sigma'\sqrt{k}\mathrm{Im}\{\tilde{x}\cdot\nabla_b u_\sigma\overline{u_\sigma}\}dS = \pm 2\sigma' r^p\int_{S_t}\eta c_1\mathrm{Im}\{\tilde{x}\cdot\nabla_b u_\sigma\overline{u_\sigma}\}dS. \tag{98}$$

This integral becomes 0 if c_1 is radial. So, in the following, we restrict c_1 to the complicated non-radial case. Then the above integral is estimated from below by

$$\geq -\int_{S_t}\frac{r^p k_1}{\sqrt{k}}\left\{(2-\epsilon)\sigma'|\tilde{x}\cdot\theta_\sigma \pm i\sqrt{k}u_\sigma|^2 + \frac{k\sigma'}{(2-\epsilon)k_1^2}(\eta c_1)^2|u_\sigma|^2\right\}dS. \tag{99}$$

Also by the Schwarz inequality

$$\begin{aligned} \mathrm{Re}\int_{S_t}&\frac{r^p k_1}{\sqrt{k}}\left(\sigma''-2\sigma'\frac{\partial_r k}{4k}-\tau\right)(\tilde{x}\cdot\theta_\sigma \pm i\sqrt{k}u_\sigma)\overline{u_\sigma}dS \\ & \geq -\int_{S_t}\frac{r^p k_1}{\sqrt{k}}\left\{\epsilon\sigma'|\tilde{x}\cdot\theta_\sigma \pm i\sqrt{k}u_\sigma|^2 + \frac{3}{\epsilon}\left(\frac{\sigma''^2}{4\sigma'}\right.\right. \\ & \left.\left. + \sigma'\left(\frac{\partial_r k}{4k}\right)^2 + \frac{\tau^2}{4\sigma'}\right)|u_\sigma|^2\right\}dS. \end{aligned} \tag{100}$$

Thus, we have for any $0 < \epsilon < 2$

$$
\begin{aligned}
J_2+&\frac{1}{2}\frac{d}{dt}\int_{S_t}\frac{r^p k_1}{\sqrt{k}}(\sigma'^2-\tau)|u_\sigma|^2 dS \\
&\geq \int_{S_t}\frac{r^p k_1}{\sqrt{k}}\bigg[(\sigma'^2-\tau)\left(\frac{p}{2r}+\frac{\partial_r k_1}{2k_1}-\frac{\partial_r k}{2k}\right)|u_\sigma|^2 \\
&\qquad -\frac{k\sigma'}{(2-\epsilon)k_1^2}(\eta c_1)^2|u_\sigma|^2-\frac{3}{\epsilon}\left(\frac{\sigma''^2}{4\sigma'}+\sigma'\left(\frac{\partial_r k}{4k}\right)^2+\frac{\tau^2}{4\sigma'}\right)|u_\sigma|^2 \\
&\qquad +\left(\sigma''\sigma'-\sigma'^2\frac{\partial_r k}{4k}+\tau\frac{\partial_r k}{4k}-\frac{\tau'}{2}\right)|u_\sigma|^2\bigg]dS.
\end{aligned}
\tag{101}
$$

Now we put

$$
\sigma(r)=\frac{m}{1-\gamma_0}r^{1-\gamma_0},\quad \tau(r)=r^{-2\gamma_0}\log r\quad (1/3<\gamma_0<1) \tag{102}
$$

where $m > 0$ is a large parameter.

Lemma 4.5 *Let p satisfy $2\gamma_0 < p < 2$. Then there exist $R_4 \geq R_3$ and $C_9, C_{10} > 0$ such that for any $m \geq 1$ and $t \geq R_4$*

$$
\begin{aligned}
J_2+&\frac{1}{2}\frac{d}{dt}\int_{S_t}\frac{r^p k_1}{\sqrt{k}}(\sigma'^2-\tau)|u_\sigma|^2 dS \\
&\geq \int_{S_t}\frac{r^p k_1}{\sqrt{k}}\left\{m^2\left(\frac{p-2\gamma_0}{2r^{1+2\gamma_0}}-o(r^{-1-2\gamma_0})\right)-o(r^{-1})k\right\}|u_\sigma|^2 dS \\
&\qquad -\int_{S_t}\frac{r^p k_1}{\sqrt{k}}\left(\frac{k}{k_1}\right)^2\frac{d_0\sigma'\sqrt{k}}{(2-\epsilon)r}|u_\sigma|^2 dS.
\end{aligned}
\tag{103}
$$

Proof In virtue of $(K2)$, $(K3)$ and (102), we have

$$
\sigma'^2\left(\frac{p}{2r}+\frac{\partial_r k_1}{2k_1}-\frac{\partial_r k}{2k}\right)+\sigma''\sigma'\geq m^2 r^{-2\gamma_0}\left(\frac{p-2\gamma_0}{2r}-C_8\mu\right), \tag{104}
$$

and

$$
\begin{aligned}
-\frac{3}{\epsilon}\left\{\frac{\sigma''^2}{4\sigma'}+\sigma'\left(\frac{\partial_r k}{4k}\right)^2\right\} &\geq -\frac{3m\gamma_0^2}{4\epsilon}r^{-2-\gamma_0}-m^2 r^{-2\gamma_0}O(\mu)-O(\mu)k, \\
&\geq -m^2 o(r^{-1-2\gamma_0})-o(r^{-1})k,
\end{aligned}
\tag{105}
$$

and

$$-\frac{3\tau^2}{4\epsilon\sigma'} - \tau\left(\frac{p}{2r} + \frac{\partial_r k_1}{2k_1} - \frac{\partial_r k}{2k}\right) - \frac{\tau'}{2}$$
$$\geq -\frac{3}{2\epsilon m} r^{-3\gamma_0}(\log r)^2 - r^{-2\gamma_0}\log r\left(\frac{p-2\gamma_0}{2r} + O(\mu) + \frac{1}{r\log r}\right)$$
$$\geq -o(r^{-1}). \tag{106}$$

Thus, we conclude the desired inequality, noting $(K6)$ and $\eta(r,\lambda) \to 1$ $(r \to \infty)$ also. □

Lemma 4.6 *For $t > R_0$ and any $C^* > 0$ we have*

$$-\int_{S_t} \frac{r^p k_1}{\sqrt{k}}\left\{o(r^{-1})k + \left(\frac{k}{k_1}\right)^2 \frac{d_0\sigma'\sqrt{k}}{(2-\epsilon)r}\right\}|u_\sigma|^2 dS.$$
$$\geq -\int_{S_t} \frac{r^p k_1}{\sqrt{k}}\left\{\frac{D^*(1+o(1))}{2r}|\tilde{x}\cdot\theta_\sigma|^2 + \frac{m^2C^*(1+o(1))}{2r^{1+2\gamma_0}}|u_\sigma|^2\right\}dS, \tag{107}$$

where $D^* = \dfrac{d_0^2}{C^*(2-\epsilon)^2}$.

Proof By means of Lemma 4.2 with $\psi = k_1$, we have

$$-\int_{S_t} \frac{r^p k_1}{\sqrt{k}} o(r^{-1})k|u_\sigma|^2 dS \geq -o(r^{-1})C_{k_1}\int_{S_t}\frac{r^p k_1}{\sqrt{k}}|\tilde{x}\cdot\theta_\sigma|^2 dS$$
$$\geq -\int_{S_t}\frac{r^p k_1}{\sqrt{k}} o(r^{-1})|\tilde{x}\cdot\theta_\sigma|^2 dS \tag{108}$$

since $C_{k_1} = 1 + o(1)$. Next note that

$$\frac{k}{k_1} = 1 + o(1) \qquad \text{and} \qquad \frac{d_0\sigma'\sqrt{k}}{(2-\epsilon)r} \leq \frac{d_0^2 k}{2C^*(2-\epsilon)^2 r} + \frac{m^2C^*}{2r^{1+2\gamma_0}}. \tag{109}$$

Then by use also of Lemma 4.2

$$-\int_{S_t}\frac{r^p k_1}{\sqrt{k}}\left(\frac{k}{k_1}\right)^2\frac{d_0\sigma'\sqrt{k}}{(2-\epsilon)r}|u_\sigma|^2 dS \geq -(1+o(1))^2$$
$$\times\left\{C_{k_1}\int_{S_t}\frac{r^p k_1}{\sqrt{k}}\frac{d_0^2}{2C^*(2-\epsilon)^2 r}|\tilde{x}\cdot\theta_\sigma|^2 dS + \int_{S_t}\frac{r^p k_1}{\sqrt{k}}\frac{m^2C^*}{2r^{1+2\gamma_0}}|u_\sigma|^2 dS\right\}. \tag{110}$$

Thus, the desired assertion holds. □

It follows from Lemmas 4.4–4.6 that for any $m \geq 1$ and $t \geq R_4$

$$\begin{aligned} \frac{d}{dt}F_{\sigma,\tau}(t) &= J_1 + J_2 + \frac{1}{2}\frac{d}{dt}\int_{S_t}\frac{r^p k_1}{\sqrt{k}}(\sigma'^2 - \tau)|u_\sigma|^2 dS \\ &\geq \int_{S_t}\frac{r^p k_1}{\sqrt{k}}\left\{\frac{p-1-\beta-(D^*/2)}{r}|\tilde{x}\cdot\theta_\sigma|^2\right. \\ &\quad \left. + \left(\frac{2-p}{2r} - o(r^{-1})\right)|\theta_\sigma|^2\right\}dS \\ &\quad + \int_{S_t}\frac{r^p k_1}{\sqrt{k}}m^2\left(\frac{p-2\gamma_0-C^*}{2r^{1+2\gamma_0}} - o(r^{-1-2\gamma_0})\right)|u_\sigma|^2 dS. \end{aligned} \tag{111}$$

Note that $d_0^2 < \left(2-\frac{2}{3}\right)2^2(2-2\beta) = \frac{32(1-\beta)}{3}$ by definition. Then we have

$$C^* < p - 2\gamma_0 \text{ and } D^* = \frac{d_0^2}{C^*(2-\epsilon)^2} < p - 2\beta \tag{112}$$

by choosing p, γ_0 and ϵ respectively very close to 2, $1/3$ and 0. Thus, we conclude

$$\frac{d}{dt}F_{\sigma,\tau}(t) \geq 0 \text{ in } t \geq R_5 \tag{113}$$

if $R_5 \geq R_4$ sufficiently large.

By assumption that the support of u is not compact, R_5 is able to satisfy

$$\int_{S_{R_5}}\frac{k_1}{\sqrt{k}}|u|^2 dS > 0. \tag{114}$$

Then as we see from (111), $F_{\sigma,\tau}(R_5)$ goes to ∞ as $m \to \infty$. We fix a large m satisfying $F_{\sigma,\tau}(R_5) > 0$. Then (113) asserts that $F_{\sigma,\tau}(t) > 0$ for $t \geq R_5$.

Now, we rewrite $F_{\sigma,\tau}(t)$ as

$$\begin{aligned} F_{\sigma,\tau}(t) &= e^{2\sigma}t^p\left\{F_1(t) + \sigma'\mathrm{Re}\int_{S_t}\frac{k_1}{\sqrt{k}}(\tilde{x}\cdot\nabla u)\bar{u}\,dS\right. \\ &\quad \left. + \int_{S_t}\frac{k_1}{\sqrt{k}}\left\{\sigma'^2 - \frac{1}{2}\tau + \sigma'\left(\frac{n-1}{2r} + \frac{\partial_r k}{4k}\right)\right\}|u|^2 dS\right\} \\ &= e^{2\sigma}t^p\left[F_1(t) + \frac{1}{2}\sigma'\frac{d}{dt}\int_{S_t}\frac{k_1}{\sqrt{k}}|u|^2 dS\right. \\ &\quad \left. + \int_{S_t}\frac{k_1}{\sqrt{k}}\left\{\sigma'^2 - \frac{1}{2}\tau - \sigma'\left(\frac{\partial_r k_1}{2k_1} - \frac{\partial_r k}{2k}\right)\right\}|u|^2 dS\right]. \end{aligned} \tag{115}$$

Note here that $F_{\sigma,\tau}(t) > 0$, $F_1(t) \leq 0$ when $t \geq R_5$. Moreover, it follows from (102) that

$$\sigma'^2 - \frac{1}{2}\tau - \sigma'\left(\frac{\partial_r k_1}{2k_1} - \frac{\partial_r k}{2k}\right) \leq 0 \;\text{ when } r \geq R_6 \tag{116}$$

if $R_6 \geq R_5$ is chosen sufficiently large. So, we conclude

$$\frac{1}{2}\sigma' \frac{d}{dt} \int_{S_t} \frac{k_1}{\sqrt{k}} |u|^2 dS > 0 \;\text{ for } t \geq R_6. \tag{117}$$

Moreover, R_6 is able to satisfy

$$\int_{S_{R_6}} \frac{k_1}{\sqrt{k}} |u|^2 dS > 0. \tag{118}$$

Then

$$\int_{S_t} \sqrt{k} |u|^2 dS \geq \min_{\tilde{x} \in S_t} \frac{k(t\tilde{x}, \lambda)}{k_1(t\tilde{x}, \lambda)} \int_{S_{R_6}} \frac{k_1}{\sqrt{k}} |u|^2 dS > 0 \tag{119}$$

for $t \geq R_6$. Since

$$\min_{\tilde{x} \in S_t} \frac{k(t\tilde{x}, \lambda)}{k_1(t\tilde{x}, \lambda)} \to 1 \;\text{ as }\; t \to \infty, \tag{120}$$

this and Lemma 4.2 with $\psi = 1$ and $\sigma = 0$ establish the conclusion of Theorem 2.1.

5 Proof of Theorem 2.2

Let $u = R(\zeta)f$ with $\{\zeta, f\} \in \Gamma_\pm \times L^2(\Omega)$. Then $u \in H^2(\Omega)$ if $c_0(r)$ is bounded. So, the proposition is obvious in this case. To show the case $c_0(r) \to -\infty$ as $r \to \infty$, we start from the following classical result of Ikebe-Saito [7]: there exists $C_{13} > 0$ depending on $\mathrm{Im}\zeta$ such that

$$\|u\| + \|(e + r)^{-\alpha/2} \nabla_b u\| \leq C_{13} \|f\|. \tag{121}$$

The Green formula gives

$$\int_{\Omega_t} \psi\{|\nabla_b u|^2 + (c - \mathrm{Re}\zeta)|u|^2\}dx$$
$$= \mathrm{Re}\int_{\Omega_t} \{\psi f\overline{u} - \psi'(\tilde{x}\cdot\nabla_b u)\overline{u}\}dx + \mathrm{Re}\int_{S_t} \psi(\tilde{x}\cdot\nabla_b u)\overline{u}dS \qquad (122)$$
$$- \int_{\partial\Omega} \psi(r)d(x)|u|^2 dS.$$

Choose here $\psi = k_0(r)^{-1} = (\lambda^* - c_0(r))^{-1}$. Then since it is bounded, it follows from (121) that

$$\liminf_{t\to\infty} \int_{S_t} \psi|\nabla_b u||u|dS = 0. \qquad (123)$$

Hence, using the Schwarz inequality and letting $t \to \infty$ in (122), we obtain

$$\frac{1}{2}\int_\Omega \frac{|\nabla_b u|^2}{k_0}dx \le \int_\Omega \left\{\frac{|c-\mathrm{Re}\zeta|}{k_0}|u|^2 + \frac{1}{k_0}|f||u| + \frac{k_0'^2}{2k_0^3}|u|^2\right\}dx$$
$$+ \int_{\partial\Omega} \psi(r)|d(x)||u|^2 dS$$
$$\le M\int_\Omega \{|u|^2 + |f||u| + \frac{1}{2}|u|^2\}dx + \int_{\partial\Omega} \psi(r)|d(x)||u|^2 dS$$
$$\le C_{14}\|f\|^2 < \infty \qquad \text{for some } C_{14} > 0, \qquad (124)$$

where

$$M = \max_{x\in\Omega}\left[\max\left\{\frac{|c-\zeta|}{k_0}, \frac{1}{k_0}, \frac{O(\mu)}{\sqrt{k_0}}\right\}\right]. \qquad (125)$$

Now, we apply it to the inequality

$$\int_{\Omega_{R,t}} \frac{\varphi_0'}{\sqrt{k_0}}|\theta|^2 dx \le 2\int_{\Omega_{R,t}} \varphi_0'\sqrt{k_0}\left\{\frac{|\nabla_b u|^2}{k_0} + \frac{|K(x,\zeta|^2}{k_0}|u|^2\right\}dx. \qquad (126)$$

Then since $\varphi_0'\sqrt{k_0}$ is bounded and

$$K(x,\zeta) = i\sqrt{k(x,\zeta)} + O(\sqrt{\mu})k_0^{1/4} \qquad (127)$$

by $(K4)$ and the equivalence of $|k|$ and k_0, we obtain letting $t \to \infty$,

$$\int_{\Omega'_R} \frac{\varphi_0'}{\sqrt{k_0}} |\theta|^2 dx < \infty. \tag{128}$$

The solution $u = R(\zeta)f$ thus satisfies our adjustment of the radiation condition.

6 Proof of Theorem 2.3

Lemma 6.1 *We have for any $R > 0$,*

$$\frac{\varphi_0'(r)}{\varphi_0(r)} = \mu_0(r)\varphi_0(r) \notin L^1([R, \infty)). \tag{129}$$

Proof By definition $\varphi_0(r) \to \infty$ as $r \to \infty$. So, the assertion holds since we have

$$\int_R^r \frac{\varphi_0'(s)}{\varphi_0(s)} ds = \log \left\{ \frac{\varphi_0(r)}{\varphi_0(R)} \right\} \to \infty \quad \text{as } r \to \infty. \tag{130}$$

Lemma 6.2 *Let u be a radiative solution of* (16).

(i) If $\operatorname{Im}\zeta \neq 0$*, then we have $u \in L^2(\Omega)$ and*

$$|\operatorname{Im}\zeta| \|u\| \leq \|f\|. \tag{131}$$

(ii) There exists $C_{15} > 0$ such that for any $R \geq R_1$ and $\zeta \in \overline{\Gamma}_\pm$,

$$\int_{\Omega'_R} \mu_0 \sqrt{k_0} |u|^2 dx$$
$$\leq C_{15}\varphi_0(R)^{-1} \left\{ \|\tilde{x} \cdot \theta\|^2_{\varphi_0' \sqrt{k_0}^{-1}, \Omega'_R} + \|u\|^2_{\mu_0 \sqrt{k_0}} + \|f\|^2_{(\mu_0 \sqrt{k_0})^{-1}} \right\}, \tag{132}$$

where $\Omega'_R = \{x \in \Omega; |x| \geq R\}$ and $k = k(x, \zeta)$.

Proof By the Green formula

$$\operatorname{Im} \int_{\Omega_r} f\bar{u} dx = -\operatorname{Im}\zeta \int_{\Omega_r} |u|^2 dx - \operatorname{Im} \int_{S_r} (\tilde{x} \cdot \nabla_b u)\bar{u} dS. \tag{133}$$

This is rewritten as

$$\mathrm{Im}\zeta \int_{\Omega_r} |u|^2 dx - \int_{S_r} \mathrm{Im}K|u|^2 dS = -\mathrm{Im}\left[\int_{S_r} \tilde{x}\cdot\theta\bar{u}dS + \int_{\Omega_r} f\bar{u}dx\right]. \tag{134}$$

Note here that $\mathrm{Im}\zeta$ and $-\mathrm{Im}K$ has the same sign in $\Gamma_\pm$ when r is large, say for $r \geq R$.

(i) First we use the inequality

$$|\mathrm{Im}\zeta| \int_{\Omega_r} |u|^2 dx \leq \int_{S_r} |\tilde{x}\cdot\theta||u|dS + \int_{\Omega_r} |f||u|dx. \tag{135}$$

The radiation condition shows

$$\int_r^\infty \mu_0\varphi_0 dt \int_{S_t} |\tilde{x}\cdot\theta||u|dS \leq \|u\|_{\mu_0\sqrt{k_0},\Omega_r'} \|\tilde{x}\cdot\theta\|_{\varphi_0'\sqrt{k_0}^{-1},\Omega_r'} < \infty. \tag{136}$$

Since $\mu_0\varphi_0 \notin L^1([R,\infty))$, this implies

$$\liminf_{t\to\infty} \int_{S_t} |\tilde{x}\cdot\theta||u|dS = 0. \tag{137}$$

Thus, letting $t \to \infty$ in (135), we conclude the assertion.

(ii) Next, we use the inequality

$$\int_{S_r} |\mathrm{Im}K||u|^2 dS \leq \int_{S_r} |\tilde{x}\cdot\theta||\bar{u}|dS + \int_{\Omega_r} |f||\bar{u}|dx \tag{138}$$

which also follows from (134). Multiplying $\mu_0(r)$ and integrating both sides over the half line (R,∞), we obtain

$$\begin{aligned}\int_{\Omega_R'} \mu_0|\mathrm{Im}K||u|^2 dx &\leq \varphi_0(R)^{-1}\left\{\int_{\Omega_R'} \mu_0\varphi_0|\tilde{x}\cdot\theta||u|dx + \int_{\Omega} |f||u|dx\right\}\\ &\leq \varphi_0(R)^{-1}\left\{\|\tilde{x}\cdot\theta\|_{\varphi_0'\sqrt{k_0}^{-1},\Omega_R'}\|u\|_{\mu_0\sqrt{k_0},\Omega_R'} + \|f\|_{(\mu_0\sqrt{k_0})^{-1}}\|u\|_{\mu_0\sqrt{k_0}}\right\}.\end{aligned} \tag{139}$$

Since $|\mathrm{Im}K| = |\mathrm{Re}\sqrt{k}| + O(\sqrt{\mu})$, the equivalence of $|k|$ and k_0 leads to the desired conclusion. □

Lemma 6.3 *Let u be a radiative solution of* (16)*. Then there exists* $C_{16} = C_{16}(\overline{\Gamma}_\pm) > 0$ *such that*

$$\int_{\Omega_{R+1}'} \frac{\varphi_0'}{\sqrt{k_0}}|\theta|^2 dx \leq C_{16}\left\{\|u\|^2_{\mu_0\sqrt{k_0},\Omega_R'} + \|f\|^2_{(\mu_0\sqrt{k_0})^{-1},\Omega_R'}\right\}. \tag{140}$$

Proof To show this lemma we use Proposition 1 with $\sigma = 0$ and $\Phi = \dfrac{\varphi_0\chi}{\sqrt{|k|}}$, where $\chi = \chi(r)$ is smooth and satisfy $\chi(r) = 0$ $(r < R)$ and $= 1$ $((r > R + 1)$:

$$\left[\int_{S_t} - \int_{S_R}\right]\frac{\varphi_0\chi}{\sqrt{|k|}}\left(|\tilde{x}\cdot\theta|^2 - \frac{1}{2}|\theta|^2\right)dS$$
$$= \int_{\Omega_{R,t}}\frac{\varphi_0\chi}{\sqrt{|k|}}\left[\frac{1}{r}(|\theta|^2 - |\tilde{x}\cdot\theta|^2) + \mathrm{Re}\left(-i\sqrt{k} + \frac{\partial_r k}{4k}\right)|\theta|^2\right.$$
$$+\left(\frac{\varphi_0'}{\varphi_0} + \frac{\chi'}{\chi} - \mathrm{Re}\frac{\partial_r k}{2k}\right)\left(|\tilde{x}\cdot\theta|^2 - \frac{1}{2}|\theta|^2\right)$$
$$\left.+\mathrm{Re}\left\{-\left(\frac{\tilde{\nabla}k}{2k}\cdot\theta\right)(\tilde{x}\cdot\overline{\theta}) + \mathcal{B}(u,\theta)\right\} - \mathrm{Re}\{f(\tilde{x}\cdot\overline{\theta})\}\right]dx. \tag{141}$$

Here $\mathrm{Re}(-i\sqrt{k}) > 0$ and by use of $(\mu_0.2)$

$$\frac{1}{r}\{|\theta|^2 - |\tilde{x}\cdot\theta|^2\} + \mathrm{Re}\frac{\partial_r k}{4k}|\theta|^2 + \left(\frac{\varphi_0'}{\varphi_0} - \mathrm{Re}\frac{\partial_r k}{2k}\right)\left(|\tilde{x}\cdot\theta|^2 - \frac{1}{2}|\theta|^2\right)$$
$$= \left(\frac{1}{r} - \frac{\varphi_0'}{\varphi_0} + \mathrm{Re}\frac{\partial_r k}{2k}\right)\{|\theta|^2 - |\tilde{x}\cdot\theta|^2\} + \frac{\varphi_0'}{2\varphi_0}|\theta|^2 \geq \frac{\varphi_0'}{2\varphi_0}|\theta|^2. \tag{142}$$

Thus, it follows from (141) that

$$\int_{S_t}\frac{\chi\varphi_0}{\sqrt{|k|}}\left(|\tilde{x}\cdot\theta|^2 - \frac{1}{2}|\theta|^2\right)dS$$
$$\geq \int_{\Omega_{R,t}}\frac{\chi\varphi_0}{\sqrt{|k|}}\left[\frac{\varphi_0'}{2\varphi_0}|\theta|^2 - O(\mu)|\theta|^2 - O(\mu)|\sqrt{k}||u||\tilde{x}\cdot\theta| - |f||\theta|\right]dx$$
$$+\int_{\Omega_{R,R+1}}\frac{\varphi_0\chi'}{\sqrt{|k|}}\left(|\tilde{x}\cdot\theta|^2 - \frac{1}{2}|\theta|^2\right)dx. \tag{143}$$

Note $(\mu_0.1)$ and the identity $\varphi_0' = \mu_0\varphi_0^2$. Then the Schwarz inequality and letting $t \to \infty$ show the inequality

$$\int_{\Omega'_{R+1}}\frac{\varphi_0'}{\sqrt{|k|}}|\theta|^2 dx \leq C_{17}\left\{\|u\|^2_{\mu_0\sqrt{|k|},\Omega'_R} + \|f\|^2_{(\mu_0\sqrt{|k|})^{-1},\Omega'_R}\right\}. \tag{144}$$

Here Lemma 6.1 is used to verify

$$\liminf_{t\to\infty} \int_{S_t} \frac{\varphi_0}{\sqrt{|k|}} \left(|\tilde{x}\cdot\theta|^2 - \frac{1}{2}|\theta|^2 \right) dS = 0. \tag{145}$$

Hence, the equivalence of $|k|$ and k_0 leads to the assertion. □

With these preparation, we are now ready to prove Theorem 2.3.

Proof of Theorem 2.3 Let $\{\zeta_j, f_j\} \subset \Gamma_\pm \times L^2_{(\mu_0\sqrt{k_0})^{-1}}$ converges to $\{\zeta_0, f_0\}$ as $j \to \infty$. Since the other case is easier, we assume that $\zeta_0 = \lambda \pm i0$, $\lambda \in I$. Let $u_j = R(\zeta_j) f_k$.

(i) As is verified by Theorem 2.2, each u_j satisfies the radiation conditions.

(ii) $\{u_k\}$ is pre-compact in $L^2_{\mu_0\sqrt{k_0}}$ if it is bounded in the same space. In fact, suppose that $\{u_j\}$ is bounded in $L^2_{\mu_0\sqrt{k_0}}(\Omega)$, then by the ellipticity of the equation we can apply Rellich's compactness criterion to show its pre-compactness in $L^2(\Omega_R)$ for any $R > R_0$. On the other hand, Lemmas 6.2 (ii) and 6.3 assert that for any $\epsilon > 0$

$$\sup_j \|u_j\|_{\mu_0\sqrt{k_0},\Omega'_R} < \epsilon \tag{146}$$

if R is chosen sufficiently large.

(iii) If $u_j \to u_0$ in $L^2_{\mu_0\sqrt{k_0}}(\Omega)$ as $j \to \infty$, then u_0 satisfies the radiation conditions with $\zeta = \zeta_0$. In fact, since $\{u_j\}$ is bounded in $L^2_{\mu_0\sqrt{k_0}}(\Omega'_R)$, we see by Lemma 6.3 that

$$\{\theta_j = \nabla_b u_j + \tilde{x}K(x,\zeta_j)u_j\} \tag{147}$$

is also bounded in $L^2_{\varphi'\sqrt{k_0}^{-1}}(\Omega'_R)$. So, $\{\theta_j\}$ has a weekly convergent subsequence in the same space. Denote the limit by w, then it follows that

$$w = \nabla_b u_0 + \tilde{x}K(x,\lambda \pm i0)u_0, \tag{148}$$

and u_0 is concluded to satisfy the radiation conditions.

(iv) The boundedness $\{u_j\}$ is proved by contradiction. In fact, assume that there exists a subsequence, which we also write $\{u_j\}$, such that $\|u_j\|_{\mu_0\sqrt{k_0}} \to \infty$ as $j \to \infty$. Put $v_j = u_j/\|u_j\|_{\mu_0\sqrt{k_0}}$. Then as is explained above, $\{\zeta_j, v_j\}$ has a convergent subsequence, and if we denote the limit by $\{\lambda \pm i0, v_0\}$, then it satisfies the eigenvalue problem (28) and also

$$\|v_0\|_{\mu_0\sqrt{k_0}} = 1,\ \|\tilde{x}\cdot\nabla v_0 + K_\pm v_0\|_{\varphi_0'\sqrt{k_0}^{-1},\Omega'_R} < \infty, \tag{149}$$

where $K_\pm = K(x, \lambda \pm i0)$. The second inequality implies

$$\liminf_{r\to\infty} \int_{S(r)} \sqrt{k_0}^{-1} |\tilde{x} \cdot \nabla v_0 + K_\pm v_0|^2 dS = 0 \tag{150}$$

since $\varphi_0'(r) \notin L^1([R, \infty))$ for any $R > 0$ by Lemma 6.1.

Comparing this with Theorem 2.1, we see that v_0 has a compact support in $x \in \mathbf{R}^n$. Hence, $v_0 \equiv 0$ by the unique continuation property for solutions to (28). But this contradicts to the first equation of (149).

(v) If we apply Theorem 2.1 once more, then $\{u_j\}$ itself is shown to converge. □

7 Examples

In the following examples each $c_0(r)$ verifies $(A0)$ with $\dfrac{c_0'(r)}{\lambda - c_0(r)} = O(r^{-1})$.

Example 7.1 Let us consider the potential (2) with

$$c_0(r) = -\{1 + \psi_0 \sin(\gamma \log r)\} r^\alpha \quad \text{and} \quad c_1(x) = 0, \tag{151}$$

where $0 \le \alpha \le 2$, $\psi_0 > 0$ and $\gamma \ne 0$ real. If $\alpha > 0$, we choose any interval $I \subset \mathbf{R}$ and $R_1 \ge R_0$ sufficiently large. Then it satisfies $(A4)$ if

$$\psi_0 < \frac{2 + \alpha}{\sqrt{(\alpha + 2)^2 + \gamma^2}}. \tag{152}$$

In fact, we have

$$\frac{-c_0'}{\lambda - c_0} = \frac{\alpha + \alpha\psi_0 \sin(\gamma \log r) + \gamma\psi_0 \cos(\gamma \log r)}{r\{1 + \psi_0 \sin(\gamma \log r)\}} \frac{-c_0}{\lambda - c_0}. \tag{153}$$

and the above condition of ψ_0 shows that

$$\frac{\alpha + \alpha\psi_0 \sin(\gamma \log r) + \gamma\psi_0 \cos(\gamma \log r)}{r\{1 + \psi_0 \sin(\gamma \log r)\}} > -\frac{2}{r}, \tag{154}$$

which implies $(A4)$ since $\dfrac{-c_0(r)}{\lambda - c_0(r)} \to 1$ as $r \to \infty$. Note that the case

$$\frac{-c_0'(r)}{\lambda - c_0(r)} \le -\frac{2}{r} \tag{155}$$

may occur if

$$\frac{2+\alpha}{\sqrt{(\alpha+2)^2+\gamma^2}} < \psi_0 < 1. \tag{156}$$

If $\alpha = 0$, then $c_0(r) = -1 - \psi_0 \sin(\gamma \log r)$ is bounded. So, $\psi_0 > 0$ may be any constant. On the other hand, we have to choose $\overline{I} = [\lambda_0, \lambda_1]$ to satisfy $\lambda_0 + 1 - \psi_0 > 0$. If we restrict $\overline{I}$ more to satisfy $\lambda_0 + 1 > \frac{1}{2}\psi_0\sqrt{4+\gamma^2}$, Then we have

$$\begin{aligned} \frac{-c_0'}{\lambda - c_0} + \frac{2}{r} &= \frac{-\psi_0\gamma\cos(\gamma\log r)}{r(\{\lambda_0 + 1 - \psi_0\sin(\gamma\log r)\}} + \frac{2}{r} \\ &\geq \frac{2(\lambda_0+1) - \psi_0\sqrt{4+\gamma^2}}{r\{\lambda_0 + 1 - \psi_0\sin(\gamma\log r)\}} > 0 \end{aligned} \tag{157}$$

showing $(A4)$.

The results are extended to $c_0(r)$ with ψ_0 replaced by $\tilde{\psi}(r)$ satisfying

$$\tilde{\psi}(r) = O(1) \text{ and } \tilde{\psi}'(r),\ \tilde{\psi}''(r) = O(r^{-1}). \tag{158}$$

In this case, $(A4)$ is satisfied if $\alpha > 0$ and

$$\psi_0 < \frac{2+\alpha}{\sqrt{(\alpha + M_\psi + 2)^2 + \gamma^2}}, \tag{159}$$

where

$$\psi_0 = \limsup_{r\to\infty} |\tilde{\psi}(r)| \text{ and } M_{\tilde{\psi}} = \limsup_{r\to\infty} \frac{r\tilde{\psi}'(r)}{\tilde{\psi}(r)}. \tag{160}$$

The corresponding extension $\lambda_0 + 1 > \frac{1}{2}\psi_0\sqrt{(2+M_{\tilde{\psi}})^2+\gamma^2}$ is also possible when $\alpha = 0$

For more general $\psi(x)$, if we put

$$\tilde{\psi}(r) = \frac{1}{|S_r|}\int_{S_r} \psi(x)dS \ \left(|S_r| = \int_{S_r} dS\right), \tag{161}$$

then the part

$$c_2(x) = -\{\psi(x) - \tilde{\psi}(r)\}\sin(\gamma\log r)r^\alpha \tag{162}$$

satisfies the assumptions on non-oscillating long-range potentials under

$$\psi(x) = O(1), \ \ \partial_r^\ell \psi(x) = O(r^{-1}), \ \ \ \partial_r^{\ell-1}\tilde{\nabla}\psi(x) = O(\mu) \qquad (\ell = 1, 2, 3). \tag{163}$$

Example 7.2 Similar results are also obtained to the potential (2) with

$$\tilde{c}(x) = -\left\{1 + \psi(x)\frac{\sin(\gamma r^\nu)}{r^\nu}\right\} r^\alpha, \tag{164}$$

where $0 \le \alpha \le 2$, $0 < \nu \le 1$, $\gamma \ne 0$ and $\psi(x)$ is a real function such that

$$\psi(x) = O(1), \quad \nabla^\ell \psi(x) = O(r^{-1}) \qquad (\ell = 1, 2). \tag{165}$$

We put

$$\psi_0 = \limsup_{r\to\infty} |\psi(x)|. \tag{166}$$

First consider the simpler case $\nu < 1$ and $\psi = \psi(r)$. In this case we choose $\tilde{c}(x) = c_0(r)$ and $c_1(x) = 0$. Then since

$$\frac{-c_0'(r)}{\lambda - c_0(r)} = \left\{\frac{\alpha - \psi(r)\nu\gamma\cos(\gamma r^\nu)}{r} + O(r^{-1-\nu})\right\}\frac{r^\alpha}{\lambda - c_0}, \tag{167}$$

noting $\dfrac{r^\alpha}{\lambda - c_0(r)} \to 1$ $(r \to \infty)$, we see that $(A4)$ holds for any $I \subset \mathbf{R}$ and R_1 large if $\alpha > 0$ and $\psi_0 < \dfrac{2+\alpha}{\nu|\gamma|}$. If $\alpha = 0$, then

$$\frac{-\partial_r c_0(r)}{\lambda - c_0(r)} = \frac{-\psi(r)\nu\gamma\cos(\gamma r^\nu) + O(r^{-\nu})}{r\{\lambda + 1 + O(r^{-\nu})\}}, \tag{168}$$

and $(A4)$ holds for I satisfying $\lambda_0 + 1 > \dfrac{1}{2}\psi_0\nu|\gamma|$.

Next in case $\nu < 1$ and $\psi = \psi(x)$ (non-radial), we need to choose $\tilde{c}(x) = c_0(r) + c_1(x)$ with $c_0(r) = -r^\alpha$ except when $\psi(x) = \tilde{\psi}(r)$+non-oscillating long-range part. In this case the restrictions $\nu \ge \dfrac{1}{2}$ and $\alpha \le 4\nu - 2$ are necessary to verify the first inequality of $(A1)$. Moreover, we have to restrict $\psi_0^2 < \sqrt{\dfrac{32(1-\beta)}{3}}$ when $\alpha = 4\nu - 2$. Under these restrictions the above argument is applied to obtain $(A4)$ with $a = 0$ for each $I \subset \mathbf{R}$ when $\alpha > 0$ and $\psi_0 < \dfrac{2+\alpha}{\nu|\gamma|}$. Also it holds for I satisfying $\lambda_0 + 1 > \dfrac{1}{2}\psi_0\nu|\gamma|$ when $\alpha = 0$.

Lastly consider the case $\nu = 1$. In this case we need to choose $c_0(r) = -r^\alpha$ even though $\psi = \psi(r)$. Inequalities of $(A1)$ is obvious for all $\alpha \le 2$ with the above restriction of ψ_0^2 when $\alpha = 2$. Under these restrictions, $(A4)$ with $a = \gamma^2$ are verified for each $I \subset \mathbf{R}$ if $\alpha > 0$ and $\psi_0 \le \dfrac{2+\alpha}{|\gamma|}$. On the other hand, if $\alpha = 0$, we have

$$\frac{-\partial_r c_1(x)}{\lambda - c_0(r) - a/4} = \frac{\psi\gamma\cos(\gamma r) + O(r^{-1})}{r\{\lambda + 1 - a/4)\}}, \quad a = \gamma^2, \tag{169}$$

and $(A4)$ holds for $I = (\lambda_0, \lambda_1)$ if

$$\lambda_0 + 1 > \frac{1}{4}\gamma^2 + \frac{1}{2}\psi_0|\gamma| \text{ or } \lambda_1 + 1 < \frac{1}{4}\gamma^2 - \frac{1}{2}\psi_0|\gamma|. \tag{170}$$

References

1. Ben-Artzi, M.: A limiting absorption principle for Schrödinger operators with spherically symmetric exploding potentials. Israel J. Math. **40**, 259–274 (1981)
2. Ben-Artzi, M.: Resolvent estimates for a certain class of Schrödinger operators with exploding potentials. J. Differ. Equ. **52**, 327–341 (1984)
3. Ben-Artzi, M.: Unitary equivalence and scattering theory for Stark-like Hamiltonians. J. Math. Phys. **25**, 951–964 (1984)
4. Ben-Artzi, M., Devinatz, A.: Spectral and scattering theory for the adiabatic oscillator and related potentials. J. Math Phys. **20**, 594–607 (1979)
5. Devinatz, A., Rejto, P.: A limiting absorption principle for Schrödinger operators with oscillating potentials I. J. Differ. Equ. **49**, 85–104 (1983)
6. Eidus, D.M.: The principle of limiting amplitude. Uspekhi Math. Nauk **24**, 91–156 (1969) (Russ. Math. Surv. **24**, 97–167 (1969))
7. Ikebe, T., Saito, Y.: Limiting absorption method and absolute continuity for the Schrödinger operator. J. Marth. Kyoto Univ. **7**, 513–542 (1972)
8. Ikebe, T., Uchiyama, J.: On the asymptotic behovior of eigenfunctions of second-order elliptic operators. J. Kyoto Univ. **11**(3), 425–448 (1971)
9. Itakura, K.: Rellich's theorem for spherically symmetric repulsive Hamiltonians. Math. Z. **291**, 1435–1449 (2019)
10. Ito, K., Skibsted, E.: Stationary scattering theory on manifolds, 1 (2016). Preprint
11. Jäger, W., Rejto, P.: Limiting absorption principle for some Schrödinger operators with exploding potentials, I, II. J. Math Anal. Appl. **91**, 192–228 (1983); **95**, 169–194 (1983)
12. Jäger, W., Rejto, P.: On a theorem of Mochizuki and Uchiyama about oscillating long range potentials. Operator Theory and Its Applications (Winnipeg, 1998). Fields Institute Communications, vol. 25, pp. 305–329. American Mathematical Society, Providence (2000)
13. Kato, T.: Growth properties of solutions of the reduced wave equation with a variable coefficient. Comm. Pure Appl. Math. **12**, 403–425 (1959)
14. Lavine, R.B.: Absolute continuity of Hamiltonian operators with repulsive potentials. Proc. Amer. Math. Soc. **22**, 55–60 (1969)
15. Mochizuki, K.: Spectral and scattering theory For second-order elliptic differential operators in an exterior domain. In: Seminar in Differential Operators in an Exterior Domain, Seminar in Differential Equations. Winter and Spring, 102pp. University of Utah, Utah (1972)

16. Mochizuki, K.: Jäger-Rejto approach on growth estimates of generalized eigenfunctions and principle of limiting absorption (Kyoto, 2000). RIMS Kokyuroku No. 1208, 38–51 (2001) (In Japanese). On the spectrum of Schrödinger operators with oscillating long-range potentials. More Progress in Analysis. In: Begehr, H.G.W., Nicolosi, F. (eds.) Proceedings of the 5th ISAAC, Catania, 2005, pp. 533–542. World Science Publishing, River Edge (2009)
17. Mochizuki, K.: Spectral and Scattering Theory for Second-Order Partial Differential Operators. Monographs and Research Notes in Mathematics, 231pp. CRC Press, Boca Raton (2017)
18. Mochizuki, K.: Growth estimates of generalized eigenfunctions and principle of limiting absorption (Kyoto, 2016). RIMS Kokyuroku No. 2045, 85–99 (2017)
19. Mochizuki, K., Uchiyama, J.: On eigenvalues in the continuum of 2-body or many- body Schrödinger operators. Nagoya Math. J. **70**, 125–141 (1978)
20. Mochizuki, K., Uchiyama, J.: Radiation conditions and spectral theory for 2-body Schrödinger operators with "oscillating" long-range potentials I. J. Math. Kyoto Univ. **18**(2), 377–408 (1978)
21. Simon, B.: On positive eigenvalues of one-body Schrödinger operators. Comm Pur Appl. Math. **22**, 531–538 (1967)
22. Uchiyama, J.: Polynomial growth or decay of eigenfunctions of second-order elliptic operators. Publ. Res. Inst. Math. Sci. **23**, 975–1006 (1987)
23. Yamada, O.: Spectral theory of magnetic Schrödinger operators with exploding potentials. J. Math. Kyoto Univ. **30**, 585–623 (1990)

Simple Proof of the Estimate of Solutions to Schrödinger Equations with Linear and Sub-linear Potentials in Modulation Spaces

Keiichi Kato

Abstract In this note, we consider the estimate of solutions to Schrödinger equations with quadratic and sub-quadratic potentials. M. Kobayashi, S. Ito and the author have shown the estimate of solutions to Schrödinger equations with quadratic potentials in modulation spaces $M^{p,p}$ and the estimate with linear and sub-linear potentials in modulation spaces $M^{p,q}$. Here we give a simple proof for the estimate of solutions to Schrödinger equations with linear and sub-linear potentials in modulation spaces $M^{p,q}$.

Keywords Schrödinger equations · Modulation spaces · Adapted a priori estimates · Wave-packet transform

1 Introduction

In this note, we study the estimate of solutions of the initial value problem of the following Schrödinger equations with quadratic potential in modulation spaces

$$\begin{cases} i\partial_t u = -\frac{1}{2}\Delta u + V(t,x)u, & (t,x) \in \mathbb{R} \times \mathbb{R}^n, \\ u(0,x) = u_0(x), & x \in \mathbb{R}^n, \end{cases} \tag{1}$$

where $u : \mathbb{R} \times \mathbb{R}^n \to \mathbb{C}$, $\Delta = \sum_{j=1}^n \frac{\partial^2}{\partial x_j^2}$, $V(t,x)$ is a real valued function and $u_0 \in L^1(\mathbb{R}^n)$. We give a simple proof for the estimate of the solutions of (1) in Modulation spaces.

K. Kato (✉)
Department of Mathematics, Tokyo University of Science, Shinjuku-ku, Tokyo, Japan
e-mail: kato@rs.tus.ac.jp

© Springer Nature Switzerland AG 2020
V. Georgiev et al. (eds.), *Advances in Harmonic Analysis and Partial Differential Equations*, Trends in Mathematics, https://doi.org/10.1007/978-3-030-58215-9_9

Assumption $V(t,x)$ is a real valued function in $C^\infty(\mathbb{R}\times\mathbb{R}^n)$ and for multi-indices α with $|\alpha|\geqslant 1$ or $|\alpha|\geqslant 2$ there exists a positive constant C_α such that

$$\left|\partial_x^\alpha V(t,x)\right| \leqslant C_\alpha \quad \text{for all } (t,x)\in\mathbb{R}\times\mathbb{R}^n. \tag{2}$$

In the context of the modulation spaces, these types of issues were initiated in the works of Bényi-Gröchenig-Okoudjou-Rogers [2], Wang-Hudzik [13] and Wang-Zhao-Guo [14], and these works have been developed by a number of authors using a large variety of methods (see, for example, Bényi-Okoudjou [1], Cordero-Nicola [3], [4], Kobayashi-Sugimoto [9], Miyachi-Nicola-Rivetti-Tabacco-Tomita [10], Tomita [11], Wang-Huang [12]).

Here we use wave packet transform defined in the following. Let $\varphi\in\mathcal{S}(\mathbb{R}^n)\backslash\{0\}$. For $f\in\mathcal{S}'(\mathbb{R}^n)$, Wave packet transform $W_\varphi f$ is defined by

$$W_\varphi f(x,\xi) = (f(\cdot),\varphi(\cdot-x)e^{i\cdot\xi}) = \int_{\mathbb{R}^n}\overline{\varphi(y-x)}f(y)e^{-iy\xi}dy, \tag{3}$$

where $(\cdot,\cdot)$ is the standard L^2-norm in $L^2(\mathbb{R}^n)$. Wave packet transform is firstly introduced by A. Córdoba and C. Fefferman [5]. This kind of transform is found in various situations and is sometimes called short time Fourier transform [6] or FBI transform.

Our idea here is to use time dependent functions as basic wave packet ϕ. To illustrate our idea, we consider the free Schrödinger case $V(t,x)\equiv 0$. Let $\varphi=\varphi(t,x)$ be a time dependent function in $C^1(\mathbb{R};\mathcal{S})$, we have by integration by parts (or the definition of derivative for tempered distribution)

$$\begin{aligned} W_{\varphi(t)}[(i\partial_t+\frac{1}{2}\triangle)u] &= \int\overline{\varphi(y-x)}(i\partial_t+\frac{1}{2}\triangle)u(t,y)e^{-iy\xi}dy \\ &= ((i\partial_t+i\xi\cdot\nabla_x-\frac{1}{2}|\xi|^2))W_{\varphi(t)}u + W_{[(i\partial_t+\frac{1}{2}\triangle)\varphi(t)]}[u]. \end{aligned} \tag{4}$$

If $(i\partial_t+\frac{1}{2}\triangle)\varphi(t)=0$, we have

$$\begin{cases} (i\partial_t+i\xi\cdot\nabla_x-\frac{1}{2}|\xi|^2)W_{\varphi(t)}u(t,x,\xi)=0, \\ W_{\varphi(0)}u(0,x,\xi)=W_{\varphi_0}u_0(x,\xi). \end{cases} \tag{5}$$

Solving (5), we have

$$W_{\varphi(t)}u(t,x,\xi) = e^{-\frac{i}{2}t|\xi|^2}W_{\varphi_0}u_0(x-\xi t,\xi). \tag{6}$$

Definition 1.1 (Modulation Spaces) Let $1\leqslant p,q\leqslant\infty$. f belongs to modulation spaces $M^{p,q}(\mathbb{R}^n)$ if $\|f\|_{M_\varphi^{p,q}} = \|\|W_\varphi f(x,\xi)\|_{L_x^p}\|_{L_\xi^q}<\infty$.

Note, that this space does not depend on φ. Then representation (6) yields:

Proposition 1.1 (Kato et al. [7])

$$\|e^{it\Delta/2}u_0\|_{M^{p,q}_{\varphi(t)}} = \|u_0\|_{M^{p,q}_{\varphi_0}}. \tag{7}$$

This equality gives the following inequality due to Bényi-Gröchenig-Okoudjou-Rogers [1]

$$\|u(t,\cdot)\|_{M^{p,q}_{\varphi_0}} \le C(1+|t|)^{n/2}\|u_0\|_{M^{p,q}_{\varphi_0}} \tag{8}$$

where $1 \leqslant p, q \leqslant \infty$, and the inequalities due to B. Wang-Hudzik [13]

$$\|u(t,\cdot)\|_{M^{p,q}_{\varphi_0}} \le C(1+|t|)^{-n(1/2-1/p)}\|u_0\|_{M^{p',q}_{\varphi_0}} \tag{9}$$

$$\|u(t,\cdot)\|_{M^{p,q}_{\varphi_0}} \le C(1+|t|)^{n(1/2-1/p)}\|u_0\|_{M^{p,q}_{\varphi_0}} \tag{10}$$

where $2 \leqslant p, q \leqslant \infty$ and p' is Hölder conjugate of p.

For the equation with a potential, we put the Taylor expansion

$$\begin{aligned} V(t,y) =& V(t, x-(y-x)) \\ =& V(t,x) + \nabla_x V(t,x)(y-x) \\ &+ \int_0^1 \sum_{|\alpha|=2} \frac{2}{\alpha!}\partial_x^\alpha V(t, x+\theta(y-x))(1-\theta)d\theta (y-x)^\alpha \end{aligned} \tag{11}$$

into

$$W_{\varphi(t)}[Vu] = \int \overline{\varphi(t, y-x)} V(t,y) u(t,y) e^{-iy\xi} dy. \tag{12}$$

Then we have

$$\begin{aligned} &(i\partial_t + i\xi\cdot\nabla_x - i\nabla_x V(t,x)\cdot\nabla_\xi)W_{\varphi(t)}u(t,x,\xi) \\ &\quad + (-\frac{1}{2}|\xi|^2 - \widetilde{V}(t,x))W_{\varphi(t)}u(t,x,\xi) = Ru(t,x,\xi), \\ &W_{\varphi(0)}u(0,x,\xi) = W_{\varphi_0}u_0(x,\xi), \end{aligned} \tag{13}$$

where $\widetilde{V}(t,x) = V(t,x) - \nabla V(t,x)\cdot x$ and

$$Ru(t,x,\xi) = \sum_{|\alpha|=2} \int \overline{\varphi(y-x)} V_\alpha(t,x,y)(y-x)^\alpha u(t,y) e^{-i\xi y} dy \tag{14}$$

with

$$V_\alpha(t,x,y) = \frac{2!}{\alpha!}\int_0^1 \partial^\alpha V(t,x+\theta(y-x))(1-\theta)d\theta. \tag{15}$$

Let $x(s;t,x,\xi)$ and $\xi(s;t,x,\xi)$ be solutions to

$$\begin{cases} \dot{x}(s) &= \xi(s),\ x(t)=x, \\ \dot{\xi}(s) &= -\nabla_x V(s,x(s)),\ \xi(t)=\xi. \end{cases} \tag{16}$$

Then we have

$$\begin{aligned} &W_{\varphi(t)}u(t,x,\xi) = \\ &\quad e^{-i\int_0^t\{\frac{1}{2}|\xi(s;t,x,\xi)|^2+\widetilde{V}(s,x(s;t,x,\xi))\}ds}W_{\varphi_0}u_0(x(0;t,x,\xi),\xi(0;t,x,\xi)) \\ &\qquad -i\int_0^t e^{-i\int_s^t\{\frac{1}{2}|\xi(s_1;t,x,\xi)|^2+\widetilde{V}(s_1,x(s_1;t,x,\xi))\}ds_1} \\ &\qquad\qquad \times Ru(s,x(s;t,x,\xi),\xi(s;t,x,\xi))ds. \end{aligned} \tag{17}$$

$1 \leqslant p \leqslant \infty$, $\varphi_0 \in \mathcal{S}(\mathbb{R}^n)\backslash\{0\}$ and $T>0$. Set $\varphi(t,x) = e^{\frac{1}{2}it\Delta}\varphi_0(x)$. The integral equation (17) yields the following theorem.

Theorem 1.1 (Kato et al. [8]) *Let* $1 \leqslant p \leqslant \infty$. *Assume that* (2) *satisfies for all* $|\alpha| \geqslant 2$. *For* $T>0$ *there exists a positive constant* C_T *such that*

$$\|u(t,\cdot)\|_{M^{p,p}_{\varphi(t,\cdot)}} \leqslant C_T\|u_0\|_{M^{p,p}_{\varphi_0}}, \tag{18}$$

for all $u_0 \in \mathcal{S}(\mathbb{R}^n)$ *and all* $t\in[-T,T]$, *where* $u(t,x)$ *denotes the solution of* (1) *in* $C(\mathbb{R};\mathcal{S}'(\mathbb{R}^n))$ *with* $u(0,x)=u_0(x)$.

Remark 1.1 Cordero-Gröchenig-Nicola-Rodino [3] have shown that for $T>0$ there exists a positive constant C_T such that

$$\|u(t,\cdot)\|_{M^{p,p}_{\varphi_0}} \leqslant C_T\|u_0\|_{M^{p,p}_{\varphi_0}}, \tag{19}$$

for all $u_0\in\mathcal{S}(\mathbb{R}^n)$ and all $t\in[-T,T]$, where $u(t,x)$ denotes the solution of (1) in $C(\mathbb{R};\mathcal{S}'(\mathbb{R}^n))$ with $u(0,x)=u_0(x)$. Theorem 1 yields this result by the same way as the proof of (8) from Proposition 1 in [7].

Here we give a sketch of the proof of Theorem 1. Put $X = x(0;t,x,\xi)$, $\Xi = \xi(0;t,x,\xi)$, then Jacobian of $(X\ \Xi)$ with respect to (x,ξ) equals 1. Taking $L^{p,p}$-norm of the both side of (17) and changing variable as $X = x(0;t,x,\xi)$, $\Xi =$

$\xi(0; t, x, \xi)$, we have

$$\|u(t,\cdot)\|_{M^{p,p}_{\varphi(t,\cdot)}} \leqslant \|u_0\|_{M^{p,p}_{\varphi_0}} + \int_0^t \|R(s, X, \Xi)\|_{L^{p,p}_{(X,\Xi)}} dt. \tag{20}$$

$\|R(s, x, \xi)\|_{L^{p,p}}$ can be estimated by $C\|u\|_{M^{p,p}_{\varphi(s)}}$, which is shown later. Gronwall's inequality completes the proof.

If (2) satisfies for $|\alpha| \geqslant 1$, we have the estimate in modulation spaces $M^{p,q}(\mathbb{R}^n)$ for all p and q with $1 \leqslant p, q \leqslant \infty$.

Theorem 1.2 (Kato et al. [8]) *Let* $1 \leqslant p, q \leqslant \infty$. *Assume that* (2) *satisfies for all* $|\alpha| \geqslant 1$. *For* $T > 0$ *there exists* $C_T > 0$ *such that*

$$\|u(t,\cdot)\|_{M^{p,q}_{\varphi(t,\cdot)}} \leqslant C_T \|u_0\|_{M^{p,q}_{\varphi_0}}, \tag{21}$$

for all $u_0 \in \mathcal{S}(\mathbb{R}^n)$ *and all* $t \in [-T, T]$, *where* $u(t, x)$ *denotes the solution of* (1) *in* $C(\mathbb{R}; \mathcal{S}'(\mathbb{R}^n))$ *with* $u(0, x) = u_0(x)$.

In [8], we have proven Theorem 2 by changing variables as $X = x(0; t, x, \xi)$, $\Xi = \xi(0; t, x, \xi)$, which is not very simple. In the next section, we give a simple proof of Theorem 2.

2 A Simple Proof

In this section, we give a simple proof for Theorem 2 different from the one in [8]. Using Taylor's expansion

$$V(t, y) = V(t, x) + \int_0^1 \nabla V(t, x + \theta(y - x))d\theta(y - x) \tag{22}$$

instead of (11), the initial value problem (1) is transformed to

$$\begin{cases} \left(i\partial_t + i\xi \cdot \nabla_x - \frac{1}{2}|\xi|^2 - V(t, x)\right) W_{\varphi(t)}u(t, x, \xi) = Ru(t, x, \xi), \\ W_{\varphi(0)}u(0, x, \xi) = W_{\varphi_0}u_0(x, \xi), \end{cases} \tag{23}$$

where

$$Ru(t, x, \xi) = \int \overline{\varphi(y - x)} V_1(t, x, y) \cdot (y - x)u(t, y)e^{-i\xi y} dy \tag{24}$$

with

$$V_1(t, x, y) = \int_0^1 \nabla_x V(t, x + \theta(y - x))d\theta. \tag{25}$$

The method of characteristic curves shows that the initial value problem (23) is equivalent to

$$W_{\varphi(t)}u(t,x,\xi) = e^{-\frac{i}{2}s|\xi|^2 - i\int_0^t V(s,x-s\xi)ds} W_{\varphi_0}u_0(x-t\xi,\xi) \\ - i\int_0^t e^{-\frac{i}{2}(t-s)|\xi|^2 - i\int_s^t V(s_1,x-s_1\xi)ds_1} \times Ru(s,x-s\xi,\xi)ds. \tag{26}$$

Taking $\|\cdot\|_{L^{p,q}}$ norm of the above equation, we have

$$\|u\|_{M^{p,q}_{\varphi(t)}} \leqslant \|u_0\|_{M^{p,q}_{\varphi_0}} + \int_0^t \|R(s,x,\xi)\|_{L^{p,q}} ds. \tag{27}$$

For the estimate of R, we use the inversion formula

$$f(x) = \frac{1}{\|\varphi\|_{L^2}} \iint \varphi(x-y) W_\varphi f(y,\xi) e^{ix\xi} dy\, d\xi, \tag{28}$$

from which we have

$$\begin{aligned} &\|R(s,x,\xi)\|_{L^{p,q}} \\ &\leqslant \| \int \overline{\varphi(s,y-x)} V_1(s,x,y)(y-x)u(s,y)e^{-i\xi y} dy \|_{L^{p,q}} \\ &\leqslant C\| \int \overline{\varphi(s,y-x)} V_1(s,x,y)(y-x) \iint \varphi(s,y-z) \\ &\qquad \times W_\varphi u(s,z,\eta) e^{iy\eta} dz\, d\eta e^{-i\xi y} dy \|_{L^{p,q}} \\ &\leqslant C\| \iint \left(\int \overline{\varphi(s,y-x)} V_1(t,x,y)(y-x)\varphi(s,y-z) e^{iy(\eta-\xi)} dy \right) \\ &\qquad \times W_\varphi u(s,z,\eta) dz\, d\eta \|_{L^{p,q}} \end{aligned} \tag{29}$$

Since $V_1(t,x,y)$ is bounded with respect to (t,x,y), we have with the fact that $(1+|\xi|^2)^{-1}(1-\Delta_y)e^{iy\xi} = e^{iy\xi}$,

$$\begin{aligned} &\left| \int \overline{\varphi(s,y-x)} V_1(t,x,y)(y-x)\varphi(s,y-z) e^{iy(\eta-\xi)} dy \right| \\ &\leqslant C(1+|x-z|)^{-n-1}(1+|\eta-\xi|)^{-n-1}, \end{aligned} \tag{30}$$

which implies

$$\|R(s,x,\xi)\|_{L^{p,q}} \leqslant C\|u\|_{M^{p,q}_{\varphi(s)}}. \tag{31}$$

Gronwall's inequality completes the proof.

Remark 2.1 Under the assumption that (2) for all $|\alpha| \geqslant 2$, the assertion of Theorem 2 is not valid in general. The case that $V(t,x) = 1/2|x|^2$ is a counter example. In this case the assertion of Theorem 2 is valid only for $p = q$.

References

1. Bényi, Á, Okoudjou, K.: Local well-posedness of nonlinear dispersive equations on modulation spaces. Bull. Lond. Math. Soc. **41**, 549–558 (2009)
2. Bényi, Á, Gröchenig, K., Okoudjou, K., Rogers, L.G.: Unimodular Fourier multipliers for modulation spaces. J. Funct. Anal. **246**, 366–384 (2007)
3. Cordero, E., Nicola, F.: Metaplectic representation on Wiener amalgam spaces and applications to the Schrödinger equation. J. Funct. Anal. **254**, 506–534 (2008)
4. Cordero, E., Nicola, F.: Strichartz estimates in Wiener amalgam spaces for the Schrödinger equation. Math. Nachr. **281**, 25–41 (2008)
5. Córdoba, A., Fefferman, C.: Wave packets and Fourier integral operators. Comm. Partial Differ. Equ. **3**, 979–1005 (1978)
6. Gröchenig, K.: Foundations of Time-Frequency Analysis. Birkhäuser, Basel (2001)
7. Kato, K., Kobayashi, M., Ito, S.: Representation of Schrödinger operator of a free particle via short time Fourier transform and its applications. Tohoku Math. J. **64**, 223–231 (2012)
8. Kato, K., Kobayashi, M., Ito, S.: Estimate on Mudulation spaces for Schrödinger evolution operators with quadratic and sub-quadratic potentials. J. Funct. Anal. **266**, 733–753 (2014)
9. Kobayashi, M., Sugimoto, M.: The inclusion relation between Sobolev and modulation spaces. J. Funct. Anal. **260**, 3189–3208 (2011)
10. Miyachi, A., Nicola, F., Rivetti, S., Tabacco, A., Tomita, N.: Estimates for unimodular Fourier multipliers on modulation spaces. Proc. Amer. Math. Soc. **137**, 3869–3883 (2009)
11. Tomita, N.: Unimodular Fourier multipliers on modulation spaces $M^{p,q}$ *for* $0 < p < 1$., Harmonic analysis and nonlinear partial differential equations. RIMS Kôkyûroku Bessatsu, vol. B18, pp. 125–131. Research Institute for Mathematical Sciences, Kyoto (2010)
12. Wang, B., Huang, C.: Frequency-uniform decomposition method for the generalized BO, KdV and NLS equations. J. Differ. Equ. **239**, 213–250 (2007)
13. Wang, B., Hudzik, H.: The global Cauchy problem for the NLS and NLKG with small rough data. J. Differ. Equ. **232**, 36–73 (2007)
14. Wang, B., Zhao, L., Guo, B.: Isometric decomposition operators, function spaces $E^{\lambda}_{p,q}$ and applications to nonlinear evolution equations. J. Funct. Anal. **233**, 1–39 (2006)

Remark on Asymptotic Order for the Energy Critical Nonlinear Damped Wave Equation to the Linear Heat Equation via the Strichartz Estimates

Takahisa Inui

Abstract We consider the damped wave equation with the energy critical power type nonlinearity. It is known that the global solution with a finite space-time norm decays to 0 as time goes to infinity. In the present paper, we give the asymptotic order that the solution goes to a solution of the linear heat equation.

Keywords Damped wave equation · Asymptotic behavior · Strichartz estimate

1 Introduction

We consider the following energy critical nonlinear damped wave equation

$$\begin{cases} \partial_t^2 u - \Delta u + \partial_t u = |u|^{\frac{4}{d-2}} u, & (t,x) \in [0,T) \times \mathbb{R}^d, \\ (u(0), \partial_t u(0)) = (u_0, u_1), & x \in \mathbb{R}^d, \end{cases} \tag{1}$$

where $d \geq 3$, (u_0, u_1) is given, and u is an unknown complex-valued function. By the Duhamel formula, the solution is given by

$$u(t,x) = \mathcal{D}(t)(u_0 + u_1) + \partial_t \mathcal{D}(t) u_0 + \int_0^t \mathcal{D}(t-s)(|u(s)|^{\frac{4}{d-2}} u(s)) ds \tag{2}$$

where $\mathcal{D}(t)$ is defined by

$$\mathcal{D}(t) := e^{-\frac{t}{2}} \mathcal{F}^{-1} L(t,\xi) \mathcal{F} \tag{3}$$

T. Inui (✉)
Department of Mathematics, Graduate School of Science, Osaka University, Toyonaka, Osaka, Japan
e-mail: inui@math.sci.osaka-u.ac.jp

© Springer Nature Switzerland AG 2020

V. Georgiev et al. (eds.), *Advances in Harmonic Analysis and Partial Differential Equations*, Trends in Mathematics, https://doi.org/10.1007/978-3-030-58215-9_10

with the Fourier transform $\mathcal{F}$, its inverse $\mathcal{F}^{-1}$, and

$$L(t,\xi) := \begin{cases} \dfrac{\sinh(t\sqrt{1/4 - |\xi|^2})}{\sqrt{1/4 - |\xi|^2}} & \text{if } |\xi| < 1/2, \\ \dfrac{\sin(t\sqrt{|\xi|^2 - 1/4})}{\sqrt{|\xi|^2 - 1/4}} & \text{if } |\xi| > 1/2. \end{cases} \tag{4}$$

The formula of $\mathcal{D}$ was obtained by Matsumura [8].

By the previous works [5] and [6] (see also [10]), it is known that (1) is locally well-posed in the energy space $H^1(\mathbb{R}^d) \times L^2(\mathbb{R}^d)$ and the global solution with a finite space-time norm decays to 0 as time goes to infinity. Namely, we have the following two propositions.

Proposition 1.1 (Local Well-Posedness in $H^1 \times L^2$) *Let $T \in (0,\infty]$. Let $(u_0, u_1) \in H^1(\mathbb{R}^d) \times L^2(\mathbb{R}^d)$ satisfy $\|(u_0,u_1)\|_{H^1\times L^2} \leq A$. Then, there exists $\delta = \delta(A) > 0$ such that if*

$$\|\mathcal{D}(t)(u_0 + u_1) + \partial_t \mathcal{D}(t) u_0\|_{L_{t,x}^{\frac{2(d+1)}{d-2}}([0,T))} \leq \delta, \tag{5}$$

then there exists a solution u to (1) *with $\|u\|_{L_{t,x}^{2(d+1)/(d-2)}([0,T))} \leq 2\delta$.*

Proposition 1.2 *If u is a global solution of* (1) *with $\|u\|_{L_{t,x}^{2(d+1)/(d-2)}([0,\infty))} < \infty$, then u satisfies*

$$\lim_{t\to\infty} \left(\|u(t)\|_{H^1} + \|\partial_t u(t)\|_{L^2}\right) = 0. \tag{6}$$

Remark 1.1 We have

$$\|\mathcal{D}(t)(u_0 + u_1) + \partial_t \mathcal{D}(t) u_0\|_{L_{t,x}^{2(d+1)/(d-2)}([0,\infty))} \lesssim \|u_0\|_{H^1} + \|u_1\|_{L^2} \tag{7}$$

by the Strichartz estimates. Therefore, if $\|u_0\|_{H^1} + \|u_1\|_{L^2}$ is sufficiently small, we can take $T = \infty$ in Proposition 1.1 and obtain $\|u\|_{L_{t,x}^{2(d+1)/(d-2)}([0,\infty))} < \infty$. Thus, the solution u decays from Proposition 1.2. Namely, we have small decaying result.

From the observation of the linear evolution operator $\mathcal{D}(t)$, we expect that the solution decays like solution to the linear heat equation $\partial_t v = \Delta v$. Our aim in this paper is to obtain the asymptotic order from the solution of (1) to a solution of the linear heat equation.

2 Main Result

Let $\mathcal{G}(t) = e^{t\Delta}$ be the solution map of the linear heat equation $\partial_t v = \Delta v$. Then, we have the following result.

Theorem 2.1 *Let* $(u_0, u_1) \in H^1(\mathbb{R}^d) \times L^2(\mathbb{R}^d)$. *Let* u *be a global solution of* (1) *with* $\|u\|_{L_{t,x}^{2(d+1)/(d-2)}([0,\infty))} < \infty$. *Then we have*

$$\lim_{t\to\infty} \{\langle t\rangle^{\alpha} \|u(t) - \mathcal{G}(t)(u_0 + u_1)\|_{H^1}\} = 0, \tag{8}$$

where $\alpha < 1/(2(d+1))$.

Remark 2.1 If $\alpha \leq 0$, the above theorem is trivial since both $\|u(t)\|_{H^1}$ and $\|\mathcal{G}(t)(u_0+u_1)\|_{H^1}$ decay to 0. Our contribution is to give the order $\alpha \in (0, 1/(2(d+1)))$. However, it is not known whether the order $1/(2(d+1))$ is optimal or not.

Remark 2.2 As stated in Remark 1.1, the assumption of Theorem 2.1 is satisfied if the initial data $\|u_0\|_{H^1} + \|u_1\|_{L^2}$ is sufficiently small.

The similar results to Theorem 2.1 were studied by many authors for the subcritical nonlinear damped wave equations. See, for instance, [1, 2, 9] and references therein. They used the L^p-L^q estimate. One of contribution in this paper is to consider the energy critical case and we use the Strichartz estimates in order to obtain the asymptotic order.

The similar results for other equations via the Strichartz estimates were investigated. See [4] for the nonlinear damped Schrödinger equation and [3] for the energy critical nonlinear wave equation with a scaling invariant damping. These results rely on the explicit time decay in front of the nonlinearity like $e^{-\beta t}\mathcal{N}(u)$ or $(1+t)^{-\gamma}\mathcal{N}(u)$. On the other hand, our equation (1) does not have the explicit time decay. We extract a time decay from the linear part.

We collect some notations.

- For the exponent p, we denote the Hölder conjugate of p by p'.
- The bracket $\langle\cdot\rangle$ is Japanese bracket, i.e., $\langle a\rangle := (1+|a|^2)^{1/2}$.
- We use $A \lesssim B$ to denote the estimate $A \leq CB$ with some constant $C > 0$.
- Let $\chi_{\leq 1} \in C_0^{\infty}(\mathbb{R})$ be a cut-off function satisfying $\chi_{\leq 1}(r) = 1$ for $|r| \leqslant 1$ and $\chi_{\leq 1}(r) = 0$ for $|r| \geqslant 2$ and let $\chi_{>1} = 1 - \chi_{\leq 1}$.
- For a function $f : \mathbb{R}^d \to \mathbb{C}$, we define the Fourier transform and the inverse Fourier transform by

$$\begin{aligned} \mathcal{F}[f](\xi) = \hat{f}(\xi) &= (2\pi)^{-d/2} \int_{\mathbb{R}^d} e^{-ix\xi} f(x)\,dx, \\ \mathcal{F}^{-1}[f](x) &= (2\pi)^{-d/2} \int_{\mathbb{R}^d} e^{ix\xi} f(\xi)\,dx. \end{aligned} \tag{9}$$

- For a measurable function $m = m(\xi)$, we denote the Fourier multiplier $m(\nabla)$ by

$$m(\nabla)f(x) = \mathcal{F}^{-1}\left[m(i\xi)\hat{f}(\xi)\right](x). \tag{10}$$

- For $s \in \mathbb{R}$ and $1 \le p \le \infty$, we denote the usual Sobolev space by

$$W^{s,p}(\mathbb{R}^d) := \left\{ f \in \mathcal{S}'(\mathbb{R}^d) : \|f\|_{W^{s,p}} = \|\langle\nabla\rangle^s f\|_{L^p} < \infty \right\}. \tag{11}$$

We write $H^s(\mathbb{R}^d) := W^{s,2}(\mathbb{R}^d)$ for simplicity.
- We define $P_{\le 1} := \mathcal{F}^{-1}\chi_{\le 1}\mathcal{F}$, $P_{>1} := \mathcal{F}^{-1}\chi_{>1}\mathcal{F}$, and

$$P_N = \mathcal{F}^{-1}\left(\chi_{\le 1}\left(\frac{\xi}{N}\right) - \chi_{\le 1}\left(\frac{2\xi}{N}\right)\right)\mathcal{F} \tag{12}$$

for $N \in 2^{\mathbb{Z}}$. For a time interval I and $F : I \times \mathbb{R}^d \to \mathbb{C}$, we set

$$\|F\|_{L^q(I:L^r)(\mathbb{R}^d)} := \left(\int_I \|F(t,\cdot)\|_{L^r(\mathbb{R}^d)}^q \, dt\right)^{1/q} \tag{13}$$

and $L^q_{t,x}(I) := L^q(I : L^q(\mathbb{R}^d))$. We sometimes use L^p_s and L^p_t to uncover time variables s and t.

3 Proof of the Main Result

We set $\|f\|_{S(I)} := \|f\|_{L^{2(d+1)/(d-2)}_{t,x}(I)}$, which appears in Propositions 1.1 and 1.2 and Theorem 2.1. And we use the auxiliary space W whose norm is defined by

$$\|f\|_{W(I)} := \|\langle\nabla\rangle^{1/2} f\|_{L^{r_*}_{t,x}(I)}, \tag{14}$$

where $r_* := 2(d+1)/(d-1)$

Lemma 3.1 (Improved Strichartz Estimates) *We have the following Strichartz estimates*

$$\begin{aligned} \left\|\langle t\rangle^{\alpha}\, \mathcal{D}(t) f\right\|_{W(I)} &\lesssim \|f\|_{L^2}, \\ \left\|\langle t\rangle^{\alpha}\, \partial_t \mathcal{D}(t) f\right\|_{W(I)} &\lesssim \|f\|_{H^1}, \end{aligned} \tag{15}$$

$$\left\| \int_0^t \langle t-s\rangle^{\alpha}\, \mathcal{D}(t-s) F(s) ds \right\|_{W(I)} \lesssim \|F\|_{L^{r_*'}_{t,x}(I)}, \tag{16}$$

where $I = [0, T)$ *is a time interval, the implicit constants are independent of* T*, and* $\alpha < 1/(2(d+1))$*,*

This is an improvement of the Strichartz estimates obtained in [5] (see also [6]).

The condition $\alpha < 1/(2(d+1))$ comes from the above Strichartz estimates. Indeed, we have

$$-\frac{d}{2}\left(\frac{1}{2}-\frac{1}{r_*}\right)+\alpha < -\frac{d-1}{2}\left(\frac{1}{2}-\frac{1}{r_*}\right). \tag{17}$$

if $\alpha < 1/(2(d+1))$. The left hand side denotes the decay order of the heat solutions with the growth order α which comes from $\langle t\rangle^{\alpha}$. The right hand side is the decay order of the wave solutions. This inequality ensures that the above Strichartz estimates with the time weight $\langle t\rangle^{\alpha}$ holds for the space W, whose spatial exponent is r_*. Namely, we extract time decay from the gap between the exponent r_* of wave and the exponent of heat for which the Strichartz estimates hold.

Proof (Lemma 3.1) By (17), the Hardy–Littlewood–Sobolev inequality, and the similar argument in [5], we get the estimates. We note that the exponent of the space W satisfies the wave admissible condition, that is, $-\frac{d-1}{2}(1/2-1/r_*) = 1/r_*$ holds. □

Lemma 3.2 *Let* u *be a global solution of* (1) *with* $\|u\|_{S([0,\infty))} < \infty$*. Then we have* $\|\langle t\rangle^{\alpha} u\|_{W([0,\infty))} < \infty$*.*

Proof By [5, Lemma 4.1], we know $\|\langle t\rangle^{\alpha} u\|_{W(K)} < C_K$ for any compact interval $K \subset [0,\infty)$. Take $0 < \tau < T < \infty$. By Duhamel's formula

$$\begin{aligned} u(t) =& \mathcal{D}(t-\tau)(u(\tau)+\partial_t u(\tau)) + \partial_t \mathcal{D}(t-\tau)u(\tau) \\ &+ \int_{\tau}^{t} \mathcal{D}(t-s)(|u|^{\frac{4}{d-2}}u)(s)ds, \end{aligned} \tag{18}$$

and the Strichartz estimates (15), we obtain

$$\begin{aligned} \left\|\langle t-\tau\rangle^{\alpha} u\right\|_{W([\tau,T))} \lesssim& \|u(\tau)\|_{H^1} + \|\partial_t u(\tau)\|_{L^2} \\ &+ \left\|\langle t-\tau\rangle^{\alpha}\int_{\tau}^{t} \mathcal{D}(t-s)(|u|^{\frac{4}{d-2}}u)(s)ds\right\|_{W([\tau,T))}. \end{aligned} \tag{19}$$

It follows from $\langle t-\tau\rangle^\alpha \lesssim \langle t-s\rangle^\alpha + \langle s-\tau\rangle^\alpha$ for $s\in[\tau,t]$ and the Strichartz estimate (16) that

$$\begin{aligned}
&\left\|\langle t-\tau\rangle^\alpha \int_\tau^t \mathcal{D}(t-s)(|u|^{\frac{4}{d-2}}u)(s)ds\right\|_{W([\tau,T))} \\
&\lesssim \left\|\int_\tau^t \langle t-s\rangle^\alpha \mathcal{D}(t-s)(|u|^{\frac{4}{d-2}}u)(s)ds\right\|_{W([\tau,T))} \\
&\quad + \left\|\int_\tau^t \mathcal{D}(t-s)(\langle\cdot-\tau\rangle^\alpha |u|^{\frac{4}{d-2}}u)(s)ds\right\|_{W([\tau,T))} \qquad (20)\\
&\lesssim \|u\|_{S([\tau,T))}^{\frac{4}{d-2}} \|u\|_{W([\tau,T))} + \|u\|_{S([\tau,T))}^{\frac{4}{d-2}} \left\|\langle t-\tau\rangle^\alpha u\right\|_{W([\tau,T))} \\
&\lesssim \|u\|_{S([\tau,T))}^{\frac{4}{d-2}} \left\|\langle t-\tau\rangle^\alpha u\right\|_{W([\tau,T))}.
\end{aligned}$$

See [5, Proof of Theorem 1.3] for the nonlinear estimate. Thus, we have

$$\begin{aligned}
\left\|\langle t-\tau\rangle^\alpha u\right\|_{W([\tau,T))} &\lesssim \|u(\tau)\|_{H^1} + \|\partial_t u(\tau)\|_{L^2} \\
&\quad + \|u\|_{S([\tau,T))}^{\frac{4}{d-2}} \left\|\langle t-\tau\rangle^\alpha u\right\|_{W([\tau,T))}.
\end{aligned} \qquad (21)$$

Since $\|u\|_{S([0,\infty))} < \infty$, we get

$$\left\|\langle t-\tau\rangle^\alpha u\right\|_{W([\tau,T])} \lesssim \|u(\tau)\|_{H^1} + \|\partial_t u(\tau)\|_{L^2} \qquad (22)$$

for large fixed τ. Thus, $\|\langle t-\tau\rangle^\alpha u\|_{W([\tau,\infty))} \lesssim \|u(\tau)\|_{H^1} + \|\partial_t u(\tau)\|_{L^2}$. It follows that

$$\begin{aligned}
\left\|\langle t\rangle^\alpha u\right\|_{W([0,\infty))} &\le \left\|\langle t\rangle^\alpha u\right\|_{W([0,\tau))} + \left\|\langle t\rangle^\alpha u\right\|_{W([\tau,\infty))} \\
&\le \langle\tau\rangle^\alpha \|u\|_{W([0,\infty))} + (2\tau^2)^{\alpha/2} \left\|\langle t-\tau\rangle^\alpha u\right\|_{W([\tau,\infty))} \qquad (23)\\
&\le C_\tau < \infty,
\end{aligned}$$

where we use $\|u\|_{W([0,\infty))} < \infty$ (see [5, Lemma 4.1]) and $\langle t\rangle \le (2\tau^2)^{1/2}\langle t-\tau\rangle$ for $t\ge\tau$. □

Lemma 3.3 *Let $d\ge 2$, $2\le r,q\le\infty$, and $0<\alpha<1/(2(d+1))$. Assume that (q,r) satisfies*

$$\frac{d}{2}\left(\frac{1}{2}-\frac{1}{r}\right) - \alpha > \frac{1}{q}. \qquad (24)$$

Then, we have

$$\left\| \langle t\rangle^{\alpha} \int_0^t \mathcal{D}(t-s)F(s)ds \right\|_{L_t^\infty H^1} \lesssim \left\| \langle t\rangle \langle \nabla\rangle^{\gamma} F \right\|_{L_t^{q'} L_x^{r'}} \tag{25}$$

where $\gamma := d(1/2 - 1/r) - 1/q$

Proof Let $\mathcal{D}_l := \mathcal{D}P_{\le 1}$ and $\mathcal{D}_h := \mathcal{D}P_{>1}$ denote the low and high frequency part of $\mathcal{D}$, respectively. The low frequency part $\mathcal{D}_l$ behaves like the heat propagator. The high frequency part $\mathcal{D}_h$ behaves like the wave propagator but it has exponential time decay. We divide $\mathcal{D}$ to $\mathcal{D}_l$ and $\mathcal{D}_h$. For the low frequency part, since $\langle t\rangle \le \sqrt{2}\langle t-s\rangle \langle s\rangle$ for $t \ge s$, we have

$$\begin{aligned} &\left\| \langle t\rangle^{\alpha} \int_0^t \mathcal{D}_l(t-s)F(s)ds \right\|_{H^1} \\ &\lesssim \left\| \int_0^t | \langle \nabla\rangle \langle t-s\rangle^{\alpha} \mathcal{D}_l(t-s) \langle s\rangle^{\alpha} F(s)| ds \right\|_{L^2} \\ &\lesssim \left\| \langle t\rangle^{\alpha} F \right\|_{L_t^{q'} L_x^{r'}} \end{aligned} \tag{26}$$

where we used $\frac{d}{2}(1/2 - 1/r) - \alpha > 1/q$, the decay order of $\mathcal{D}_l$, and the Young inequality. See the argument in [5, Lemma 2.3]. For the high frequency part, it is enough to consider the frequency localized operator

$$\mathcal{W}_N^{\pm}(t) := e^{\pm it\sqrt{-\Delta - 1/4}} P_{>1} P_N, \tag{27}$$

where $N \in 2^{\mathbb{N}}$. We have

$$\left\| \int_0^t | \langle t-s\rangle^{\alpha} e^{-\frac{t-s}{2}} \mathcal{W}_N^{\pm}(t-s)F(s)| ds \right\|_{L_t^q L_x^r} \lesssim N^{2\gamma} \left\| P_N F \right\|_{L_t^{q'} L_x^{r'}}. \tag{28}$$

Indeed, this follows from the argument in [5, Lemma 2.6]. Now, we have

$$\begin{aligned} &\left\| \langle t\rangle^{\alpha} \int_0^t e^{-\frac{t-s}{2}} \mathcal{W}_N^{\pm}(t-s)F(s)ds \right\|_{L^2}^2 \\ &= \left\langle \langle t\rangle^{\alpha} \int_0^t e^{-\frac{t-s}{2}} \mathcal{W}_N^{\pm}(t-s)F(s)ds, \langle t\rangle^{\alpha} \int_0^t e^{-\frac{t-\tau}{2}} \mathcal{W}_N^{\pm}(t-\tau)F(\tau)d\tau \right\rangle_{L^2} \\ &= \int_0^t \int_0^s \left\langle \langle t\rangle^{\alpha} e^{-\frac{t-s}{2}} \mathcal{W}_N^{\pm}(t-s)F(s), \langle t\rangle^{\alpha} e^{-\frac{t-\tau}{2}} \mathcal{W}_N^{\pm}(t-\tau)F(\tau) \right\rangle_{L^2} d\tau ds \\ &\quad + \int_0^t \int_0^\tau \left\langle \langle t\rangle^{\alpha} e^{-\frac{t-s}{2}} \mathcal{W}_N^{\pm}(t-s)F(s), \langle t\rangle^{\alpha} e^{-\frac{t-\tau}{2}} \mathcal{W}_N^{\pm}(t-\tau)F(\tau) \right\rangle_{L^2} ds d\tau \\ &=: I + II. \end{aligned} \tag{29}$$

It is enough to estimate I since II is estimated in the same way.

$$|I| \le \int_0^t \int_0^s e^{-(t-s)} \langle t\rangle^{2\alpha} \left\langle |P_N F(s)|, \left| e^{-\frac{s-\tau}{2}} \mathcal{W}_N^{\pm}(s-\tau)F(\tau)\right|\right\rangle_{L^2} d\tau ds$$
$$\lesssim \int_0^t \int_0^s e^{-(t-s)} \left(\frac{\langle t\rangle}{\langle s\rangle}\right)^{2\alpha}$$
$$\times \left\langle \langle s\rangle^{\alpha} |P_N F(s)|, \left| \langle s-\tau\rangle^{\alpha} e^{-\frac{s-\tau}{2}} \mathcal{W}_N^{\pm}(s-\tau) \langle \tau\rangle^{\alpha} F(\tau)\right|\right\rangle_{L^2} d\tau ds \tag{30}$$

where we use $1 \le \sqrt{2}\langle s\rangle^{-1} \langle s-\tau\rangle \langle \tau\rangle$ for $s \ge \tau$. Thus, since $\langle t\rangle \langle s\rangle^{-1} \le \sqrt{2}\langle t-s\rangle$ for $t \ge s$, we get

$$(R.H.S)$$
$$\lesssim \int_0^t \int_0^s e^{-(t-s)} \langle t-s\rangle^{2\alpha}$$
$$\times \left\langle \langle s\rangle^{\alpha} |P_N F(s)|, \left| \langle s-\tau\rangle^{\alpha} e^{-\frac{s-\tau}{2}} \mathcal{W}_N^{\pm}(s-\tau) \langle \tau\rangle^{\alpha} F(\tau)\right|\right\rangle_{L^2} d\tau ds$$
$$\lesssim \int_0^t \left\langle \langle s\rangle^{\alpha} |P_N F(s)|, \int_0^s \left| \langle s-\tau\rangle^{\alpha} e^{-\frac{s-\tau}{2}} \mathcal{W}_N^{\pm}(s-\tau) \langle \tau\rangle^{\alpha} F(\tau)\right| d\tau \right\rangle_{L^2} ds. \tag{31}$$

By the Hölder inequality, it holds that

$$(R.H.S)$$
$$\lesssim \left\| \langle t\rangle^{\alpha} P_N F\right\|_{L_t^{q'} L_x^{r'}} \left\| \int_0^s \left| \langle s-\tau\rangle^{\alpha} e^{-\frac{s-\tau}{2}} \mathcal{W}_N^{\pm}(s-\tau) \langle \tau\rangle^{\alpha} F(\tau)\right| d\tau \right\|_{L_s^q L_x^r}$$
$$\lesssim N^{2\gamma} \left\| \langle t\rangle^{\alpha} P_N F\right\|_{L_t^{q'} L_x^{r'}}^2 \tag{32}$$

by the estimate (28). This implies that

$$\left\| \langle t\rangle^{\alpha} \int_0^t \mathcal{D}_h(t-s)F(s)ds \right\|_{H^1} \lesssim \left\| \langle t\rangle^{\alpha} \langle \nabla\rangle^{\gamma} F\right\|_{L_t^{q'} L_x^{r'}}. \tag{33}$$

Combining this and the estimate of the low frequency, we get the desired estimate.
□

Proof *(Theorem 2.1)* By the Duhamel formula, we have

$$\begin{aligned}
&\langle t\rangle^{\alpha} \|u(t) - \mathcal{G}(t)(u_0 + u_1)\|_{H^1} \\
&\leq \langle t\rangle^{\alpha} \|\mathcal{D}(t)(u_0 + u_1) + \partial_t \mathcal{D}(t)u_1 - \mathcal{G}(t)(u_0 + u_1)\|_{H^1} \\
&\quad + \langle t\rangle^{\alpha} \left\| \int_0^t \mathcal{D}(t-s)(|u|^{\frac{4}{d-2}})u(s)ds \right\|_{H^1} \\
&=: I + II.
\end{aligned} \tag{34}$$

We calculate I.

$$I \leq \langle t\rangle^{\alpha} \|\mathcal{D}(t)(u_0 + u_1) - \mathcal{G}(t)(u_0 + u_1)\|_{H^1} + \langle t\rangle^{\alpha} \|\partial_t \mathcal{D}(t)u_0\|_{H^1} \tag{35}$$

It is known that (see [2] and also [1, 7, 9])

$$\begin{aligned}
&\|\mathcal{D}(t)f - \mathcal{G}(t)f\|_{H^1} \lesssim \langle t\rangle^{-1} \|f\|_{L^2} + e^{-t/4} \|f\|_{L^2}, \\
&\|\partial_t \mathcal{D}(t)f\|_{H^1} \lesssim \langle t\rangle^{-1} \|f\|_{L^2} + e^{-t/4} \|f\|_{H^1}.
\end{aligned} \tag{36}$$

Thus, since $\alpha < 1/(2(d+1)) < 1$, we obtain

$$I \to 0 \text{ as } t \to \infty. \tag{37}$$

Next, we calculate II. Take small $\varepsilon > 0$ such that $\alpha + \varepsilon < 1/(2(d+1))$.

$$\begin{aligned}
&\langle t\rangle^{\alpha} \left\| \int_0^t \mathcal{D}(t-s)(|u|^{\frac{4}{d-2}}u)(s)ds \right\|_{H^1} \\
&= \langle t\rangle^{-\varepsilon} \left\| \langle t\rangle^{\alpha+\varepsilon} \int_0^t \mathcal{D}(t-s)(|u|^{\frac{4}{d-2}}u)(s)ds \right\|_{H^1}.
\end{aligned} \tag{38}$$

By Lemma 3.3, where we note that the exponent of the space W satisfies the condition of Lemma 3.3 and $\gamma = 1/2$, and Lemma 3.2, we get

$$\left\| \langle t\rangle^{\alpha+\varepsilon} \int_0^t \mathcal{D}(t-s)\mathcal{N}(u(s))ds \right\|_{H^1} \lesssim \|u\|_{S([0,\infty))}^{\frac{4}{d-2}} \left\| \langle t\rangle^{\alpha+\varepsilon} u \right\|_{W([0,\infty))} < \infty. \tag{39}$$

Thus, $II \to 0$ as $t \to \infty$. □

Acknowledgments The author would like to express deep appreciation to Professor Satoshi Masaki, Professor Yoshihiro Ueda, and, Professor Yuta Wakasugi for encouragement. Moreover, he also thanks to Professor Vladimir Georgiev, Professor Tohru Ozawa, Professor Michael Ruzhansky, and, Professor Jens Wirth for inviting me the session "Harmonic Analysis and Partial Differential Equations" in 12th ISAAC Congress. The author is supported by JSPS KAKENHI Grant-in-Aid for Early-Career Scientists JP18K13444.

References

1. Hosono, T., Ogawa, T.: Large time behavior and L^p-L^q estimate of solutions of 2-dimensional nonlinear damped wave equations. J. Differ. Equ. **203**(1), 82–118 (2004)
2. Ikeda, M., Inui, T., Okamoto, M., Wakasugi, Y.: L^p-L^q estimates for the damped wave equation and the critical exponent for the nonlinear problem with slowly decaying data. Commun. Pure Appl. Anal. **18**(4), 1967–2008 (2019)
3. Inui, T.: Asymptotic behavior of the energy critical nonlinear wave equation with a special scale invariant damping. Adv. Stud. Pure Math. To appear
4. Inui, T.: Asymptotic behavior of the nonlinear damped Schrödinger equation. Proc. Amer. Math. Soc. **147**(2), 763–773 (2019)
5. Inui, T.: The Strichartz estimates for the damped wave equation and the behavior of solutions for the energy critical nonlinear equation. NoDEA Nonlin. Differ. Equ. Appl. **26**(6), Paper No. 50 (2019)
6. Inui, T., Wakasugi, Y.: Endpoint Strichartz estimate for the damped wave equation and its application (2019). Preprint. arXiv:1903.05891. https://arxiv.org/abs/1903.05891
7. Marcati, P., Nishihara, K.: The L^p-L^q estimates of solutions to one-dimensional damped wave equations and their application to the compressible flow through porous media. J. Differ. Equ. **191**, 445–469 (2003)
8. Matsumura, A.: On the asymptotic behavior of solutions of semi-linear wave equations. Publ. Res. Inst. Math. Sci. **12**(1), 169–189 (1976/77)
9. Narazaki, T., Nishihara, K.: Asymptotic behavior of solutions for the damped wave equation with slowly decaying data. J. Math. Anal. Appl. **338**, 803–819 (2008)
10. Watanabe, T.: Strichartz type estimates for the damped wave equation and their application. RIMS Kôkyûroku Bessatsu **B63**, 77–101 (2017)

On Uniqueness for the Generalized Choquard Equation

Vladimir Georgiev and George Venkov

Abstract We consider the generalized Choquard equation describing trapped electron gas in three dimensional case. The study of orbital stability of the energy minimizers (known as ground states) depends essentially in the local uniqueness of these minimizers. The uniqueness of the minimizers for the case $p = 2$, i.e. for the case of Hartree–Choquard is well known. The main difficulty for the case $p \neq 2$ is connected with possible lack of control on the L^p norm of the minimizers. Our main result treats the local uniqueness of radial positive minimizers for $p \in (5/3, 7/3)$.

Keywords Choquard equation · Ground states · Local uniqueness

1 Main Results

In this work we study the existence, uniqueness and orbital stability of ground states associated with the generalized Choquard equation

$$\begin{aligned} & i\partial_t u + \Delta u + I(|u|^p)|u|^{p-2}u = 0, \quad (t,x) \in \mathbb{R}_+ \times \mathbb{R}^3, \\ & u(0,x) = u_0(x) \end{aligned} \tag{1}$$

V. Georgiev
Dipartimento di Matematica, Università di Pisa, Pisa, Italy

Faculty of Science and Engineering, Waseda University, Shinjuku-ku, Tokyo, Japan

Institute of Mathematics and Informatics at BAS, Sofia, Bulgaria
e-mail: georgiev@dm.unipi.it

G. Venkov (✉)
Faculty of Applied Mathematics and Informatics, Technical University of Sofia, Sofia, Bulgaria
e-mail: gvenkov@tu-sofia.bg

© Springer Nature Switzerland AG 2020
V. Georgiev et al. (eds.), *Advances in Harmonic Analysis and Partial Differential Equations*, Trends in Mathematics, https://doi.org/10.1007/978-3-030-58215-9_11

assuming $5/3 < p < 7/3$. Here and below $I(f)$ is the Riesz potential defined by

$$I(f)(x) = (-\Delta)^{-1} f(x) = \frac{1}{4\pi} \int_{\mathbb{R}^3} \frac{f(y)dy}{|x-y|}. \tag{2}$$

In general the existence of ground state is studied in [4] and [13], while the decay and scattering properties is considered in [12, 15]. A detailed classification result for linearized stability properties of the standing waves is obtained in [5]. Considering linearization of (1) around standing waves, one can apply the classification results from [5, 9] and deduce that linearized orbital stability holds for $p \in (5/3, 7/3)$, while linearized orbital instability is fulfilled for $p \in [7/3, 5)$.

The ground states are defined as solutions to the energy minimization problem

$$\mathcal{E}_\sigma = \inf_{u \in H^1,\ \|u\|_{L^2}^2 = \sigma} E_p(u). \tag{3}$$

Here and below

$$E_p(u) = \frac{1}{2}\|\nabla u\|_{L^2}^2 - \frac{1}{2p} D(|u|^p, |u|^p), \tag{4}$$

where

$$D(|u|^p, |u|^p) = \langle I(|u|^p), |u|^p \rangle_{L^2} = \left\| (-\Delta)^{-1/2} |u|^p \right\|_{L^2}^2. \tag{5}$$

In general the orbital stability property concerns the set

$$\mathfrak{G} = \{Q \in H^1(\mathbb{R}^3);\ E_p(Q) = \mathcal{E}_\sigma,\ \|Q\|_{L^2}^2 = \sigma\}. \tag{6}$$

The fact that the energy functional $E_p(u)$ is invariant under the group of rotations and transform $u \to -u$ implies that one can study the asymptotic properties of solutions to (1) with radial initial data close to the set of radial positive ground states

$$\mathfrak{G}_{rad} = \{Q \in H^1_{rad}(\mathbb{R}^3);\ E_p(Q) = \mathcal{E}_\sigma,\ \|Q\|_{L^2}^2 = \sigma,\ Q > 0\}, \tag{7}$$

where

$$H^1_{rad} = \{u \in H^1(\mathbb{R}^3);\ u(x) = u(|x|)\} \tag{8}$$

The well-known result of Lieb [10] asserts that $\mathfrak{G}_{rad}$ consists of unique element for $p = 2$. Despite of the variety of results on existence, classification and asymptotic behavior of solution to the minimization problem (3), much less is known about the structure (cardinality, discreteness) of the set $\mathfrak{G}_{rad}$ for $p \in (5/3, 7/3) \setminus \{2\}$. Some partial result cover the cases $p \in (2, 2+\delta)$ with $\delta > 0$

small enough (see [17]) or $p \in (2, 7/3)$ (see [7]). In both cases local uniqueness of ground states is established thanks to different kinds of nondegeneracy properties of linearized operators associated with the energy functional.

The main goal of this work is to show that the set $\mathfrak{G}_{rad}$ consists of a finite number of positive radial minimizers for the whole range $p \in (5/3, 7/3)$.

Our main result treats the local uniqueness of minimizers Q of (3).

Theorem 1.1 *Assume* $5/3 < p < 7/3$. *Then one can find* $\varepsilon > 0$ *so that for any two radial positive minimizers* $Q_1, Q_2 \in H^1_{rad}$ *of* (3)*, satisfying*

$$\|Q_1 - Q_2\|_{L^2} \leq \varepsilon, \tag{9}$$

we have $Q_1 = Q_2$.

Any minimizer of (4) satisfies the Euler–Lagrange equation

$$-\Delta u + \omega u = I(|u|^p)|u|^{p-2}u, \tag{10}$$

where $\omega > 0$ is the Lagrange multiplier. Using the rescaling properties of the energy functional and the Euler–Lagrange equation, we can choose $\sigma > 0$ so that the Lagrange multiplier in (10) is $\omega = 1$.

The classical case $p = 2$ has been studied in [10], the approach is based on shooting method and the fact that the Riesz potential behaves like

$$I(|u|^2)(x) = \frac{\|u\|_{L^2}^2}{4\pi|x|} + o\left(|x|^{-1}\right), \quad x \to \infty. \tag{11}$$

Indeed, taking any two solutions u_1, u_2, then from (11) and using the Pohozaev normalization conditions (see for example [7]) one can deduce that

$$I(|u_1|^2)(x) - I(|u_2|^2)(x) = o\left(|x|^{-1}\right), \quad x \to \infty. \tag{12}$$

This gives the possibility to apply Sturm argument and following shooting method to obtain uniqueness. If $p \neq 2$, then (11) becomes

$$I(|u|^p)(x) = \frac{1}{4\pi}\frac{\|u\|_{L^p}^p}{|x|} + o\left(|x|^{-1}\right), \quad x \to \infty \tag{13}$$

and obviously we loose the control on the asymptotics of Riesz potential at infinity, since in this case the L^p norm is not presented in Pohozaev normalization conditions (one can compare with the case of p-Choquard model [6]).

There are different method to prove the uniqueness of positive radial minimizes of nonlinear elliptic equations with local type nonlinearities. The method of McLeod and Serrin [11] and the subsequent refinements due to Kwong [8] are also based on Sturm oscillation argument and therefore they work effectively for local type nonlinearities. In our case the nonlinearities involve the nonlocal Riesz potential and consequently we have met essential difficulties to follow this strategy.

Alternative method to show uniqueness of minimizer for Weinstein functionals have been proposed in [3] for the case of local type nonlinearity by studying

$$\frac{\|u\|_{L^2}^{5-p}\|\nabla u\|_{L^2}^{3p-5}}{D(|u|^p, |u|^p)}. \tag{14}$$

Performing the substitution of u by $Q + \varepsilon h$ and making a Taylor expansion of the above quotient near $\varepsilon = 0$, one can reduce the local existence result to the proof that the operator

$$L_+ = -\Delta + \omega - pI(Q^{p-1}\cdot)Q^{p-1} - (p-1)I(Q^p)Q^{p-2}, \tag{15}$$

has a unique negative eigenvalue and a kernel of dimension not greater than 2. However, the lack of Sturm comparison argument for nonlocal ODE causes essential difficulties to show the non-degeneracy of L_+, i.e. to check that the kernel of L_+ on H^1_{rad} is trivial. Our approach to obtain the local uniqueness of the minimizer might allow degeneracy of L_+, but the local uniqueness is based on the appropriate analytic continuation $K(z)$ of the function

$$K : \varepsilon \to E_p\left(\sqrt{\sigma}\frac{Q+\varepsilon h}{\|Q+\varepsilon h\|_{L^2}}\right), \tag{16}$$

where $h \in H^1_{rad}$ is a nontrivial element in the kernel of L_+. The crucial point is to show the identity $K(z) = K(0)$ for z in the domain of analyticity of $K(z)$ and to find a suitable curve $z = z(R)$, $R > 0$ in this domain so that

$$\lim_{R\to\infty} K(z(R)) = E_p(\sqrt{\sigma}\, h). \tag{17}$$

2 Analytic Extension of the Energy Functional

Our goal is to show the uniqueness of the minimizer Q, associated to the minimization problem

$$\mathcal{E}_\sigma = \inf_{u\in H^1,\ \|u\|_{L^2}^2=\sigma} E_p(u), \tag{18}$$

where E_p is defined in (4).

Let us assume that Q is a positive radial minimizer normalized by $\|Q\|_{L^2}^2 = \sigma$ and take $h \perp Q$ with $\|h\|_{L^2} = 1$. Consider the vector

$$Q(\lambda) = \sqrt{\sigma}\frac{Q+\lambda h}{\|Q+\lambda h\|_{L^2}} = \sqrt{\frac{\sigma}{\sigma+\lambda^2}}(Q+\lambda h) \tag{19}$$

for $\lambda \in [0, 1)$. Our goal is to show that the function

$$K : (\lambda, h) \to E_p\left(Q(\lambda)\right). \tag{20}$$

has no minimum for $\lambda > 0$ small enough.

Let us recall the following definition from [7].

Definition 2.1 If

$$A \subseteq S_Q(H^1_{rad}) = \{g \in H^1_{rad};\ \|g\|_{L^2} = 1, g \perp Q\}, \tag{21}$$

then we shall say that local uniqueness of the minimizer Q holds on A if there exists $\varepsilon_0 = \varepsilon_0(A) > 0$, so that

$$E_p\left(\sqrt{\sigma}\frac{Q+\varepsilon h}{\|Q+\varepsilon h\|_{L^2}}\right) > E_p(Q) \tag{22}$$

is fulfilled for any $h \in A$ and for any $\varepsilon \in (0, \varepsilon_0]$.

It is not difficult to restrict the proof of the local uniqueness to the subspace $\mathrm{Ker}L_+$, where

$$L_+h = -\Delta h + h - pQ^{p-1}(-\Delta)^{-1}(Q^{p-1}h) - (p-1)Q^{p-2}h(-\Delta)^{-1}(Q^p). \tag{23}$$

Lemma 2.1 *If A is a closed subset in*[1] $S_Q(H^1_{rad})$, *such that* $A \cap \mathrm{Ker}L_+ = \emptyset$, *then the local uniqueness of the minimizer Q holds on A.*

The proof can be found in the Appendix.

The local uniqueness on $\mathrm{Ker}L_+$ needs appropriate analytic extension of the energy function K. First we deal with the nonlocal term in the energy functional.

Lemma 2.2 *If Q is a positive radial minimizer of E_p, and*

$$h \in S_Q(H^1_{rad}) \cap \mathrm{Ker}L_+, \tag{24}$$

[1] Recall that $S_Q(H^1_{rad})$ is defined in (21).

then

- *there exists a constant* $C > 0$, *so that*

$$\frac{|h(r)|}{Q(r)} \leq C; \tag{25}$$

- *for any* $\delta \in (0, 1/C)$ *the function*

$$z \to D((Q + zh)^p, (Q + zh)^p) \tag{26}$$

is a complex analytic function in the complement $\Sigma_\delta^c = \mathbb{C} \setminus \Sigma_\delta$ *of the set*

$$\Sigma_\delta = \{z \in \mathbb{C}; d(z, \Sigma) \leq \delta\}, \tag{27}$$

where

$$\Sigma = \{z \in \mathbb{C}; \mathrm{Im} z = 0, |\mathrm{Re} z| > 1/C\}, \tag{28}$$

and $d(z_1, z_2) = |z_1 - z_2|$ *is the standard distance in the complex plane.*

Proof First, we check (25). Since $h \perp Q$ satisfies

$$\begin{aligned} (1 - \Delta)h &= pBQ^{p-1} + (p-1)AQ^{p-2}h, \\ -\Delta B &= Q^{p-1}h, \end{aligned} \tag{29}$$

we can use the argument from [7] and we can find $r_* > 0$, so that $h(r) \neq 0$ for $r > r_*$ and the following estimate holds

$$|h(r)| \lesssim \frac{e^{-r}}{r}, \tag{30}$$

for $r > r_*$.

The positive radial ground state Q satisfies the system

$$\begin{aligned} (1 - \Delta)Q &= AQ^{p-1}, \\ -\Delta A &= Q^p. \end{aligned} \tag{31}$$

By using the arguments in [1, 2, 7, 14], we have the following asymptotic expansions of Q and A as $r \to \infty$

$$Q(r) = \frac{e^{-r}}{r}\left(c_0 + O\left(\frac{1}{r}\right)\right), \tag{32}$$

with $c_0 > 0$. Thus, from (30) and (32) we get (25).

Next we turn to the proof of the analyticity of $D((Q+zh)^p, (Q+zh)^p)$, where

$$D(f, g) = \iint_{\mathbb{R}^3 \times \mathbb{R}^3} \frac{f(x)g(y)}{|x-y|} dxdy. \tag{33}$$

Obviously, we have to find domain in the complex plane where $\mathrm{Arg}(1+zh(r)/Q(r))$ is analytic. Set

$$w(z, r) = 1 + z\frac{h(r)}{Q(r)}. \tag{34}$$

The cut of the Arg—function is determined by

$$\mathrm{Im} w = 0, \ \mathrm{Re} w < 0. \tag{35}$$

Therefore,

$$\mathrm{Arg}(w(z, r)) = \mathrm{Arg}\left(1 + z\frac{h(r)}{Q(r)}\right) \tag{36}$$

is analytic function in domain without intersection with

$$\{z \in \mathbb{C}; \mathrm{Im} w = 0, \mathrm{Re} w < 0\} \tag{37}$$

This forbidden set is included in the set Σ defined in (28). Therefore, for any $\delta \in (0, 1)$ the function

$$\mathrm{Arg}(w(z, r)) = \mathrm{Arg}\left(1 + z\frac{h(r)}{Q(r)}\right) \tag{38}$$

is analytic in Σ_δ^c defined in (27). Since this domain is independent of r we can assert that

$$\begin{aligned} D((Q+zh)^p, (Q+zh)^p) &= \iint_{\mathbb{R}^3 \times \mathbb{R}^3} \frac{(Q+zh)^p(x)(Q+zh)^p(y)}{|x-y|} dxdy \\ &= \iint_{\mathbb{R}^3 \times \mathbb{R}^3} \left(1 + z\frac{h(x)}{Q(x)}\right)^p \left(1 + z\frac{h(y)}{Q(y)}\right)^p \frac{Q(x)^p Q(y)^p}{|x-y|} dxdy \end{aligned} \tag{39}$$

is also analytic in this domain. This completes the proof. □

We can make the following simple observation. For any real B and for any $\delta \in (0, \sqrt{\sigma})$ the function

$$z \to \left(\|Q\|_{L^2}^2 + z^2 \|h\|_{L^2}^2\right)^{B/2} = \left(\sigma + z^2\right)^{B/2} \tag{40}$$

is analytic in the complement $\Lambda_\delta^c = \mathbb{C} \setminus \Lambda_\delta$ of the set

$$\Lambda_\delta = \{z \in \mathbb{C}; d(z, \Lambda) \leq \delta\}, \tag{41}$$

where

$$\Lambda = \{z \in \mathbb{C}; \mathrm{Re} z = 0, |\mathrm{Im} z| > \sqrt{\sigma}\}, \tag{42}$$

Now we are in position to unify previous results and verify the analyticity of

$$\begin{aligned} K(z,h) &= E_p\left(\sqrt{\sigma}\frac{Q+zh}{\|Q+zh\|_{L^2}}\right) \\ &= \frac{\sigma(\|\nabla Q\|_{L^2}^2 + z^2\|\nabla h\|_{L^2}^2)}{2(\sigma+z^2)} - \frac{\sigma^p D(|Q+zh|^p, |Q+zh|^p)}{2p(\sigma+z^2)^p}. \end{aligned} \tag{43}$$

Corollary 2.1 *For any $\delta > 0$ satisfying*

$$\delta < \min\left(\frac{1}{C}, \sqrt{\sigma}\right), \tag{44}$$

the function

$$K : z \to \frac{\sigma(\|\nabla Q\|_{L^2}^2 + z^2\|\nabla h\|_{L^2}^2)}{2(\sigma+z^2)} - -\frac{\sigma^p D(|Q+zh|^p, |Q+zh|^p)}{2p(\sigma+z^2)^p} \tag{45}$$

is analytic in the region

$$\Sigma_\delta^c \cap \Lambda_\delta^c. \tag{46}$$

(see Fig. 1).

3 Proof of the Main Result

We now give the proof of Theorem 1.1. We take Q to be positive, radial minimizer associated to the problem

$$\mathcal{E}_\sigma = \inf_{u\in H^1, \|u\|_{L^2}^2=\sigma} E_p(u). \tag{47}$$

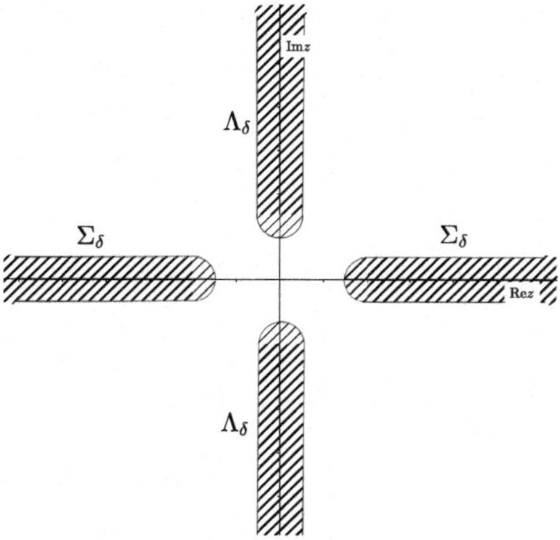

Fig. 1 The domain of analyticity of $K(z)$, here depicted by the complement of the shaded region $\Sigma_\delta \cup \Lambda_\delta$

Using the Definition 2.1 and Lemma 2.1 we have reduced the proof to the local uniqueness on small neighborhood of $\mathrm{Ker}L_+$ in $S_Q(H^1_{rad})$. For any $h \in \mathrm{Ker}L_+ \cap S_Q(H^1_{rad})$ we define $Q(\lambda)$ as in (19).

If

$$h \in \mathrm{Ker}L_+, \tag{48}$$

then Lemma 4.1 implies that h is a linear combination of at most two vectors e_1, e_2 forming the basis of the kernel of L_+. Therefore, there exists $\theta_0 \in [0, 2\pi)$, so that

$$h = \cos(\theta_0)e_1 + \sin(\theta_0)e_2 \tag{49}$$

and for a fixed θ_0 we can use the Taylor expansion of the function

$$K : \lambda \to E_p\left(Q(\lambda)\right) \tag{50}$$

near $\lambda = 0$

$$K(\lambda) = \sum_{\ell=0}^{\infty} \frac{\varepsilon^\ell}{\ell!} \left(\frac{d}{d\lambda}\right)^\ell K\Big|_{\lambda=0}. \tag{51}$$

Since

$$K(0) = E_p(Q) \tag{52}$$

and Q is a minimizer, we have two possibilities:

Case (A) there exists even ℓ such that

$$\left(\frac{d}{d\lambda}\right)^{\ell} K\Big|_{\lambda=0} > 0; \tag{53}$$

Case (B) all derivatives of order $\ell \geq 1$ in the Taylor expansion are identically zero.

In the Case (A) we have local uniqueness at h. Actually, the function $K(\lambda)$ depends on the parameter θ_0 and considering

$$h(\theta) = \cos(\theta)e_1 + \sin(\theta)e_2, \tag{54}$$

we have the function

$$K(\lambda, \theta) = E_p\left(\sqrt{\frac{\sigma}{\sigma + \lambda^2}}(Q + \lambda h(\theta))\right). \tag{55}$$

defined in a small neighborhood of $\lambda = 0, \theta = \theta_0$. In the Case (A) we can use the continuous dependence of

$$\left(\frac{d}{d\lambda}\right)^{\ell} K(\lambda, \theta)\Big|_{\lambda=0} \tag{56}$$

in θ close to θ_0 and see that we have local uniqueness in a small neighborhood of $\mathrm{Ker}L_+$ in $S_Q(H^1_{rad})$.

The proof will be completed if we can show that Case (B) is impossible, i.e. we assume that if $h \in \mathrm{Ker}L_+ \cap S_Q(H^1_{rad})$ is such that

$$\sqrt{\frac{\sigma}{\sigma + \lambda^2}}(Q + \lambda h(\theta)) \tag{57}$$

is a second minimizer and all derivatives of $K(\lambda)$ are identically zero, then we shall arrive at a contradiction.

The function K has an analytic extension in the region $\Sigma^c_\delta \cap \Lambda^c_\delta$, due to Corollary 2.1. Therefore, in the Case (B) we have immediately

$$K(z) = K(0). \tag{58}$$

Choosing $z(R) = R + iR$, it is obvious that the line $z(R)$ belongs to the complement of the shaded region of Fig. 1. Now we take $R \to \infty$, use the relations

$$\frac{z(R)}{\sqrt{\sigma + z(R)^2}} \to \frac{1+i}{\sqrt{2i}} = \frac{1+i}{1+i} = 1, \tag{59}$$

$$\left(\frac{1 + z(R)h(|x|)/Q(|x|)}{\sqrt{\sigma + z(R)^2}}\right)^p \to \left(\frac{|h(|x|)|}{Q(|x|)}\right)^p e^{\mathrm{i}\pi p(1-\mathrm{sgn}h)/2}, \tag{60}$$

and hence

$$\begin{aligned} 2\mathrm{Re}&\frac{D((Q+zh)^p,(Q+zh)^p)}{(\sigma + zR^2)^p} \\ &= \int_{h(x)>0}\int_{h(y)>0} \mathrm{Re}\frac{(Q+zh)^p(x)(Q+zh)^p(y)}{|x-y|(\sigma+zR^2)^p}dxdy \\ &\quad + 2\int_{h(x)<0}\int_{h(y)>0} \mathrm{Re}\frac{(Q+zh)^p(x)(Q+zh)^p(y)}{|x-y|(\sigma+zR^2)^p}dxdy \\ &\quad + \int_{h(x)<0}\int_{h(y)<0} \mathrm{Re}\frac{(Q+zh)^p(x)(Q+zh)^p(y)}{|x-y|(\sigma+zR^2)^p}dxdy \\ &\to \int_{h(x)>0}\int_{h(y)>0} \frac{|h(x)|^p|h(y)|^p}{|x-y|}dxdy \\ &\quad + 2\int_{h(x)>0}\int_{h(y)<0} \frac{|h(x)|^p|h(y)|^p}{|x-y|}\cos(p\pi)dxdy \\ &\quad + \int_{h(x)<0}\int_{h(y)<0} \frac{|h(x)|^p|h(y)|^p}{|x-y|}\cos^2(p\pi)dxdy. \end{aligned} \tag{61}$$

Therefore, we see that

$$\begin{aligned} K(0) =&\mathrm{Re}K(0) = \lim_{R\to\infty} \mathrm{Re}K(z(R)) \\ =&\frac{\sigma(\|\nabla h\|_{L^2}^2)}{2} - \frac{\sigma^p}{2p}\int_{h(x)>0}\int_{h(y)>0} \frac{|h(x)|^p|h(y)|^p}{|x-y|}dxdy \\ &- \frac{2\sigma^p}{2p}\int_{h(x)>0}\int_{h(y)<0} \frac{|h(x)|^p|h(y)|^p}{|x-y|}\cos(p\pi)dxdy \\ &- \frac{\sigma^p}{2p}\int_{h(x)<0}\int_{h(y)<0} \frac{|h(x)|^p|h(y)|^p}{|x-y|}\cos(p\pi)dxdy. \end{aligned} \tag{62}$$

Now we take into account the simple estimate

$$\begin{aligned}
&\frac{\sigma(\|\nabla h\|_{L^2}^2)}{2} - \frac{\sigma^p}{2p}\int_{h(x)>0}\int_{h(y)>0}\frac{|h(x)|^p|h(y)|^p}{|x-y|}dxdy \\
&- \frac{2\sigma^p}{2p}\int_{h(x)>0}\int_{h(y)<0}\frac{|h(x)|^p|h(y)|^p}{|x-y|}\cos(p\pi)dxdy \\
&- \frac{\sigma^p}{2p}\int_{h(x)<0}\int_{h(y)<0}\frac{|h(x)|^p|h(y)|^p}{|x-y|}\cos(p\pi)dxdy \\
&> E_p(\sqrt{\sigma}h)
\end{aligned} \tag{63}$$

fulfilled, when the set $\{h(x) < 0\}$ has nonzero Lebesgue measure in view of the fact that $\cos(p\pi) < 1$ for $p \neq 2$, $p \in (5/3, 7/3)$. The last observation easily implies that

$$E_p(Q) > E_p(\sqrt{\sigma}h) \tag{64}$$

and this contradicts the fact that Q is a minimizer of $E_p(u)$ under the constraint condition $\|u\|_{L^2}^2 = \sigma$. Therefore, the unique remaining possibility is that the set $\{h(x) < 0\}$ has zero Lebesgue measure and hence $h(x) \geq 0$ almost everywhere. This contradicts the orthogonality condition $h \perp Q$ and completes the proof that the Case B) is not possible.

Therefore, the proof of the Theorem is completed.

4 Simple ODE Lemmas

In this section we shall discuss the dimension of the kernel of L_+.

Lemma 4.1 *If $5/3 < p < 7/3$, then we have*

$$\dim(\mathrm{Ker} L_+) \leq 2. \tag{65}$$

Proof Any positive radial solution w to the equation $L_+ w = 0$ is a solution of the ordinary differential equation

$$-r^{-2}\partial_r(r^2\partial_r w(r)) + \omega w = pI(Q^{p-1}w)Q^{p-1} + (p-1)I(Q^p)Q^{p-2}w. \tag{66}$$

Then the couple of w and $B = I(Q^{p-1}w)$ satisfies the system of nonlinear second order differential equations

$$\begin{aligned}
&w''(r) + \frac{2}{r}w'(r) = \omega w(r) - pBQ^{p-1} - (p-1)I(Q^p)Q^{p-2}w, \\
&B''(r) + \frac{2}{r}B'(r) = -Q^{p-1}w.
\end{aligned} \tag{67}$$

subject to initial data

$$
\begin{aligned}
w(0) &= w_0 \neq 0, \quad B(0) = B_0 \neq 0, \\
w'(0) &= 0, \quad B'(0) = 0.
\end{aligned} \tag{68}
$$

Making the substitution

$$
w = \frac{\tilde{w}}{r}, \quad B = \frac{\tilde{B}}{r} \tag{69}
$$

we find the standard Cauchy problem

$$
\begin{aligned}
\tilde{w}''(r) &= \omega \tilde{w}(r) - p\tilde{B} Q^{p-1} - (p-1) I(Q^p) Q^{p-2} \tilde{w}, \\
\tilde{B}''(r) &= -Q^{p-1} \tilde{w}.
\end{aligned} \tag{70}
$$

subject to initial data

$$
\begin{aligned}
\tilde{w}(0) &= 0, \quad \tilde{B}(0) = 0, \\
\tilde{w}'(0) &= w_0 \neq 0, \quad \tilde{B}'(0) = B_0 \neq 0.
\end{aligned} \tag{71}
$$

The linear space of the solutions obviously has dimension at most 2. This completes the proof of the Lemma. □

Appendix: Proof of Lemma 2.1

The Taylor expansion for ε close to 0 gives

$$
K(\varepsilon, h) = K(0, h) + \partial_\varepsilon K(0, h)\varepsilon + \frac{1}{2}\partial_\varepsilon^2 K(0, h)\varepsilon^2 + o(\varepsilon^2), \tag{72}
$$

where

$$
\partial_\varepsilon K(0, h) = \langle Q, L_- h \rangle_{L^2} \tag{73}
$$

and

$$
L_- h = -\Delta h + h - (-\Delta)^{-1}(Q^p) Q^{p-2} h. \tag{74}
$$

The fact that Q is minimizer satisfying the corresponding Euler–Lagrange equation (10) (with $\omega = 1$) implies $L_- Q = 0$, so we have

$$\partial_\varepsilon K(0, h) = \langle Q, L_- h \rangle_{L^2} = 0. \tag{75}$$

We have also

$$\partial_\varepsilon^2 K(0, h) = \langle L_+ h, h \rangle_{L^2}, \tag{76}$$

where

$$L_+ h = -\Delta h + h - pQ^{p-1}(-\Delta)^{-1}(Q^{p-1}h) - (p-1)Q^{p-2}h(-\Delta)^{-1}(Q^p). \tag{77}$$

In general, the operator L_+ satisfies (see (3.9) in [5])

$$\langle L_+ h, h \rangle_{L^2} \geq 0, \quad h \perp Q. \tag{78}$$

Moreover, one can verify the property

$$\{h \in H^1_{rad};\, h \perp Q, \langle L_+ h, h \rangle_{L^2} = 0\} = \{h \in H^1_{rad};\, h \perp Q, L_+ h = 0\}, \tag{79}$$

that follows from the stronger coerciveness property

$$\inf_{h \in H^1_{rad}, \|h\|_{L^2} = 1, h \perp Q, h \perp \mathrm{Ker}\ L_+} \langle L_+ h, h \rangle_{L^2} \geq C > 0. \tag{80}$$

To check this we follow [3], [7] and [16]. More precisely, we assume

$$\inf_{h \in H^1_{rad}, \|h\|_{L^2} = 1, h \perp Q, h \perp \mathrm{Ker}\ L_+} \langle L_+ h, h \rangle_{L^2} = 0 \tag{81}$$

and equip the space

$$\mathrm{Ker} L_+ \cap H^1_{rad} \cap \{Q\}^\perp = \{h \in H^1_{rad};\, h \perp Q, L_+(h) = 0\} \tag{82}$$

with orthonormal basis. Since this space has maximal dimension $k \leq 2$ (see Lemma 4.1), we consider only the case $k = 2$, since the cases $k = 1$ or $k = 0$ are similar. For this, let e_1, e_2 form orthonormal basis of the space (82). The minimization problem (81) has a minimization sequence $\{h_k\}_{k \in \mathbb{N}}$, satisfying all constraints. On the other hand, we have $L_+ = -\Delta + 1 + \mathfrak{K}$, where the operator $\mathfrak{K}$ is represented by

$$\mathfrak{K}(h) = -pQ^{p-1}(-\Delta)^{-1}(Q^{p-1}h) - (p-1)Q^{p-2}h(-\Delta)^{-1}(Q^p). \tag{83}$$

One can easily show that it is a compact operator in H^1_{rad}. In conclusion, taking a subsequence of $\{h_k\}$ we prove its convergence in H^1_{rad} to some $h^* \in H^1_{rad}$ satisfying $\|h^*\|_{L^2} = 1$, $h^* \perp Q$, $h^* \perp \mathrm{Ker} L_+$ and

$$L_+ h^* = \lambda Q + \lambda_1 e_1 + \lambda_2 e_2. \tag{84}$$

Multiplying by e_1 and e_2, we get $\lambda_1 = \lambda_2 = 0$. Hence, we have

$$L_+ h^* = \lambda Q. \tag{85}$$

Next, we have to show that $\lambda = 0$ and this step uses the crucial relation established in [7] (see also [16]).

Lemma 4.2 *If $S = x \cdot \nabla_x$ is the scaling operator in $\mathbb{R}^3$, then*

$$L_+ (SQ) = -2Q. \tag{86}$$

Then we can write

$$\big\langle L_+ h^*, (SQ)\big\rangle_{L^2} = \big\langle h^*, L_+ (SQ)\big\rangle_{L^2} = -2\langle h^*, Q\rangle_{L^2} = 0. \tag{87}$$

Multiplying Eq. (85) by SQ, we arrive at

$$0 = \lambda \langle Q, SQ\rangle_{L^2} = -\lambda \frac{3}{2} \|Q\|^2_{L^2}. \tag{88}$$

Thus we see that $\lambda = 0$, so $h^* \in \mathrm{Ker} L^+$ and this contradicts the properties $h^* \perp \mathrm{Ker} L^+$ and $\|h^*\|_{L^2} = 1$. Therefore, we have the coercive estimate

$$\langle L_+ h, h\rangle_{L^2} \geq C \|h\|^2_{L^2},\ \forall\, h \in H^1_{rad},\ h \perp Q,\ h \perp \mathrm{Ker} L_+. \tag{89}$$

Acknowledgments The first author was supported in part by INDAM, GNAMPA—Gruppo Nazionale per l'Analisi Matematica, la Probabilità e le loro Applicazioni, by Institute of Mathematics and Informatics, Bulgarian Academy of Sciences, by Top Global University Project, Waseda University, the Project PRA 2018 49 of University of Pisa and project "Dinamica di equazioni nonlineari dispersive", "Fondazione di Sardegna", 2016.

The second author was supported in part by Research and Development Sector of the Technical University of Sofia.

References

1. Berestycki, H., Lions, P.-L.: Nonlinear scalar field equations. I. Existence of a ground state. Arch. Rational Mech. Anal. **82**(4), 313–345 (1983)
2. Berestycki, H., Lions, P.-L.: Nonlinear scalar field equations. II. Exis-tence of infinitely many solutions. Arch. Rational Mech. Anal. **82**(4), 347–375 (1983)

3. Chang, S.M., Gustafson, S., Nakanishi, K., Tsai, T.P.: Spectra of linearized operators for NLS solitary waves. SIAM J. Math. Anal. **39**(4), 1070–1111 (2007)
4. Genev, H., Venkov, G.: Soliton and blow-up solutions to the time-dependent Schrödinger-Hartree equation. Discrete Contin. Dyn. Syst. Ser. S **5**(5), 903–923 (2012)
5. Georgiev, V., Stefanov, A.: On the classification of the spectrally stable standing waves of the Hartree problem. Physica D **370**, 29–39 (2018)
6. Georgiev, V., Tarulli, M., Venkov, G.: Existence and uniqueness of ground states for p-Choquard model. Nonlin. Anal. **179**, 131–145 (2019)
7. Georgiev, V., Tarulli, M., Venkov, G.: Orbital stability of solitary waves for the generalized Choquard model (2019), e-print. arXiv:1908.08106
8. Kwong, M.K.: Uniqueness of positive solutions of $\Delta u - u + u^p = 0$. Arch. Rational Mech. Anal. **105**(3), 243–266 (1989)
9. Lenzmann, E.: Uniqueness of ground states for pseudorelativistic Hartree equations. Anal. PDE **2**(1), 1–27 (2009)
10. Lieb, E.H.: Existence and uniqueness of the minimizing solution of Choquard's nonlinear equation. Stud. Appl. Math. **57**(2), 93–105 (1976/1977)
11. McLeod, K., Serrin, J.: Uniqueness of positive radial solutions of $\Delta u + f(u) = 0$ in $\mathbb{R}^N$. Arch. Ration. Mech. Anal. **99**(2), 115–145 (1987)
12. Moroz, V., Van Schaftingen, J.: Groundstates of nonlinear Choquard equations: existence, qualitative properties and decay asymptotics. J. Funct. Anal. **265**(2), 153–184 (2013)
13. Moroz, V., Van Schaftingen, J.: Existence of ground states for a class of nonlinear Choquard equations. Trans. Amer. Math. Soc. **367**(9), 6557–6579 (2015)
14. Strauss, W.: Existence of solitary waves in higher dimension. Comm. Math. Phys. **55**, 149–162 (1977)
15. Tarulli, M., Venkov, G.: Decay and Scattering in energy space for the solution of weakly coupled Choquard and Hartree-Fock equations (2019), e-print. arXiv:1904.10364
16. Weinstein, M.: Modulational stability of ground states of nonlinear Schrodinger equations. SIAM J. Math. Anal. **16**(3), 472–491 (1985)
17. Xiang, Ch.-L.: Uniqueness and nondegeneracy of ground states for Choquard equations in three dimensions. Calc. Var. Partial Differ. Equ. **55**(6), 134 (2016)

Characterization of the Ground State to the Intercritical NLS with a Linear Potential by the Virial Functional

Masaru Hamano and Masahiro Ikeda

Abstract This paper is a sequel to the paper (Hamano and Ikeda, Proc. Amer. Math. Soc. To appear). In this paper, we consider the nonlinear Schrödinger equation with a real-valued linear potential and p-th order gauge invariant nonlinearity. Our aim of this paper is to characterize the ground state of the elliptic equation corresponding to the time-dependent problem by the virial functional (see Definition 1.2). In order to study the ground state, we consider attainability of a minimization problem about the elliptic equation. As its application, we give a sufficient condition on the initial data, under which the energy solution can be extended globally in time. We note that the condition is defined by the minimum value of the minimization problem.

Keywords Nonlinear Schrödinger equation · Linear potential · Ground state · Virial functional.

1 Introduction

1.1 Background

In this paper, we consider the Cauchy problem of the following nonlinear Schrödinger equation with a linear potential:

$$\begin{cases} i\partial_t u + \Delta_V u = -|u|^{p-1}u, & (t,x) \in \mathbb{R}\times\mathbb{R}^d, \\ u(0,x) = u_0(x) \in H^1(\mathbb{R}^d), \end{cases} \tag{1}$$

M. Hamano (✉)
Department of Mathematics, Graduate School of Science and Engineering, Saitama University, Saitama-shi, Saitama, Japan
e-mail: m.hamano.733@ms.saitama-u.ac.jp

M. Ikeda
Center for Advanced Intelligence Project, RIKEN, Chuo City, Tokyo, Japan

Department of Mathematics, Faculty of Science and Technology, Keio University, Yokohama, Kanagawa, Japan
e-mail: masahiro.ikeda@riken.jp; masahiro.ikeda@keio.jp

© Springer Nature Switzerland AG 2020
V. Georgiev et al. (eds.), *Advances in Harmonic Analysis and Partial Differential Equations*, Trends in Mathematics, https://doi.org/10.1007/978-3-030-58215-9_12

where $p > 1$,

$$-\Delta_V := -\Delta + V = -\sum_{j=1}^{d} \frac{\partial^2}{\partial x_j^2} + V \tag{2}$$

is the Schrödinger operator, $u = u(t,x)$ is a complex-valued unknown function, $u_0 = u_0(x)$ is a complex-valued given function, $V = V(x)$ is a real-valued given function.

Equation (1) has a physical background as follows. Equation (1) with $V \in L^\infty(\mathbb{R}^d)$ is a model proposed to describe the local dynamics at a nucleation site (see [20]).

Hong [12] proved that if $d = 3$ and $\|V_-\|_{\mathcal{K}} < 4\pi$, then the Schrödinger operators $-\Delta_V$ and $1 - \Delta_V$ satisfy

$$\langle -\Delta_V f, f\rangle_{L^2} = \|\nabla f\|_{L^2}^2 + \int_{\mathbb{R}^3} V(x)|f(x)|^2 dx \geq \left(1 - \frac{\|V_-\|_{\mathcal{K}}}{4\pi}\right)\|\nabla f\|_{L^2}^2 > 0, \tag{3}$$

$$\langle (1-\Delta_V) f, f\rangle_{L^2} \geq \|f\|_{L^2}^2 + \left(1 - \frac{\|V_-\|_{\mathcal{K}}}{4\pi}\right)\|\nabla f\|_{L^2}^2 > 0 \tag{4}$$

for any $f \in H^1(\mathbb{R}^3)\setminus\{0\}$, that is, they are non-negative operators in $H^1(\mathbb{R}^3)$, where $V_-(x) := \min\{V(x), 0\}$ and $\|\cdot\|_{\mathcal{K}}$ is called a global Kato norm and is given by

$$\|V\|_{\mathcal{K}} := \sup_{x\in\mathbb{R}^3} \int_{\mathbb{R}^3} \frac{|V(y)|}{|x-y|} dy. \tag{5}$$

Therefore, the fractional operators $(-\Delta_V)^{\frac{1}{2}}$ and $(1-\Delta_V)^{\frac{1}{2}}$ are well defined on the domain $H_V^1(\mathbb{R}^3)$, whose norm is defined by

$$\|f\|_{H_V^1}^2 := \|f\|_{L^2}^2 + \|(-\Delta_V)^{\frac{1}{2}} f\|_{L^2}^2 = \|f\|_{L^2}^2 + \|\nabla f\|_{L^2}^2 + \int_{\mathbb{R}^3} V(x)|f(x)|^2 dx. \tag{6}$$

Moreover, "equivalence of norms between $\|(a-\Delta)^{\frac{s}{2}} f\|_{L_x^r}$ and $\|(a-\Delta_V)^{\frac{s}{2}} f\|_{L_x^r}$ for $a = 0, 1$, $0 \leq s \leq 2$, $1 < r < \frac{3}{s}$ (Lemma 2.1)" and Strichartz estimates for $\{e^{it\Delta_V}\}_{t\in\mathbb{R}}$ (Theorem 2.1) hold under the conditions $d = 3$, $V \in L^{\frac{3}{2}}(\mathbb{R}^3) \cap \mathcal{K}_0(\mathbb{R}^3)$, and $\|V_-\|_{\mathcal{K}} < 4\pi$. Here, $\mathcal{K}_0(\mathbb{R}^3)$ is the closure of the function space composed of bounded and compactly supported functions with respect to the global Kato norm, that is, the space $\mathcal{K}_0(\mathbb{R}^3)$ is defined as

$$\mathcal{K}_0(\mathbb{R}^3) := \overline{\left\{f \in L^\infty(\mathbb{R}^3) : \operatorname{supp} f \text{ is compact.}\right\}}^{\|\cdot\|_{\mathcal{K}}}. \tag{7}$$

As a consequence of these properties, we see that (1) is locally well-posed in the energy space H^1 under the assumptions (see Theorem 5.1). The local solution in time given in Theorem 5.1 conserves the following mass and energy with respect to time $t \in (T_{\min}, T_{\max})$, where $(T_{\min}, T_{\max})$ denotes the maximal existence time interval:

$$\text{(Mass)} \quad M[u(t)] := \|u(t)\|_{L^2}^2, \tag{8}$$

$$\text{(Energy)} \quad E_V[u(t)] := \frac{1}{2}\|(-\Delta_V)^{\frac{1}{2}}u(t)\|_{L^2}^2 - \frac{1}{p+1}\|u(t)\|_{L^{p+1}}^{p+1}. \tag{9}$$

A solution to (1) has various kinds of behaviors in time by the choice of initial data and a potential V. For example, we can consider the following behaviors in time.

Definition 1.1 (Scattering, Blow-up, Grow-up, and Standing Wave)

- (Scattering)
  We say that the solution u to (1) scatters in positive time (resp. negative time) if $T_{\max} = \infty$ (resp. $T_{\min} = -\infty$) and there exists $\psi_+ \in H^1(\mathbb{R}^d)$ (resp. $\psi_- \in H^1(\mathbb{R}^d)$) such that

$$\lim_{t\to+\infty} \|u(t) - e^{it\Delta_V}\psi_+\|_{H^1} = 0 \quad \left(\text{resp.} \lim_{t\to-\infty} \|u(t) - e^{it\Delta_V}\psi_-\|_{H^1} = 0\right), \tag{10}$$

  which means that the nonlinear solution $u(t)$ to (1) approaches to a linear solution $e^{it\Delta_V}\psi_+$ (resp. $e^{it\Delta_V}\psi_-$) as $t \to \infty$ (resp. $t \to -\infty$).
- (Blow-up)
  We say that the solution u to (1) blows up in positive time (resp. negative time) if $T_{\max} < \infty$ (resp. $T_{\min} > -\infty$).
- (Grow-up)
  We say that the solution u to (1) grows up in positive time (resp. negative time) if $T_{\max} = \infty$ (resp. $T_{\min} = -\infty$) and

$$\limsup_{t\to\infty} \|u(t)\|_{H^1} = \infty, \quad \left(\text{resp.} \limsup_{t\to-\infty} \|u(t)\|_{H^1} = \infty\right). \tag{11}$$

- (Standing wave)
  We say that the solution u to (1) is a standing wave if $u = e^{i\omega t}Q_{\omega,V}$ for some $\omega \in \mathbb{R}$, where $Q_{\omega,V}$ satisfies the elliptic equation

$$-\omega Q_{\omega,V} + \Delta_V Q_{\omega,V} = -|Q_{\omega,V}|^{p-1}Q_{\omega,V}. \tag{12}$$

1.2 Known Results

After the work by Kenig and Merle [16], global dynamics in time for focusing cases without smallness assumption on data have been studied. More precisely, Kenig and Merle [16] proved under $d = 3, 4, 5$ and energy-critical case ($p = 1 + \frac{4}{d-2}$) that if $u_0 \in \dot{H}^1_{\mathrm{rad}}(\mathbb{R}^d)$ satisfies

$$E_0[u_0] < E_0[W], \qquad \|\nabla u_0\|_{L^2} < \|\nabla W\|_{L^2}, \tag{13}$$

then the solution u to (1) with $V = 0$ scatters in both time directions and if $u_0 \in \dot{H}^1_{\mathrm{rad}}(\mathbb{R}^d) \cap |x|^{-1}L^2(\mathbb{R}^d)$ satisfies

$$E_0[u_0] < E_0[W], \qquad \|\nabla u_0\|_{L^2} > \|\nabla W\|_{L^2}, \tag{14}$$

then the solution u to (1) with $V = 0$ blows up. Here, W is the Talenti function:

$$W(x) = \left(1 + \frac{|x|^2}{d(d-2)}\right)^{-\frac{d-2}{2}} \in \dot{H}^1_{\mathrm{rad}}(\mathbb{R}^d), \tag{15}$$

solves (12) with $\omega = 0$ and $V = 0$, and satisfies the sharp Sobolev inequality:

$$\|f\|_{L^{\frac{2d}{d-2}}} \leq \frac{\|W\|_{L^{\frac{2d}{d-2}}}}{\|\nabla W\|_{L^2}} \|\nabla f\|_{L^2} \tag{16}$$

for any $f \in \dot{H}^1(\mathbb{R}^d)$. They made use of this inequality to give the conditions (13) and (14) on the data. Holmer and Roudenko [11] showed an analogous result of Kenig and Merle [16] under the mass-supercritical case and the energy-subcritical case ($p = 3$ and $u_0 \in H^1_{\mathrm{rad}}(\mathbb{R}^3)$) by applying Kenig–Merle's argument. In order to give the similar condition on the initial data as (13) and (14), they used the following characterization of the ground state to (12) with $\omega = 1$ and $V = 0$. The ground state $Q_{1,0} \in H^1(\mathbb{R}^d)$ satisfies the sharp Gagliardo–Nirenberg inequality:

$$\|f\|_{L^{p+1}}^{p+1} \leq \frac{\|Q_{1,0}\|_{L^{p+1}}^{p+1}}{\|Q_{1,0}\|_{L^2}^{p+1-\frac{d(p-1)}{2}} \|\nabla Q_{1,0}\|_{L^2}^{\frac{d(p-1)}{2}}} \|f\|_{L^2}^{p+1-\frac{d(p-1)}{2}} \|\nabla f\|_{L^2}^{\frac{d(p-1)}{2}} \tag{17}$$

for any $f \in H^1(\mathbb{R}^d)$. Duyckaerts et al. [6] extended the argument [11] to the non-radial case. Fang et al. [7] showed a scattering result, and Akahori and Nawa [1] showed a scattering result, a blow-up result, and a blow-up or grow-up result for the mass-supercritical and energy-subcritical (1) with $V = 0$ under more general nonlinear power and space dimensions. Fang et al. [7] used a characterization of the ground state to (12) with $\omega = 1$ and $V = 0$ by the Gagliardo–Nirenberg inequality (17). Akahori and Nawa [1] utilized not only its characterization by the sharp Gagliardo–Nirenberg inequality but also other characterizations of the ground state to (12) with $\omega > 0$ and $V = 0$. For example, the ground state $Q_{\omega,0}$ to (12) with $\omega > 0$ and $V = 0$ attains a minimization problem $m^{d,2}_{\omega,0}$, where

$$m^{\alpha,\beta}_{\omega,V} := \inf\left\{S_{\omega,V}(f) : f \in H^1(\mathbb{R}^d) \setminus \{0\},\ K^{\alpha,\beta}_{\omega,V}(f) = 0\right\}, \tag{18}$$

$$S_{\omega,V}(f) := \frac{\omega}{2}M[f] + E_V[f], \tag{19}$$

and

$$K^{\alpha,\beta}_{\omega,V}(f) := \mathcal{L}^{\alpha,\beta}S_{\omega,V}(f) := \frac{\partial}{\partial\lambda}S_{\omega,V}(e^{\alpha\lambda}f(e^{\beta\lambda}\cdot))\Big|_{\lambda=0} \tag{20}$$

for $(\alpha, \beta) \in \mathbb{R}^2$. Du et al. [5] showed a blow-up or grow-up result for (1) with $V = 0$ under the mass-supercritical case and energy-subcritical case. Dodson and Murphy [4] gave an alternative proof for the scattering result under the same setting as [11].

Next, we recall several related results with $V \neq 0$. Hong [12] gave the following sufficient condition of scattering for (1) with $d = 3$, $V \neq 0$, $p = 3$, and $u_0 \in H^1(\mathbb{R}^3)$.

Theorem 1.1 (Hong [12, Theorem 1.7]) *Let $d = 3$, $p = 3$, and $u_0 \in H^1(\mathbb{R}^3)$. Suppose that V satisfies $V \in L^{\frac{3}{2}}(\mathbb{R}^3) \cap \mathcal{K}_0(\mathbb{R}^3)$, $V \geq 0$, $x \cdot \nabla V \in L^{\frac{3}{2}}(\mathbb{R}^3)$, and $x \cdot \nabla V \leq 0$. We also assume that*

$$\begin{aligned} M[u_0]E_V[u_0] &< M[Q_{1,0}]E_0[Q_{1,0}] \\ \|u_0\|_{L^2}\|(-\Delta_V)^{\frac{1}{2}}u_0\|_{L^2} &< \|Q_{1,0}\|_{L^2}\|\nabla Q_{1,0}\|_{L^2}. \end{aligned} \tag{21}$$

Then, a solution u to (1) satisfies

$$\|u(t)\|_{L^2}\|(-\Delta_V)^{\frac{1}{2}}u(t)\|_{L^2} < \|Q_{1,0}\|_{L^2}\|\nabla Q_{1,0}\|_{L^2} \tag{22}$$

for any $t \in \mathbb{R}$ and u scatters.

The authors [9] improved the sufficient condition of scattering in [12]. More precisely, the authors extended a range of the exponent p for the nonlinearity and replaced $\|(-\Delta_V)^{\frac{1}{2}}u_0\|_{L^2}$ with $\|\nabla u_0\|_{L^2}$. However, the setting is restricted to radial solutions. Furthermore, the authors gave a "sufficient condition of blow-up or grow-up" and a sufficient condition of blow-up.

Theorem 1.2 (Hamano and Ikeda [9, Theorem 1.2]) *Let $d = 3$, $\frac{7}{3} < p < 5$, and $u_0 \in H^1(\mathbb{R}^3)$. Suppose that V satisfies $V \geq 0$ and $x \cdot \nabla V \in L^{\frac{3}{2}}(\mathbb{R}^3)$. We also* *assume that*

$$M[u_0]^{1-s_c}E_V[u_0]^{s_c} < M[Q_{1,0}]^{1-s_c}E_0[Q_{1,0}]^{s_c}, \tag{23}$$

where $s_c = \frac{3}{2} - \frac{2}{p-1}$.

(1) (Scattering)
If $V \in L^{\frac{3}{2}}(\mathbb{R}^3) \cap \mathcal{K}_0(\mathbb{R}^3)$, $x \cdot \nabla V \le 0$, *and*

$$\|u_0\|_{L^2}^{1-s_c}\|\nabla u_0\|_{L^2}^{s_c} < \|Q_{1,0}\|_{L^2}^{1-s_c}\|\nabla Q_{1,0}\|_{L^2}^{s_c}, \tag{24}$$

then $(T_{min}, T_{max}) = \mathbb{R}$, *that is, u exists globally in time and*

$$\|u(t)\|_{L^2}^{1-s_c}\|\nabla u(t)\|_{L^2}^{s_c} < \|Q_{1,0}\|_{L^2}^{1-s_c}\|\nabla Q_{1,0}\|_{L^2}^{s_c} \tag{25}$$

for any $t \in \mathbb{R}$. *Moreover, if* u_0 *and* V *are radially symmetric, then u scatters.*

(2) (Blow-up or grow-up)
If "$V \in L^{\frac{3}{2}}(\mathbb{R}^3) \cap \mathcal{K}_0(\mathbb{R}^3)$ *or* $V \in L^{\sigma}(\mathbb{R}^3)$ *for some* $\frac{3}{2} < \sigma \le \infty$", $2V + x \cdot \nabla V \ge 0$, *and*

$$\|u_0\|_{L^2}^{1-s_c}\|(-\Delta_V)^{\frac{1}{2}}u_0\|_{L^2}^{s_c} > \|Q_{1,0}\|_{L^2}^{1-s_c}\|\nabla Q_{1,0}\|_{L^2}^{s_c}, \tag{26}$$

then

$$\|u(t)\|_{L^2}^{1-s_c}\|(-\Delta_V)^{\frac{1}{2}}u(t)\|_{L^2}^{s_c} > \|Q_{1,0}\|_{L^2}^{1-s_c}\|\nabla Q_{1,0}\|_{L^2}^{s_c} \tag{27}$$

for any $t \in (T_{min}, T_{max})$ *and u blows up or grows up. Furthermore, if the following (i) or (ii) holds:*

(i) "u_0 *and* V *are radially symmetric",* $x \cdot \nabla V \ge 0$, *and* $V \in L^{\infty}(\mathbb{R}^3)$,
(ii) $xu_0 \in L^2(\mathbb{R}^3)$,

then u blows up.

Remark 1.1 s_c comes from the scale invariance of (1) with $V = 0$. If a function u is a solution to (1) with $V = 0$, then $u_{[\lambda]}$ is also a solution to (1) with $V = 0$ for $\lambda > 0$, where $u_{[\lambda]}(t, x) := \lambda^{\frac{2}{p-1}}u(\lambda^2 t, \lambda x)$. When $t = 0$, the transform becomes $(u_0)_{\{\lambda\}}(x) := \lambda^{\frac{2}{p-1}}u_0(\lambda x)$. Then, the following identity holds:

$$\|(u_0)_{\{\lambda\}}\|_{\dot{H}^{s_c}} = \|u_0\|_{\dot{H}^{s_c}}, \tag{28}$$

where s_c is defined as

$$s_c := \frac{d}{2} - \frac{2}{p-1} \tag{29}$$

in $\mathbb{R}^d$.

Remark 1.2 Mizutani [19] proved that for any $\psi \in H^1$, there exists $\phi_\pm \in H^1(\mathbb{R}^3)$ such that

$$\lim_{t \to \pm\infty} \|e^{it\Delta_V}\psi - e^{it\Delta}\phi_\pm\|_{H^1} = 0 \tag{30}$$

under the assumptions $V \in L^{\frac{3}{2}}(\mathbb{R}^3)$ and $V \geq 0$, where the double-sign corresponds. Combining this limit and Theorem 1.2 (1) (or Theorem 1.1), we can see that the nonlinear solution u to (1) approaches to a free solution $e^{it\Delta}\phi_\pm$ as $t \to \pm\infty$ for some $\phi_\pm \in H^1(\mathbb{R}^3)$.

In the proofs of Theorems 1.1 and 1.2, they made use of a characterization of the ground state $Q_{1,0}$ to (12) with $\omega = 1$ and $V = 0$ by the sharp Gagliardo–Nirenberg inequality. Our aim is to classify initial data by using the ground state $Q_{\omega,V}$ to (12) with $\omega > 0$ and more abstract potential $V(\neq 0)$ than previous works [14], [18]. As the result, we replace the threshold $M[Q_{1,0}]E_0[Q_{1,0}]$ in Theorem 1.1 and $M[Q_{1,0}]^{1-s_c}E_0[Q_{1,0}]^{s_c}$ in Theorem 1.2 with the quantity used $Q_{\omega,V}$. The similar studies for a specific potential is seen in Ikeda and Inui [14] (a repulsive delta potential $V = \gamma\delta$ for $\gamma > 0$ in $\mathbb{R}$) and Killip-Murphy et al. [18] (an inverse square potential $V = \frac{a}{|x|^2}$ for $a > 0$ in $\mathbb{R}^3$). As the first step of replacing the threshold $M[Q_{1,0}]^{1-s_c}E_0[Q_{1,0}]^{s_c}$, we study (12). For studying (12), we consider the minimization problems $m_{\omega,V}^{\alpha,\beta}$ (see (18) for the definition) and

$$r_{\omega,V}^{\alpha,\beta} := \inf\left\{S_{\omega,V}(f) : f \in H^1_{\text{rad}}(\mathbb{R}^d) \setminus \{0\},\ K_{\omega,V}^{\alpha,\beta}(f) = 0\right\}, \tag{31}$$

where $S_{\omega,V}$ is defined as (19), $K_{\omega,V}^{\alpha,\beta}$ is defined as (20), and $(\alpha, \beta) \in \mathbb{R}^2$ satisfies

$$\alpha > 0,\ \ \beta \geq 0,\ \ \underline{\mu} := 2\alpha - d\beta \geq 0. \tag{32}$$

Remark 1.3 $K_{\omega,V}^{\alpha,\beta}$ can be written as follows:

$$\begin{aligned} K_{\omega,V}^{\alpha,\beta}(f) = {} & \frac{\omega(2\alpha - d\beta)}{2}\|f\|_{L^2}^2 + \frac{2\alpha - (d-2)\beta}{2}\|\nabla f\|_{L^2}^2 \\ & + \frac{2\alpha - d\beta}{2}\int_{\mathbb{R}^d} V(x)|f(x)|^2 dx - \frac{\beta}{2}\int_{\mathbb{R}^d} (x \cdot \nabla V)|f(x)|^2 dx \\ & - \frac{(p+1)\alpha - d\beta}{p+1}\|f\|_{L^{p+1}}^{p+1} \end{aligned} \tag{33}$$

$$= \frac{\omega\{2\alpha - (d-2)\beta\}}{2}\|f\|_{L^2}^2 + \frac{2\alpha - (d-2)\beta}{2}\|(-\Delta_V)^{\frac{1}{2}} f\|_{L^2}^2$$
$$- \frac{\beta}{2}\int_{\mathbb{R}^d}(2\omega + 2V + x \cdot \nabla V)|f(x)|^2 dx \tag{34}$$
$$- \frac{(p+1)\alpha - d\beta}{p+1}\|f\|_{L^{p+1}}^{p+1}.$$

Definition 1.2 The Nehari functional and the virial functional are defined as

- (Nehari functional)

$$K_{\omega,V}^{1,0}(f) = \mathcal{N}_{\omega,V}(f) = \omega\|f\|_{L^2}^2 + \|(-\Delta_V)^{\frac{1}{2}} f\|_{L^2}^2 - \|f\|_{L^{p+1}}^{p+1}, \tag{35}$$

- (virial functional)

$$K_{\omega,V}^{d,2}(f) = K_V(f) = 2\|\nabla f\|_{L^2}^2 - \int_{\mathbb{R}^d}(x \cdot \nabla V)|f(x)|^2 dx - \frac{d(p-1)}{p+1}\|f\|_{L^{p+1}}^{p+1}. \tag{36}$$

Remark 1.4 It is well known that $m_{\omega,0}^{\alpha,\beta}$ and $r_{\omega,0}^{\alpha,\beta}$ are attained by the ground state of (12) with $V = 0$. Therefore, $m_{\omega,0}^{\alpha,\beta}$ and $r_{\omega,0}^{\alpha,\beta}$ are independent of (α, β) and the identity $m_{\omega,0}^{\alpha,\beta} = r_{\omega,0}^{\alpha,\beta}$ holds.

Remark 1.5 The relation (32) deduces

$$\overline{\mu} := 2\alpha - (d-2)\beta > 0 \quad \text{and} \quad (p+1)\alpha - d\beta > (p-1)\alpha - 2\beta > 0, \tag{37}$$

which and (32) are used in the proof of Theorem 1.5.

The authors [10] compared the minimization problem $m_{\omega,V}^{1,0}$ with the minimization problem $r_{\omega,V}^{1,0}$ and showed existence of a ground state of (12).

Theorem 1.3 (Existence of a Ground State to (1), **[10])** *Let* $d = 3$, $1 < p < 5$, $\omega > 0$, $V \geq 0$.

- *(Non-radial case) Let* $V \in L^{\frac{3}{2}}(\mathbb{R}^3) + L^{\sigma}(\mathbb{R}^3)$ *for some* $\frac{3}{2} < \sigma < \infty$ *and* $V > 0$. *Then,* $m_{\omega,V} = m_{\omega,0}$ *and* $m_{\omega,V}$ *is never attained.*
- *(Radial case) Let* $V \in L_{rad}^{\frac{3}{2}}(\mathbb{R}^3)$. *Then,* $r_{\omega,V}^{1,0}$ *is attained. Moreover,* $\mathcal{M}_{\omega,V,rad}^{1,0} = \mathcal{G}_{\omega,V,rad}$ *holds, where*

$$\mathcal{M}_{\omega,V,\,rad}^{\alpha,\beta} = \left\{\phi \in H_{rad}^1(\mathbb{R}^d) \setminus \{0\} : S_{\omega,V}(\phi) = r_{\omega,V}^{\alpha,\beta},\ K_{\omega,V}^{\alpha,\beta}(\phi) = 0\right\}, \tag{38}$$

$$\mathcal{G}_{\omega,V,\,rad} = \left\{\phi \in \mathcal{A}_{\omega,V,\,rad} : S_{\omega,V}(\phi) \le S_{\omega,V}(\psi)\ \textit{for any}\ \psi \in \mathcal{A}_{\omega,V,\,rad}\right\}, \tag{39}$$

$$\mathcal{A}_{\omega,V,\,rad} = \left\{\psi \in H^1_{rad}(\mathbb{R}^d) \setminus \{0\} : S'_{\omega,V}(\psi) = 0\right\}. \tag{40}$$

Then, we also investigated global existence of radial solution to (1) below the ground state $Q_{\omega,V}$ to (12) with $\omega > 0$.

Theorem 1.4 (Global Well-Posedness Below the Ground State, [10]) *Let $d = 3$ and $1 < p < 5$. Let $V \in L^{\frac{3}{2}}_{rad}(\mathbb{R}^3) \cap \mathcal{K}_0(\mathbb{R}^3)$, $V \ge 0$, and $u_0 \in H^1_{rad}(\mathbb{R}^3)$. We also assume that there exists $\omega > 0$ such that*

$$S_{\omega,V}(u_0) < S_{\omega,V}(Q_{\omega,V})\ (= r^{1,0}_{\omega,V}) \quad \textit{and} \quad \mathcal{N}_{\omega,V}(u_0) \ge 0. \tag{41}$$

Then, the solution u to (1) satisfies

$$\mathcal{N}_{\omega,V}(u(t)) \ge 0 \tag{42}$$

for any $t \in (T_{min}, T_{max})$ and u exists globally in time.

However, it is not expected that one can prove scattering result by using the above characterization (Theorem 1.3) of the ground state $Q_{\omega,V}$. In fact, the uniform positivity of the functional $K^{3,2}_{\omega,V}$ is not clear in our potential V. In this paper, we study the extended minimization problem $r^{\alpha,\beta}_{\omega,V}$ with (α, β) satisfying (32), which contains not only $r^{1,0}_{\omega,V}$ but also $r^{3,2}_{\omega,V}$.

1.3 Main Results

To state main results, we define the following quantity:

$$\omega_0 := -\frac{1}{2} \operatorname*{ess\,inf}_{x \in \mathbb{R}^d}(2V + x \cdot \nabla V). \tag{43}$$

Here, we state our main results. First, we give properties of $m^{\alpha,\beta}_{\omega,V}$, attainability of $r^{\alpha,\beta}_{\omega,V}$, and prove that the set of ground states to (12) coincides with the set of minimizers to $r^{\alpha,\beta}_{\omega,V}$.

Theorem 1.5 *Let $d \ge 3$ and $1 + \frac{4}{d} < p < 1 + \frac{4}{d-2}$.*

- *(Non-radial case) Let $V, x \cdot \nabla V \in L^{\frac{d}{2}}(\mathbb{R}^d) + L^{\sigma}(\mathbb{R}^d)$ for some $\frac{d}{2} \le \sigma < \infty$, $V \ge 0$, $x \cdot \nabla V < 0$, and $2V + x \cdot \nabla V \ge 0$. Then, for each (α, β) with (32) and*

$\omega > 0$, $m^{\alpha,\beta}_{\omega,V} = m^{\alpha,\beta}_{\omega,0}$ holds and $m^{\alpha,\beta}_{\omega,V}$ is never attained. As a result, a value of $m^{\alpha,\beta}_{\omega,V}$ is independent of (α, β).

- *(Radial case) Let V be radially symmetric. Let $V \in L^{\frac{d}{2}}(\mathbb{R}^d) + L^\infty(\mathbb{R}^d)$, $V \geq 0$, $x \cdot \nabla V \in L^{\frac{d}{2}}(\mathbb{R}^d) + L^\infty(\mathbb{R}^d)$, $x \cdot \nabla V \leq 0$, and $\omega_0 < \infty$. Then, $r^{\alpha,\beta}_{\omega,V}$ is attained for each (α, β) with (32) and $\omega > 0$ satisfying $\omega \geq \omega_0$. Moreover, if V satisfies $x\nabla^2 V x^T \in L^{\frac{d}{2}}(\mathbb{R}^d) + L^\infty(\mathbb{R}^d)$ and $3x \cdot \nabla V + x\nabla^2 V x^T \leq 0$, then any minimizer of $r^{\alpha,\beta}_{\omega,V}$ is a solution to (12), where $\nabla^2 V$ denotes the Hessian matrix of V. In particular, $\mathcal{M}^{\alpha,\beta}_{\omega,V,rad} = \mathcal{G}_{\omega,V,rad}$ holds, where $\mathcal{M}^{\alpha,\beta}_{\omega,V,rad}$ and $\mathcal{G}_{\omega,V,rad}$ are defined as (38) and (39) respectively.*

The inequality $m^{\alpha,\beta}_{\omega,V} \leq r^{\alpha,\beta}_{\omega,V}$ holds by their definitions. When the potential V satisfy assumptions of Non-radial case and Radial case in Theorem 1.5, a possibility of $m^{\alpha,\beta}_{\omega,V} = r^{\alpha,\beta}_{\omega,V}$ is eliminated. That is, a difference occurs exactly between non-radial case and radial case.

Corollary 1.1 *Let $d \geq 3$ and $1 + \frac{4}{d} < p < 1 + \frac{4}{d-2}$. Let V be radially symmetric. Let V, $x \cdot \nabla V \in L^{\frac{d}{2}}(\mathbb{R}^d) + L^\sigma(\mathbb{R}^d)$ for some $\frac{d}{2} < \sigma < \infty$, $V \geq 0$, $x \cdot \nabla V < 0$, and $2V + x \cdot \nabla V \geq 0$. Then, it follows for each (α, β) with (32) and $\omega > 0$ that*

$$m^{\alpha,\beta}_{\omega,V} < r^{\alpha,\beta}_{\omega,V}. \tag{44}$$

$m^{\alpha,\beta}_{\omega,V}$ is not attained and $r^{\alpha,\beta}_{\omega,V}$ is attained under the assumptions of Corollary 1.1, so the corollary holds.

The condition (23) can be replaced with $m^{\alpha,\beta}_{\omega,V}$ by the following proposition.

Proposition 1.1 *Let $d \geq 3$, $1 + \frac{4}{d} < p < 1 + \frac{4}{d-2}$, $u_0 \in H^1(\mathbb{R}^d)$, $V \in L^\gamma(\mathbb{R}^d) + L^\sigma(\mathbb{R}^d)$, $x \cdot \nabla V \in L^{\frac{d}{2}}(\mathbb{R}^d) + L^\sigma(\mathbb{R}^d)$ for some $\frac{d}{2} < \gamma \leq \sigma < \infty$, $V \geq 0$, $x \cdot \nabla V \leq 0$, and $2V + x \cdot \nabla V \geq 0$. Then, the following two conditions are equivalent.*

(1) $M[u_0]^{1-s_c} E_V[u_0]^{s_c} < M[Q_{1,0}]^{1-s_c} E_0[Q_{1,0}]^{s_c}$,
(2) There exists $\omega > 0$ and (α, β) with (32) such that $S_{\omega,V}(u_0) < m^{\alpha,\beta}_{\omega,V}$.

Then, we investigate global existence of a solution to time-dependent (1).

Theorem 1.6 (Global Well-Posedness in H^1) *Let $1 + \frac{4}{d} < p < 1 + \frac{4}{d-2}$.*

- *(Non-radial case) Let $u_0 \in H^1(\mathbb{R}^d)$, $V \in L^\gamma(\mathbb{R}^d) + L^\sigma(\mathbb{R}^d)$, $x \cdot \nabla V \in L^{\frac{d}{2}}(\mathbb{R}^d) + L^\sigma(\mathbb{R}^d)$ for some $\frac{d}{2} < \gamma \leq \sigma < \infty$, $V \geq 0$, $x \cdot \nabla V \leq 0$, and $2V + x \cdot \nabla V \geq 0$. We also assume that there exist (α, β) satisfying (32) and $\omega > 0$ such that*

$$M[u_0]^{1-s_c} E_V[u_0]^{s_c} < M[Q_{1,0}]^{1-s_c} E_0[Q_{1,0}]^{s_c}, \quad K^{\alpha,\beta}_{\omega,V}(u_0) \geq 0. \tag{45}$$

Then, the solution u to (1) *satisfies*

$$K_{\omega,V}^{\alpha,\beta}(u(t)) \geq 0 \tag{46}$$

for any $t \in (T_{min}, T_{max})$ and exists globally in time.

- *(Radial case) Let $u_0 \in H^1_{rad}(\mathbb{R}^d)$, $x \cdot \nabla V \in L^{\frac{d}{2}}(\mathbb{R}^d) + L^\infty(\mathbb{R}^d)$, $V \geq 0$, $x \cdot \nabla V \leq 0$, and $\omega_0 < \infty$. Let a radial function V satisfy the following (i) or (ii):*

  (i) $d = 3$, $V \in L^{\frac{3}{2}}(\mathbb{R}^3) \cap \mathcal{K}_0(\mathbb{R}^3)$,
  (ii) $d \geq 3$, $V \in L^\gamma(\mathbb{R}^d) + L^\infty(\mathbb{R}^d)$ for some $\frac{d}{2} < \gamma < \infty$.

  We also assume that there exist (α, β) with (32) *and $\omega > 0$ satisfying $\omega \geq \omega_0$ such that*

$$S_{\omega,V}(u_0) < r_{\omega,V}^{\alpha,\beta}, \quad K_{\omega,V}^{\alpha,\beta}(u_0) \geq 0. \tag{47}$$

  Then, the solution u to (1) *satisfies*

$$K_{\omega,V}^{\alpha,\beta}(u(t)) \geq 0 \tag{48}$$

  for any $t \in (T_{min}, T_{max})$ and u exists globally in time.

Remark 1.6 When $\beta \neq 0$, the term $x \cdot \nabla V$ appears in $K_{\omega,V}^{\alpha,\beta}$. Thus, we impose the assumption $x \cdot \nabla V \in L^{\frac{d}{2}}(\mathbb{R}^d) + L^\infty(\mathbb{R}^d)$ in Theorem 1.5, which assures that $K_{\omega,V}^{\alpha,\beta}$ is well defined. The repulsive condition $x \cdot \nabla V \leq 0$ is used to control the fourth term in (33). The condition of the frequency $\omega \geq \omega_0$ implies

$$2\omega + 2V + x \cdot \nabla V \geq 0 \tag{49}$$

and is used to control the third term in (34). On the other hand, when $\beta = 0$, we do not need the assumptions on $x \cdot \nabla V$ and can replace $\omega \geq \omega_0$ with $\omega > 0$.

1.4 Organization of the Paper

The organization of the rest of this paper is as follows. In Sect. 2, we collect some tools used throughout this paper. In Sect. 3, we consider properties of $m_{\omega,V}^{\alpha,\beta}$ (Non-radial case in Theorem 1.5 and Proposition 1.1). In Sect. 4, we consider attainability of $r_{\omega,V}^{\alpha,\beta}$ and prove that the set of the ground state to (12) corresponds with the set of minimizers to $r_{\omega,V}^{\alpha,\beta}$ (Radial case in Theorem 1.5). In Sect. 5, we prove global well-posedness in H^1 for (1) (Theorem 1.6).

2 Preliminaries

In this section, we define some notations and collect some tools, which are used throughout this paper.

2.1 *Notation and Definition*

For nonnegative X and Y, we write $X \lesssim Y$ to denote $X \leq CY$ for some $C > 0$. If $X \lesssim Y \lesssim X$ holds, we write $X \sim Y$. The dependence of implicit constants on parameters will be indicated by subscripts, e.g. $X \lesssim_u Y$ denotes $X \leq CY$ for some $C = C(u)$. We write $a' \in [1, \infty]$ to denote the Hölder dual exponent to $a \in [1, \infty]$, that is, the solution $\frac{1}{a} + \frac{1}{a'} = 1$.

For $1 \leq p \leq \infty$, $L^p = L^p(\mathbb{R}^d)$ denotes the usual Lebesgue space. For a Banach space X, we use $L^q(I; X)$ to denote the Banach space of functions $f : I \times \mathbb{R}^3 \longrightarrow \mathbb{C}$ whose norm is

$$\|f\|_{L^q(I;X)} := \left(\int_I \|f(t)\|_X^q dt \right)^{\frac{1}{q}} < \infty, \tag{50}$$

with the usual modificafion when $q = \infty$. We extend our notation as follows: If a time interval is not spacified, then the t-norm is evaluated over $(-\infty, \infty)$. To indicate a restriction to a time subinterval $I \subset (-\infty, \infty)$, we will write as $L^q(I)$.

We define the Fourier transform of f on $\mathbb{R}^d$ as

$$\mathcal{F}f(\xi) = \widehat{f}(\xi) := \int_{\mathbb{R}^d} e^{-2\pi i x \cdot \xi} f(x) dx \tag{51}$$

and define the inverse Fourier transform of f on $\mathbb{R}^d$ as

$$\mathcal{F}^{-1} f(x) = \check{f}(x) := \int_{\mathbb{R}^d} e^{2\pi i x \cdot \xi} f(\xi) d\xi, \tag{52}$$

where $x \cdot \xi$ denotes the usual inner product of x and ξ on $\mathbb{R}^d$.

$W^{s,p} = (1 - \Delta)^{-\frac{s}{2}} L^p$ and $\dot{W}^{s,p} = (-\Delta)^{-\frac{s}{2}} L^p$ are the inhomogeneous Sobolev space and the homogeneous Sobolev space, respectively for $s \in \mathbb{R}$ and $p \in [1, \infty]$, where $(1 - \Delta)^{\frac{s}{2}} = \mathcal{F}^{-1}(1 + 4\pi^2|\xi|^2)^{\frac{s}{2}} \mathcal{F}$ and $(-\Delta)^{\frac{s}{2}} = \mathcal{F}^{-1}(2\pi|\xi|)^s \mathcal{F}$, respectively. When $p = 2$, we express $W^{s,2} = H^s$ and $\dot{W}^{s,2} = \dot{H}^s$. We also define the Sobolev spaces with a potential V by $W_V^{s,p} = (1 - \Delta_V)^{-\frac{s}{2}} L^p$ and $\dot{W}_V^{s,p} = (-\Delta_V)^{-\frac{s}{2}} L^p$ for $s \in \mathbb{R}$ and $p \in [1, \infty]$. When $p = 2$, we express $W_V^{s,2} = H_V^s$ and $\dot{W}_V^{s,2} = \dot{H}_V^s$.

2.2 Some Tools

Lemma 2.1 (Norm Equivalence [12]) *If $d = 3$, $V \in L^{\frac{3}{2}}(\mathbb{R}^3) \cap \mathcal{K}_0(\mathbb{R}^3)$, and $\|V_-\|_{\mathcal{K}} < 4\pi$, then*

$$\|(-\Delta_V)^{\frac{s}{2}} f\|_{L^r} \sim \|f\|_{\dot{W}^{s,r}}, \quad \|(1-\Delta_V)^{\frac{s}{2}} f\|_{L^r} \sim \|f\|_{W^{s,r}}, \tag{53}$$

where $1 < r < \frac{3}{s}$ and $0 \le s \le 2$.

Definition 2.1 (L^2-Admissible Pair and Strichartz Norm) We say that a pair of exponents (q, r) is called L^2-admissible in three space dimensions if $2 \le q, r \le \infty$ and

$$\frac{2}{q} + \frac{3}{r} = \frac{3}{2}. \tag{54}$$

We define the Strichartz norm by

$$\|u\|_{S(L^2)} := \sup_{\substack{(q,r):L^2\text{admissible} \\ 2\le q\le\infty, 2\le r\le 6}} \|u\|_{L_t^q L_x^r} \tag{55}$$

and its dual norm by

$$\|u\|_{S'(L^2)} := \inf_{\substack{(q,r):L^2\text{admissible} \\ 2\le q\le\infty, 2\le r\le 6}} \|u\|_{L_t^{q'} L_x^{r'}}. \tag{56}$$

Theorem 2.1 (Strichartz Estimate [8], [15]) *If $d = 3$, $V \in L^{\frac{3}{2}}(\mathbb{R}^3) \cap \mathcal{K}_0(\mathbb{R}^3)$, and $\|V_-\|_{\mathcal{K}} < 4\pi$, then the following estimates hold.*

- *(Homogeneous estimates)*

$$\|e^{it\Delta_V} f\|_{S(L^2)} \lesssim \|f\|_{L_x^2}. \tag{57}$$

- *(Inhomogeneous estimates)*

$$\left\| \int_0^t e^{i(t-s)\Delta_V} F(\cdot, s) ds \right\|_{S(L^2)} \lesssim \|F\|_{S'(L^2)}. \tag{58}$$

where implicit constants are independent of f and F. Even if time is restricted, Theorem 2.1 still holds.

Lemma 2.2 (Fractional Calculus [17]) *Suppose $G \in C^1(\mathbb{C})$ and $s \in (0, 1]$. Let $1 < r, r_2 < \infty$ and $1 < r_1 \le \infty$ satisfying $\frac{1}{r} = \frac{1}{r_1} + \frac{1}{r_2}$. Then, we have*

$$\||\nabla|^s G(u)\|_{L^r} \lesssim \|G'(u)\|_{L^{r_1}} \||\nabla|^s u\|_{L^{r_2}}. \tag{59}$$

Proposition 2.1 (Pohozaev Identity Without a Potential) *Let $d \ge 3$ and $1 < p < 1 + \frac{4}{d-2}$. The ground state $Q_{\omega,0}$ for (12) with $V = 0$ satisfies the following Pohozaev identities:*

$$\begin{aligned} \|Q_{\omega,0}\|_{L^{p+1}}^{p+1} &= \frac{2(p+1)}{d+2-(d-2)p}\omega\|Q_{\omega,0}\|_{L^2}^2, \\ \|Q_{\omega,0}\|_{L^{p+1}}^{p+1} &= \frac{2(p+1)}{d(p-1)}\|\nabla Q_{\omega,0}\|_{L^2}^2. \end{aligned} \tag{60}$$

For the proof of this proposition, see [3, Lemma 8.1.2]. This proposition deduces the following relations:

$$\begin{aligned} E_0[Q_{\omega,0}] &= \frac{dp-(d+4)}{2d(p-1)}\|\nabla Q_{\omega,0}\|_{L^2}^2, \\ C_{\mathrm{GN}} &= \frac{2(p+1)}{d(p-1)} \frac{1}{\|Q_{1,0}\|_{L^2}^{\frac{d+2-(d-2)p}{2}} \|\nabla Q_{1,0}\|_{L^2}^{\frac{dp-(d+4)}{2}}}, \end{aligned} \tag{61}$$

where C_{GN} denotes the best constant of the Gagliardo-Nirenberg inequality (17).

Lemma 2.3 (Brezis and Lieb [2]) *The following holds.*

(1) Let $1 < q < \infty$ and $d \ge 1$. We assume that $\{u_n\}$ is bounded in $L^q(\mathbb{R}^d)$ and satisfies $u_n \longrightarrow u$ a.e. in $\mathbb{R}^d$. Then, we have $u \in L^q(\mathbb{R}^d)$ and

$$\lim_{n\to\infty} \left(\|u_n\|_{L^q}^q - \|u_n - u\|_{L^q}^q\right) = \|u\|_{L^q}^q. \tag{62}$$

(2) Let X be a Hilbert space. If $u_n \rightharpoonup u$ in X, then we have

$$\lim_{n\to\infty} \left(\|u_n\|_X^2 - \|u_n - u\|_X^2\right) = \|u\|_X^2. \tag{63}$$

3 Non-radial Case

In this section, we investigate properties of $m_{\omega,V}$, which is defined as (18). The argument is based on [14].

We define the following functional:

$$\begin{aligned} T_{\omega,V}^{\alpha,\beta}(f) &:= S_{\omega,V}(f) - \frac{1}{2\alpha-(d-2)\beta} K_{\omega,V}^{\alpha,\beta}(f) \\ &= \frac{\beta}{2\{2\alpha-(d-2)\beta\}} \int_{\mathbb{R}^d} (2\omega + 2V + x\cdot\nabla V)|f(x)|^2 dx \\ &\qquad + \frac{(p-1)\alpha - 2\beta}{(p+1)\{2\alpha-(d-2)\beta\}} \|f\|_{L^{p+1}}^{p+1}. \end{aligned} \tag{64}$$

Lemma 3.1 *Let $d \geq 3$, $1+\frac{4}{d} < p < 1+\frac{4}{d-2}$, and let (α,β) satisfy* (32). *Let $V \in L^{\frac{d}{2}}(\mathbb{R}^d) + L^\infty(\mathbb{R}^d)$, $V \geq 0$, $x\cdot\nabla V \in L^{\frac{d}{2}}(\mathbb{R}^d) + L^\infty(\mathbb{R}^d)$, $x\cdot\nabla V \leq 0$, and $\omega_0 < \infty$. If $\omega > 0$ satisfies $\omega \geq \omega_0$, then*

$$m_{\omega,V}^{\alpha,\beta} = \inf\left\{T_{\omega,V}^{\alpha,\beta}(f) : f \in H^1(\mathbb{R}^d)\setminus\{0\},\ K_{\omega,V}^{\alpha,\beta}(f) \leq 0\right\} \tag{65}$$

holds.

Proof When f satisfies $K_{\omega,V}^{\alpha,\beta}(f) \leq 0$, there exists $0 < \lambda \leq 1$ such that $K_{\omega,V}^{\alpha,\beta}(\lambda f) = 0$. For such $0 < \lambda \leq 1$, we have

$$m_{\omega,V}^{\alpha,\beta} \leq S_{\omega,V}(\lambda f) = T_{\omega,V}^{\alpha,\beta}(\lambda f) \leq T_{\omega,V}^{\alpha,\beta}(f) \tag{66}$$

by $\omega \geq \omega_0$. Therefore, we obtain

$$\begin{aligned} m_{\omega,V}^{\alpha,\beta} &= \inf\left\{T_{\omega,V}^{\alpha,\beta}(f) : f \in H^1(\mathbb{R}^d)\setminus\{0\},\ K_{\omega,V}^{\alpha,\beta}(f) = 0\right\} \\ &\geq \inf\left\{T_{\omega,V}^{\alpha,\beta}(f) : f \in H^1(\mathbb{R}^d)\setminus\{0\},\ K_{\omega,V}^{\alpha,\beta}(f) \leq 0\right\} \geq m_{\omega,V}^{\alpha,\beta}. \end{aligned} \tag{67}$$

Proposition 3.1 *Let $d \geq 3$, $1+\frac{4}{d} < p < 1+\frac{4}{d-2}$, $\omega > 0$, and let (α,β) satisfy* (32). *Let $V, x\cdot\nabla V \in L^{\frac{d}{2}}(\mathbb{R}^d) + L^\sigma(\mathbb{R}^d)$ for some $\frac{d}{2} \leq \sigma < \infty$, $V \geq 0$, $x\cdot\nabla V \leq 0$, and $2V + x\cdot\nabla V \geq 0$. Then, the identity*

$$m_{\omega,V}^{\alpha,\beta} = m_{\omega,0}^{\alpha,\beta} \tag{68}$$

holds.

Proof When $V = 0$, this proposition is trivial. We assume $V \neq 0$. First, we prove $m_{\omega,V}^{\alpha,\beta} \geq m_{\omega,0}^{\alpha,\beta}$. We take any $f \in H^1\setminus\{0\}$ satisfying $K_{\omega,V}^{\alpha,\beta}(f) = 0$. Since $K_{\omega,0}^{\alpha,\beta}(f) \leq K_{\omega,V}^{\alpha,\beta}(f) = 0$ by $V \geq 0$, using Lemma 3.1 and $2V + x\cdot\nabla V \geq 0$, we have

$$m_{\omega,0}^{\alpha,\beta} \leq T_{\omega,0}^{\alpha,\beta}(f) \leq T_{\omega,V}^{\alpha,\beta}(f) = S_{\omega,V}(f), \tag{69}$$

which implies

$$m_{\omega,0}^{\alpha,\beta} \le m_{\omega,V}^{\alpha,\beta}. \tag{70}$$

Next, we prove $m_{\omega,0}^{\alpha,\beta} \ge m_{\omega,V}^{\alpha,\beta}$. We note that $Q_{\omega,0}$ attains $m_{\omega,0}$. For any $\{y_n\}$ satisfying $|y_n| \to \infty$, it follows that

$$S_{\omega,V}(Q_{\omega,0}(\cdot - y_n)) \longrightarrow S_{\omega,0}(Q_{\omega,0}) = m_{\omega,0}^{\alpha,\beta} \quad \text{as} \quad n \to \infty \tag{71}$$

by $V \in L^{\frac{d}{2}}(\mathbb{R}^d)+L^{\sigma}(\mathbb{R}^d)$ for some $\frac{d}{2} \le \sigma < \infty$. For any $\{y_n\}$ satisfying $|y_n| \to \infty$, we also have

$$K_{\omega,V}^{\alpha,\beta}(Q_{\omega,0}(\cdot - y_n)) > K_{\omega,0}^{\alpha,\beta}(Q_{\omega,0}(\cdot - y_n)) = K_{\omega,0}^{\alpha,\beta}(Q_{\omega,0}) = 0 \tag{72}$$

by $V \ge 0$ and $x \cdot \nabla V \le 0$. Since $K_{\omega,V}^{\alpha,\beta}(Q_{\omega,0}(\cdot - y_n)) > 0$ and $K_{\omega,V}^{\alpha,\beta}(\lambda Q_{\omega,0}(\cdot - y_n)) < 0$ for a sufficiently large $\lambda > 1$, we can take $\lambda_n > 1$ with $K_{\omega,V}^{\alpha,\beta}(\lambda_n Q_{\omega,0}(\cdot - y_n)) = 0$. For this sequence $\{\lambda_n\}$, we have $\lambda_n \longrightarrow 1$ as $n \to \infty$. Indeed, $K_{\omega,V}^{\alpha,\beta}(\lambda_n Q_{\omega,0}(\cdot - y_n)) = 0$ and $K_{\omega,0}^{\alpha,\beta}(Q_{\omega,0}) = 0$ deduces

$$\begin{aligned} 0 = &\frac{(p+1)\alpha - d\beta}{p+1}(1-\lambda_n^{p-1})\|Q_{\omega,0}\|_{L^{p+1}}^{p+1} \\ &+ \frac{2\alpha - d\beta}{2}\int_{\mathbb{R}^d} V(x)Q_{\omega,0}(x-y_n)^2 dx \\ &- \frac{\beta}{2}\int_{\mathbb{R}^d}(x \cdot \nabla V)Q_{\omega,0}(x-y_n)^2 dx. \end{aligned} \tag{73}$$

Thus, we have $\lambda_n \longrightarrow 1$ as $n \to \infty$. Hence, $S_{\omega,V}(\lambda_n Q_{\omega,0}(\cdot - y_n)) \longrightarrow S_{\omega,0}(Q_{\omega,0}) = m_{\omega,0}^{\alpha,\beta}$ as $n \to \infty$ and $K_{\omega,V}^{\alpha,\beta}(\lambda_n Q_{\omega,V}(\cdot - y_n)) = 0$ for each $n \in \mathbb{N}$. This implies $m_{\omega,0}^{\alpha,\beta} \ge m_{\omega,V}^{\alpha,\beta}$. □

Proposition 3.2 *Besides the assumptions of Proposition 3.1, we assume that $x \cdot \nabla V < 0$. Then, $m_{\omega,V}$ is not attained.*

Proof We note that $V \ge 0$ and $x \cdot \nabla V < 0$ imply $V > 0$. We assume for contradiction that there exists $\phi \in H^1(\mathbb{R}^d)$ such that ϕ attains $m_{\omega,V}^{\alpha,\beta}$. We take $y \in \mathbb{R}^d$ satisfying

$$K_{\omega,V}^{\alpha,\beta}(\phi(\cdot - y)) < K_{\omega,V}^{\alpha,\beta}(\phi) = 0 \tag{74}$$

and

$$\int_{\mathbb{R}^d}(2V + x \cdot \nabla V)|\phi(x-y)|^2 dx \le \int_{\mathbb{R}^d}(2V + x \cdot \nabla V)|\phi(x)|^2 dx. \tag{75}$$

We note that if we take $y \in \mathbb{R}^d$ with sufficiently large $|y|$, then this inequality holds by $V,\ x \cdot \nabla V \in L^{\frac{d}{2}} + L^{\sigma}$ for some $\frac{d}{2} \le \sigma < \infty$, $V > 0$, and $x \cdot \nabla V < 0$. Since $K^{\alpha,\beta}_{\omega,V}(\lambda\phi(\cdot - y)) > 0$ for some $0 < \lambda < 1$, combining this inequality and (74), there exists $0 < \lambda_0 < 1$ such that

$$K^{\alpha,\beta}_{\omega,V}(\lambda_0\phi(\cdot - y)) = 0. \tag{76}$$

Therefore, the definition of $m^{\alpha,\beta}_{\omega,V}$, (74), and (75) imply

$$\begin{aligned} m^{\alpha,\beta}_{\omega,V} &\le S_{\omega,V}(\lambda_0\phi(\cdot - y)) = T_{\omega,V}(\lambda_0\phi(\cdot - y)) \\ &< T_{\omega,V}(\phi(\cdot - y)) \le T_{\omega,V}(\phi) = S_{\omega,V}(\phi) = m^{\alpha,\beta}_{\omega,V}. \end{aligned} \tag{77}$$

This is contradiction. □

Finally, we prove Proposition 1.1.

Proof For $(\alpha, \beta) \in \mathbb{R}^2$, we set $Q_{\omega,0} = \omega^{\alpha} Q_{1,0}(\omega^{\beta}\,\cdot)$, where (α, β) will be chosen later. By the equation (12) with $V = 0$, it follows that

$$-\omega^{\alpha+1}Q_{1,0}(\omega^{\beta}\,\cdot) + \omega^{\alpha+2\beta}(\Delta Q_{1,0})(\omega^{\beta}\,\cdot) + \omega^{p\alpha}|Q_{1,0}(\omega^{\beta}\,\cdot)|^{p-1}Q_{1,0}(\omega^{\beta}\,\cdot) = 0 \tag{78}$$

for any $\omega > 0$. Hence, the identities $\alpha + 1 = \alpha + 2\beta = p\alpha$ hold, that is, $Q_{\omega,0} = \omega^{\frac{1}{p-1}} Q_{1,0}(\omega^{\frac{1}{2}}\,\cdot)$ hold. Then,

$$S_{\omega,0}(Q_{\omega,0}) = \omega^{\frac{d+2-(d-2)p}{2(p-1)}} S_{1,0}(Q_{1,0}) \tag{79}$$

holds. Thus, the condition (2) in Proposition 1.1 is equivalent to

$$S_{\omega,V}(u_0) < \omega^{\frac{d+2-(d-2)p}{2(p-1)}} S_{1,0}(Q_{1,0}) \tag{80}$$

by using Proposition 3.1. Here, we define a function $f(\omega) := \omega^{\frac{d+2-(d-2)p}{2(p-1)}} S_{1,0}(Q_{1,0}) - S_{\omega,V}(u_0)$ on $\omega \in (0, \infty)$. Since

$$f'(\omega) = \frac{d+2-(d-2)p}{2(p-1)}\omega^{\frac{d+4-dp}{2(p-1)}} S_{1,0}(Q_{1,0}) - \frac{1}{2}M[u_0], \tag{81}$$

solving $f'(\omega_0) = 0$, we get

$$\omega_0 = \left\{ \frac{p-1}{d+2-(d-2)p} \cdot \frac{M[u_0]}{S_{1,0}(Q_{1,0})} \right\}^{\frac{2(p-1)}{d+4-dp}}. \tag{82}$$

The function f has a maximum value at $\omega = \omega_0$ by $p > 1 + \frac{4}{d}$. Therefore, if there exists $\omega > 0$ such that $S_{\omega,V}(u_0) < S_{\omega,0}(Q_{\omega,0})$, then $f(\omega_0) > 0$ holds. Since

$$\begin{aligned} &f(\omega_0)\\ &= \left\{\frac{d+2-(d-2)p}{p-1}\right\}^{\frac{2(p-1)}{dp-(d+4)}} \frac{dp-(d+4)}{2\{d+2-(d-2)p\}} \frac{S_{1,0}(Q_{1,0})^{\frac{2(p-1)}{dp-(d+4)}}}{M[u_0]^{\frac{d+2-(d-2)p}{dp-(d+4)}}}\\ &\quad - E_V[u_0], \end{aligned} \tag{83}$$

$f(\omega_0) > 0$ implies

$$\begin{aligned} \frac{d+2-(d-2)p}{p-1} \left[\frac{dp-(d+4)}{2\{d+2-(d-2)p\}}\right]^{\frac{dp-(d+4)}{2(p-1)}} S_{1,0}(Q_{1,0})& \\ > M[u_0]^{1-s_c} E_V[u_0]^{s_c}.& \end{aligned} \tag{84}$$

Here, we calculate $S_{1,0}(Q_{1,0})$ by using (61).

$$S_{1,0}(Q_{1,0}) = \frac{p-1}{d+2-(d-2)p} \left[\frac{2\{d+2-(d-2)p\}}{dp-(d+4)}\right]^{s_c} M[Q_{1,0}]^{1-s_c} E_0[Q_{1,0}]^{s_c}. \tag{85}$$

Therefore, we obtain

$$M[Q_{1,0}]^{1-s_c} E_0[Q_{1,0}]^{s_c} > M[u_0]^{1-s_c} E_V[u_0]^{s_c}. \tag{86}$$

4 Radial Case

In this section, we prove attainability of $r^{\alpha,\beta}_{\omega,V}$ defined as (31) and that the set of the ground state to (12) corresponds with the set of minimizers to $r^{\alpha,\beta}_{\omega,V}$. The argument is based on [13].

Proposition 4.1 (Positivity of $T^{\alpha,\beta}_{\omega,V}$) *Let $d \geq 3$, $1 + \frac{4}{d} < p < 1 + \frac{4}{d-2}$, and let (α, β) satisfy (32). We assume that $V \in L^{\frac{d}{2}}(\mathbb{R}^d) + L^\infty(\mathbb{R}^d)$, $V \geq 0$, $x \cdot \nabla V \in L^{\frac{d}{2}}(\mathbb{R}^d) + L^\infty(\mathbb{R}^d)$, and $\omega_0 < \infty$. If $\omega > 0$ satisfies $\omega \geq \omega_0$ and $f \in H^1(\mathbb{R}^d) \setminus \{0\}$*

satisfies $K_{\omega,V}^{\alpha,\beta}(f) = 0$, *then we have*

$$T_{\omega,V}^{\alpha,\beta}(f) > 0. \tag{87}$$

Proof This proposition follows from (32) and $2\omega + 2V + x \cdot \nabla V \geq 0$ a.e. $x \in \mathbb{R}^d$. □

Proposition 4.2 (Equivalence of H^1-Norm and $S_{\omega,V}$) *Let $d \geq 3$, $1 + \frac{4}{d} < p < 1 + \frac{4}{d-2}$, and let (α, β) satisfy (32). Let $V \in L^{\frac{d}{2}}(\mathbb{R}^d) + L^\infty(\mathbb{R}^d)$, $V \geq 0$, $x \cdot \nabla V \in L^{\frac{d}{2}}(\mathbb{R}^d) + L^\infty(\mathbb{R}^d)$, and $\omega_0 < \infty$. We assume that $\omega > 0$ satisfies $\omega \geq \omega_0$ and $f \in H^1(\mathbb{R}^d)$ satisfies $K_{\omega,V}^{\alpha,\beta}(f) \geq 0$. Then,*

$$\begin{aligned} \{(p-1)\alpha - 2\beta\} S_{\omega,V}(f) &\leq \frac{(p-1)\alpha - 2\beta}{2} J_{\omega,V}(f) \\ &\leq \{(p+1)\alpha - d\beta\} S_{\omega,V}(f) \end{aligned} \tag{88}$$

holds, where $J_{\omega,V}(f) := \omega \|f\|_{L^2}^2 + \|(-\Delta_V)^{\frac{1}{2}} f\|_{L^2}^2$.

Proof The first inequality holds clearly. We prove the second inequality.

$$\begin{aligned} &\frac{(p-1)\alpha - 2\beta}{2} J_{\omega,V}(f) \leq \frac{(p-1)\alpha - 2\beta}{2} J_{\omega,V}(f) + K_{\omega,V}^{\alpha,\beta}(f) \\ &= \{(p+1)\alpha - d\beta\} S_{\omega,V}(f) - \frac{\beta}{2} \int_{\mathbb{R}^d} (2\omega + 2V + x \cdot \nabla V)|f(x)|^2 dx \\ &\leq \{(p+1)\alpha - d\beta\} S_{\omega,V}(f). \end{aligned} \tag{89}$$

Lemma 4.1 (Positivity of $K_{\omega,V}^{\alpha,\beta}$) *Let $d \geq 3$ and $1 + \frac{4}{d} < p < 1 + \frac{4}{d-2}$. Let $V \in L^{\frac{d}{2}}(\mathbb{R}^d) + L^\infty(\mathbb{R}^d)$, $V \geq 0$, $x \cdot \nabla V \in L^{\frac{d}{2}}(\mathbb{R}^d) + L^\infty(\mathbb{R}^d)$, and $x \cdot \nabla V \leq 0$. Let (α, β) satisfy (32). Suppose that $\{f_n\}$ is a bounded sequence in $H^1(\mathbb{R}^d) \setminus \{0\}$ and satisfies $\|\nabla f_n\|_{L^2} \longrightarrow 0$ as $n \to \infty$. Then, there exists $n_0 \in \mathbb{N}$ such that*

$$K_{\omega,V}^{\alpha,\beta}(f_n) > 0 \tag{90}$$

for any $n \geq n_0$.

Proof We take a positive constant C such as $\sup_{n\in\mathbb{N}} \|f_n\|_{L^2} \le C$. Applying the Gagliardo–Nirenberg inequality (17), we have

$$\begin{aligned} K^{\alpha,\beta}_{\omega,V}(f_n) \ge& \frac{\omega(2\alpha - d\beta)}{2}\|f_n\|^2_{L^2} + \frac{2\alpha-(d-2)\beta}{2}\|\nabla f_n\|^2_{L^2} \\ &+ \frac{2\alpha - d\beta}{2}\int_{\mathbb{R}^d} V(x)|f_n(x)|^2dx - \frac{\beta}{2}\int_{\mathbb{R}^d}(x\cdot\nabla V)|f_n(x)|^2dx \\ &- \frac{(p+1)\alpha - d\beta}{p+1}C_{\mathrm{GN}}\|\nabla f_n\|_{L^2}^{\frac{d(p-1)}{2}}\|f_n\|_{L^2}^{\frac{d+2-(d-2)p}{2}} \\ \ge& \left(\frac{2\alpha-(d-2)\beta}{2} - \frac{(p+1)\alpha-d\beta}{p+1}C_{\mathrm{GN}}C^{\frac{d+2-(d-2)p}{2}}\|\nabla f_n\|_{L^2}^{\frac{dp-(d+4)}{2}}\right)\|\nabla f_n\|^2_{L^2}. \end{aligned} \tag{91}$$

When $\|\nabla f_n\|_{L^2} \neq 0$ is sufficiently small, we obtain $K^{\alpha,\beta}_{\omega,V}(f_n) > 0$. □

Lemma 4.2 (Positivity of $r^{\alpha,\beta}_{\omega,V}$) *Let $d \ge 3$, $1+\frac{4}{d} < p < 1+\frac{4}{d-2}$, and let (α,β) satisfy* (32). *Let $V \in L^{\frac{d}{2}}(\mathbb{R}^d) + L^\infty(\mathbb{R}^d)$, $V \ge 0$, $x\cdot\nabla V \in L^{\frac{d}{2}}(\mathbb{R}^d)+L^\infty(\mathbb{R}^d)$, $x\cdot\nabla V \le 0$, and $\omega_0 < \infty$. If $\omega \ge \omega_0$, then we have*

$$0 < r^{\alpha,\beta}_{\omega,V}. \tag{92}$$

Proof We take any $f \in H^1_{\mathrm{rad}}(\mathbb{R}^d)\setminus\{0\}$ satisfying $K^{\alpha,\beta}_{\omega,V}(f) = 0$. Then,

$$\begin{aligned} &\frac{2\alpha-(d-2)\beta}{2}\|\nabla f\|^2_{L^2} \\ &\le \frac{\omega(2\alpha-d\beta)}{2}\|f\|^2_{L^2} + \frac{2\alpha-(d-2)\beta}{2}\|\nabla f\|^2_{L^2} \\ &\quad + \frac{2\alpha-d\beta}{2}\int_{\mathbb{R}^d}V(x)|f(x)|^2dx - \frac{\beta}{2}\int_{\mathbb{R}^d}(x\cdot\nabla V)|f(x)|^2dx \\ &= \frac{(p+1)\alpha-d\beta}{p+1}\|f\|^{p+1}_{L^{p+1}} \\ &\le \frac{(p+1)\alpha-d\beta}{p+1}C_{\mathrm{GN}}\|f\|_{L^2}^{\frac{d+2-(d-2)p}{2}}\|\nabla f\|_{L^2}^{\frac{d(p-1)}{2}} \end{aligned} \tag{93}$$

by $K^{\alpha,\beta}_{\omega,V}(f) = 0$ and the Gagliardo–Nirenberg inequality (17). Using this inequality and Proposition 4.2, we have

$$1 \lesssim \|f\|_{L^2}^{\frac{d+2-(d-2)p}{2}}\|\nabla f\|_{L^2}^{\frac{dp-(d+4)}{2}} \le \|f\|^{p-1}_{H^1} \lesssim J_{\omega,V}(f)^{\frac{p-1}{2}} \lesssim S_{\omega,V}(f)^{\frac{p-1}{2}}. \tag{94}$$

Lemma 4.3 *Let $d \geq 3$, $1 + \frac{4}{d} < p < 1 + \frac{4}{d-2}$, and let (α, β) satisfy (32). Let $V \in L^{\frac{d}{2}}(\mathbb{R}^d) + L^\infty(\mathbb{R}^d)$, $V \geq 0$, $x \cdot \nabla V \in L^{\frac{d}{2}}(\mathbb{R}^d) + L^\infty(\mathbb{R}^d)$, $x \cdot \nabla V \leq 0$, and $\omega_0 < \infty$. If $\omega > 0$ satisfies $\omega \geq \omega_0$ and $f \in H^1_{rad}(\mathbb{R}^d) \setminus \{0\}$ satisfies $K^{\alpha,\beta}_{\omega,V}(f) \leq 0$, then there exists $0 < \lambda \leq 1$ such that*

$$K^{\alpha,\beta}_{\omega,V}(\lambda f) = 0 \quad \text{and} \quad r^{\alpha,\beta}_{\omega,V} \leq S_{\omega,V}(\lambda f) = T^{\alpha,\beta}_{\omega,V}(\lambda f) \leq T^{\alpha,\beta}_{\omega,V}(f). \tag{95}$$

In particular,

$$r^{\alpha,\beta}_{\omega,V} = \inf\left\{T^{\alpha,\beta}_{\omega,V}(f) : f \in H^1_{rad}(\mathbb{R}^d) \setminus \{0\},\ K^{\alpha,\beta}_{\omega,V}(f) \leq 0\right\} \tag{96}$$

holds.

Proof This lemma follows from the same argument as Lemma 3.1. □

The following theorem is attainability part in Radial case of Theorem 1.5.

Theorem 4.1 *Let $d \geq 3$, $1 + \frac{4}{d} < p < 1 + \frac{4}{d-2}$, and let (α, β) satisfies (32). Let $V \in L^{\frac{d}{2}}(\mathbb{R}^d) + L^\infty(\mathbb{R}^d)$, $V \geq 0$, $x \cdot \nabla V \in L^{\frac{d}{2}}(\mathbb{R}^d) + L^\infty(\mathbb{R}^d)$, $x \cdot \nabla V \leq 0$, and $\omega_0 < \infty$. If $\omega > 0$ satisfies $\omega \geq \omega_0$, then $r^{\alpha,\beta}_{\omega,V}$ is attained.*

Proof We take a minimizing sequence $\{f_n\} \subset H^1_{\text{rad}}(\mathbb{R}^d) \setminus \{0\}$ satisfying

$$K^{\alpha,\beta}_{\omega,V}(f_n) = 0 \quad \text{for any} \quad n \in \mathbb{N} \tag{97}$$

and

$$S_{\omega,V}(f_n) = T^{\alpha,\beta}_{\omega,V}(f_n) \searrow r^{\alpha,\beta}_{\omega,V} \quad \text{as} \quad n \to \infty. \tag{98}$$

We see that $\{f_n\}$ is a bounded sequence in $L^2(\mathbb{R}^d)$ and $\dot{H}^1_V(\mathbb{R}^d)$ by Proposition 4.2. From $V \geq 0$, $\{f_n\}$ is also a bounded sequence in $H^1(\mathbb{R}^d)$. Thus, we can take a subsequence of $\{f_n\}$ satisfying $f_n \rightharpoonup f_\infty$ in $H^1(\mathbb{R}^d)$, which is still denoted by $\{f_n\}$. Since $H^1(\mathbb{R}^d) \subset L^{p+1}(\mathbb{R}^d)$ is compact for $1 < p < 1 + \frac{4}{d-2}$, we can take a subsequence of $\{f_n\}$ satisfying $f_n \longrightarrow f_\infty$ in $L^{p+1}(\mathbb{R}^d)$, which is still denoted by $\{f_n\}$. Therefore, we get

$$\|f_\infty\|_{L^2} \leq \liminf_{n\to\infty} \|f_n\|_{L^2}, \tag{99}$$

$$\|(-\Delta_V)^{\frac{1}{2}} f_\infty\|_{L^2} \leq \liminf_{n\to\infty} \|(-\Delta_V)^{\frac{1}{2}} f_n\|_{L^2}, \tag{100}$$

$$\|f_\infty\|^{p+1}_{L^{p+1}} = \lim_{n\to\infty} \|f_n\|^{p+1}_{L^{p+1}}. \tag{101}$$

To prove (100), we used the following inequality

$$\int_{\mathbb{R}^d} V(x)|f_\infty(x)|^2 dx \le \liminf_{n\to\infty} \int_{\mathbb{R}^d} V(x)|f_n(x)|^2 dx, \tag{102}$$

which holds by the following argument. Using the Hölder inequality,

$$\left|\int_{\mathbb{R}^d} V(x) f_n(x)\overline{f_\infty(x)} dx\right| \le \left(\int_{\mathbb{R}^d} V(x)|f_n(x)|^2 dx\right)^{\frac{1}{2}} \left(\int_{\mathbb{R}^d} V(x)|f_\infty(x)|^2 dx\right)^{\frac{1}{2}} \tag{103}$$

and taking liminf of the both sides, we get the desired inequality (102). The relations (99), (100), and (101) deduce

$$S_{\omega,V}(f_\infty) \le \liminf_{n\to\infty} S_{\omega,V}(f_n) = r_{\omega,V}^{\alpha,\beta}. \tag{104}$$

First, we prove that f_∞ is not trivial. We assume $f_\infty = 0$ for contradiction. Then, we have

$$\begin{aligned} 0 &= \frac{(p+1)\alpha - d\beta}{p+1} \lim_{n\to\infty} \|f_n\|_{L^{p+1}}^{p+1} \\ &= \lim_{n\to\infty} \left\{ \frac{\omega(2\alpha - d\beta)}{2} \|f_n\|_{L^2}^2 + \frac{2\alpha - (d-2)\beta}{2} \|\nabla f_n\|_{L^2}^2 \right. \\ &\quad \left. + \frac{2\alpha - d\beta}{2} \int_{\mathbb{R}^d} V(x)|f_n(x)|^2 dx - \frac{\beta}{2} \int_{\mathbb{R}^d} (x\cdot\nabla V)|f_n(x)|^2 dx \right\} \\ &\ge \frac{2\alpha - (d-2)\beta}{2} \liminf_{n\to\infty} \|\nabla f_n\|_{L^2}^2 \ge 0 \end{aligned} \tag{105}$$

by (101) and (97), that is, $\liminf_{n\to\infty} \|\nabla f_n\|_{L^2} = 0$. From Lemma 4.1, we get $K_{\omega,V}^{\alpha,\beta}(f_n) > 0$ for sufficiently large n. This is contradiction to (97). Next, we prove that there exists a function $Q_{\omega,V} \in H^1 \setminus \{0\}$ such that $S_{\omega,V}(Q_{\omega,V}) = r_{\omega,V}^{\alpha,\beta}$ and $K_{\omega,V}^{\alpha,\beta}(Q_{\omega,V}) = 0$. From the weak limit $f_n \rightharpoonup f_\infty$ in $L^2(\mathbb{R}^d)$ as $n \to \infty$, we can see that

$$\begin{aligned} &\int_{\mathbb{R}^d} V(x)|f_n(x)|^2 dx - \int_{\mathbb{R}^d} V(x)|f_n(x) - f_\infty(x)|^2 dx \\ &\qquad = 2\mathrm{Re} \int_{\mathbb{R}^d} V(x) f_n(x)\overline{f_\infty(x)} dx - \int_{\mathbb{R}^d} V(x)|f_\infty(x)|^2 dx \\ &\qquad \longrightarrow \int_{\mathbb{R}^d} V(x)|f_\infty(x)|^2 dx \quad \text{as} \quad n \to \infty \end{aligned} \tag{106}$$

and

$$\int_{\mathbb{R}^d} (x \cdot \nabla V)|f_n(x)|^2 dx - \int_{\mathbb{R}^d} (x \cdot \nabla V)|f_n(x) - f_\infty(x)|^2 dx \\ \longrightarrow \int_{\mathbb{R}^d} (x \cdot \nabla V)|f_\infty(x)|^2 dx \quad \text{as} \quad n \to \infty. \tag{107}$$

Combining (97), (106), (107), and Lemma 2.3, we have

$$S_{\omega,V}(f_n) - S_{\omega,V}(f_n - f_\infty) \longrightarrow S_{\omega,V}(f_\infty) \quad \text{as} \quad n \to \infty, \tag{108}$$

$$T^{\alpha,\beta}_{\omega,V}(f_n) - T^{\alpha,\beta}_{\omega,V}(f_n - f_\infty) \longrightarrow T^{\alpha,\beta}_{\omega,V}(f_\infty) \quad \text{as} \quad n \to \infty, \tag{109}$$

$$-K^{\alpha,\beta}_{\omega,V}(f_n - f_\infty) \longrightarrow K^{\alpha,\beta}_{\omega,V}(f_\infty) \quad \text{as} \quad n \to \infty. \tag{110}$$

The inequality

$$\|(2\omega + 2V + x \cdot \nabla V)^{\frac{1}{2}} f_\infty\|_{L^2} \leq \liminf_{n\to\infty} \|(2\omega + 2V + x \cdot \nabla V)^{\frac{1}{2}} f_n\|_{L^2} \tag{111}$$

follows from the same argument as (102). This inequality deduces

$$T^{\alpha,\beta}_{\omega,V}(f_\infty) \leq \liminf_{n\to\infty} T^{\alpha,\beta}_{\omega,V}(f_n) = r^{\alpha,\beta}_{\omega,V}. \tag{112}$$

Since

$$\lim_{n\to\infty} T^{\alpha,\beta}_{\omega,V}(f_n - f_\infty) = \lim_{n\to\infty} T^{\alpha,\beta}_{\omega,V}(f_n) - T^{\alpha,\beta}_{\omega,V}(f_\infty) \\ = r^{\alpha,\beta}_{\omega,V} - T^{\alpha,\beta}_{\omega,V}(f_\infty) < r^{\alpha,\beta}_{\omega,V} \tag{113}$$

by Proposition 4.1, there exists $n_0 \in \mathbb{N}$ such that

$$K^{\alpha,\beta}_{\omega,V}(f_n - f_\infty) > 0 \tag{114}$$

for any $n \geq n_0$ by Lemma 4.3. That is, we have

$$K^{\alpha,\beta}_{\omega,V}(f_\infty) \leq 0 \tag{115}$$

and

$$r^{\alpha,\beta}_{\omega,V} \leq T^{\alpha,\beta}_{\omega,V}(f_\infty). \tag{116}$$

Combining (112) and (116), we have

$$r^{\alpha,\beta}_{\omega,V} = T^{\alpha,\beta}_{\omega,V}(f_\infty). \tag{117}$$

Applying Lemma 4.3, there exists $0 < \lambda \leq 1$ such that

$$\begin{aligned} &K^{\alpha,\beta}_{\omega,V}(\lambda f_\infty) = 0, \text{ and} \\ &r^{\alpha,\beta}_{\omega,V} \leq S_{\omega,V}(\lambda f_\infty) = T^{\alpha,\beta}_{\omega,V}(\lambda f_\infty) \leq T^{\alpha,\beta}_{\omega,V}(f_\infty) = r^{\alpha,\beta}_{\omega,V}. \end{aligned} \tag{118}$$

If we set $Q_{\omega,V} = \lambda f_\infty$, then

$$K^{\alpha,\beta}_{\omega,V}(Q_{\omega,V}) = 0 \quad \text{and} \quad S_{\omega,V}(Q_{\omega,V}) = r^{\alpha,\beta}_{\omega,V}. \tag{119}$$

From now on, we prove the latter half in Radial case of Theorem 1.5.

Lemma 4.4 *Let $d \geq 3$, $1 + \frac{4}{d} < p < 1 + \frac{4}{d-2}$, V, $x \cdot \nabla V$, $x\nabla^2 V x^T \in L^{\frac{d}{2}}(\mathbb{R}^d) + L^\infty(\mathbb{R}^d)$, $3x \cdot \nabla V + x\nabla^2 V x^T \leq 0$, and let (α, β) satisfy (32).*

$$(\mathcal{L}^{\alpha,\beta} - \overline{\mu})(\mathcal{L}^{\alpha,\beta} - \underline{\mu})S_{\omega,V}(f) \leq -\frac{(p-1)\alpha\{(p-1)\alpha - 2\beta\}}{p+1}\|f\|^{p+1}_{L^{p+1}} \tag{120}$$

holds for any $f \in H^1(\mathbb{R}^d)$. In particular, if f satisfies $K^{\alpha,\beta}_{\omega,V}(f) = 0$, then it follows that

$$\mathcal{L}^{\alpha,\beta}K^{\alpha,\beta}_{\omega,V}(f) \leq -\overline{\mu}\underline{\mu}T^{\alpha,\beta}_{\omega,V}(f) - \frac{(p-1)\alpha\{(p-1)\alpha - 2\beta\}}{p+1}\|f\|^{p+1}_{L^{p+1}}. \tag{121}$$

Proof By the simple calculation, we have

$$\mathcal{L}^{\alpha,\beta}\|f\|^2_{L^2} = \underline{\mu}\|f\|^2_{L^2}, \tag{122}$$

$$\mathcal{L}^{\alpha,\beta}\|\nabla f\|^2_{L^2} = \overline{\mu}\|\nabla f\|^2_{L^2}, \tag{123}$$

$$\mathcal{L}^{\alpha,\beta}\|f\|^{p+1}_{L^{p+1}} = \{(p+1)\alpha - d\beta\}\|f\|^{p+1}_{L^{p+1}}, \tag{124}$$

$$\begin{aligned} \mathcal{L}^{\alpha,\beta}\int_{\mathbb{R}^d} V(x)|f(x)|^2 dx = \underline{\mu}\int_{\mathbb{R}^d} V(x)|f(x)|^2 dx \\ - \beta\int_{\mathbb{R}^d}(x \cdot \nabla V)|f(x)|^2 dx, \end{aligned} \tag{125}$$

and

$$\mathcal{L}^{\alpha,\beta}\int_{\mathbb{R}^d}(x\cdot\nabla V)|f(x)|^2dx=\{2\alpha-(d+1)\beta\}\int_{\mathbb{R}^d}(x\cdot\nabla V)|f(x)|^2dx$$
$$-\beta\int_{\mathbb{R}^d}(x\nabla^2Vx^T)|f(x)|^2dx. \tag{126}$$

These identities and $3x\cdot\nabla V+x\nabla^2Vx^T\le 0$ imply desired results. □

Proposition 4.3 *Besides the assumptions of Lemma 4.4, we assume that V be radially symmetric. Then,*

$$\mathcal{M}^{\alpha,\beta}_{\omega,V,rad}\subset\mathcal{G}_{\omega,V,rad} \tag{127}$$

holds.

Proof We take $\phi\in\mathcal{M}^{\alpha,\beta}_{\omega,V,\mathrm{rad}}$. From Lemma 4.4, we have

$$\begin{aligned}\left\langle\left(K^{\alpha,\beta}_{\omega,V}\right)'(\phi),\mathcal{L}^{\alpha,\beta}\phi\right\rangle&=\mathcal{L}^{\alpha,\beta}K^{\alpha,\beta}_{\omega,V}(\phi)\\&\le-\overline{\mu}\underline{\mu}T^{\alpha,\beta}_{\omega,V}(\phi)-\frac{(p-1)\alpha\{(p-1)\alpha-2\beta\}}{p+1}\|\phi\|^{p+1}_{L^{p+1}}<0.\end{aligned} \tag{128}$$

Thus, there exists the Lagrange multiplier $\eta\in\mathbb{R}$ such that

$$S'_{\omega,V}(\phi)=\eta\left(K^{\alpha,\beta}_{\omega,V}\right)'(\phi). \tag{129}$$

This identity deduces

$$\begin{aligned}0&=K^{\alpha,\beta}_{\omega,V}(\phi)=\mathcal{L}^{\alpha,\beta}S_{\omega,V}(\phi)\\&=\left\langle S'_{\omega,V}(\phi),\mathcal{L}^{\alpha,\beta}\phi\right\rangle=\eta\left\langle\left(K^{\alpha,\beta}_{\omega,V}\right)'(\phi),\mathcal{L}^{\alpha,\beta}\phi\right\rangle.\end{aligned} \tag{130}$$

This implies $\eta=0$ by (128). Therefore, we obtain $S'_{\omega,V}(\phi)=0$ by (129). We take any $\psi\in\mathcal{A}_{\omega,V,\mathrm{rad}}$. Then, we have

$$K^{\alpha,\beta}_{\omega,V}(\psi)=\left\langle S'_{\omega,V}(\psi),\mathcal{L}^{\alpha,\beta}(\psi)\right\rangle=0. \tag{131}$$

Therefore, we obtain $S_{\omega,V}(\phi)\le S_{\omega,V}(\psi)$, that is, $\phi\in\mathcal{G}_{\omega,V,\mathrm{rad}}$. □

Proposition 4.4 *We assume the same conditions with Proposition 4.3. If $\mathcal{M}^{\alpha,\beta}_{\omega,V,rad}$ is not empty, then*

$$\mathcal{G}_{\omega,V,rad}\subset\mathcal{M}^{\alpha,\beta}_{\omega,V,rad} \tag{132}$$

holds.

Proof We take $\phi \in \mathcal{G}_{\omega,V,\mathrm{rad}}$. We take a $\psi \in \mathcal{M}^{\alpha,\beta}_{\omega,V,\mathrm{rad}} \subset \mathcal{G}_{\omega,V,\mathrm{rad}}$, where the last inclusion holds by Proposition 4.3. Let $v \in H^1_{\mathrm{rad}} \setminus \{0\}$ satisfy $K^{\alpha,\beta}_{\omega,V}(v) = 0$. Then, it follows that

$$S_{\omega,V}(\phi) = S_{\omega,V}(\psi) \le S_{\omega,V}(v). \tag{133}$$

In addition, we have $K^{\alpha,\beta}_{\omega,V}(\phi) = \left\langle S'_{\omega,V}(\phi), \mathcal{L}^{\alpha,\beta}\phi \right\rangle = 0$. Therefore, we obtain $\phi \in \mathcal{M}^{\alpha,\beta}_{\omega,V,\mathrm{rad}}$. □

The identity

$$\mathcal{M}^{\alpha,\beta}_{\omega,V,\mathrm{rad}} = \mathcal{G}_{\omega,V,\mathrm{rad}} \tag{134}$$

in Theorem 1.5 holds by Theorem 4.1, Proposition 4.3, and Proposition 4.4.

5 Well-Posedness in H^1 for Time-Dependent NLS

5.1 Local Well-Posedness in H^1

First, we recall local well-posedness to (1) in H^1 and $d = 3$.

Theorem 5.1 (Local Well-Posedness in H^1 (I)) *Let $1 < p < 5$, $V \in L^{\frac{3}{2}}(\mathbb{R}^3) \cap \mathcal{K}_0(\mathbb{R}^3)$, and $\|V_-\|_{\mathcal{K}} < 4\pi$. For any $u_0 \in H^1(\mathbb{R}^3)$, there exist $T = T(\|u_0\|_{H^1}, V) > 0$ and a unique solution u to* (1) *on a time interval $[-T, T]$ with*

$$u \in C([-T,T]; H^1(\mathbb{R}^3)) \cap L^q(-T,T; W^{1,r}_V(\mathbb{R}^3)) \tag{135}$$

for any L^2-admissible pair (q,r). Moreover, the solution u conserves its mass and energy defined as (8) *and* (9) *respectively.*

Proof This result is deduced by a contraction argument by using Lemma 2.1, Theorem 2.1, and Lemma 2.2. We refer to [12] and [9] for the detail. □

The following local well-posedness holds by Cazenave [3, Theorem 4.3.1].

Theorem 5.2 (Local Well-Posedness in H^1 (II)) *Let $d \ge 3$, $1 < p < 1 + \frac{4}{d-2}$, $V \in L^\gamma(\mathbb{R}^d) + L^\infty(\mathbb{R}^d)$ for some $\frac{d}{2} < \gamma \le \infty$. For any $u_0 \in H^1(\mathbb{R}^d)$, there exist $T = T(u_0) > 0$ and a unique solution u to* (1) *on a time interval $[-T, T]$. The solution u conserves its mass and energy defined as* (8) *and* (9) *respectively. Moreover, the solution u has the following blow-up alternative: If $T_{max} < \infty$, then $\lim_{t \nearrow T_{max}} \|u(t)\|_{H^1} = \infty$ (resp. if $T_{min} > -\infty$, then $\lim_{t \searrow T_{min}} \|u(t)\|_{H^1} = \infty$).*

5.2 *Global Well-Posedness in H^1*

In this subsection, we prove Theorem 1.6. Non-radial case in Theorem 1.6 is deduced by the same argument with Radial case in Theorem 1.6 and Proposition 1.1. Hence, we only prove Radial case. If $u_0 \in H^1_{\mathrm{rad}}(\mathbb{R}^d)$ satisfies the assumptions of Theorem 1.6 and $K^{\alpha,\beta}_{\omega,V}(u_0) = 0$, then $u_0(x) \equiv 0$ by the definition of $r^{\alpha,\beta}_{\omega,V}$. Thus, $u(t,x) \equiv 0$ holds. From now on, we only consider the case $K^{\alpha,\beta}_{\omega,V}(u_0) > 0$.

Lemma 5.1 *Let $d \geq 3$, $1 + \frac{4}{d} < p < 1 + \frac{4}{d-2}$, $\omega_0 < \infty$, $x \cdot \nabla V \in L^{\frac{d}{2}}(\mathbb{R}^d) + L^\infty(\mathbb{R}^d)$, $V \geq 0$, $x \cdot \nabla V \leq 0$. Let the radial potential V satisfy the following (i) or (ii):*

(i) $d = 3$, $V \in L^{\frac{3}{2}}(\mathbb{R}^3) \cap \mathcal{K}_0(\mathbb{R}^3)$,
(ii) $d \geq 3$, $V \in L^\gamma(\mathbb{R}^d) + L^\infty(\mathbb{R}^d)$ for some $\frac{d}{2} < \gamma \leq \infty$.

We assume that a radial datum $u_0 \in H^1_{rad}(\mathbb{R}^d)$ satisfies

$$S_{\omega,V}(u_0) < r^{\alpha,\beta}_{\omega,V} \quad \text{and} \quad K^{\alpha,\beta}_{\omega,V}(u_0) > 0 \tag{136}$$

for some (α, β) satisfying (32) and $\omega \geq \omega_0$. Then, the solution u to (1) satisfies

$$K^{\alpha,\beta}_{\omega,V}(u(t)) > 0 \tag{137}$$

for any $t \in (T_{min}, T_{max})$.

Proof If the conclusion does not hold, then there exists $t_0 \in (T_{\min}, T_{\max})$ such that $K^{\alpha,\beta}_{\omega,V}(u(t_0)) = 0$. For such t_0, we have

$$r^{\alpha,\beta}_{\omega,V} \leq S_{\omega,V}(u(t_0)) \tag{138}$$

by the definition of $r^{\alpha,\beta}_{\omega,V}$. On the other hand, the conservation laws implies

$$S_{\omega,V}(u(t_0)) = S_{\omega,V}(u_0) < r^{\alpha,\beta}_{\omega,V}. \tag{139}$$

This is contradiction. □

Theorem 5.3 (Global Well-Posedness in H^1) *We assume the same assumptions with Lemma 5.1. Then, the solution u to (1) with the initial data u_0 exists globally in time.*

Proof Since $\frac{1}{(p+1)\alpha-d\beta}K^{\alpha,\beta}_{\omega,V}(u(t)) > 0$ by Lemma 5.1, we have

$$
\begin{aligned}
-\frac{1}{p+1}\|u(t)\|^{p+1}_{L^{p+1}} &> -\frac{2\alpha-(d-2)\beta}{2\{(p+1)\alpha-d\beta\}}J_{\omega,V}(f) \\
&\quad + \frac{\beta}{2\{(p+1)\alpha-d\beta\}}\int_{\mathbb{R}^d}(2\omega+2V+x\cdot\nabla V)|u(t,x)|^2dx \\
&\geq -\frac{2\alpha-(d-2)\beta}{2\{(p+1)\alpha-d\beta\}}J_{\omega,V}(f).
\end{aligned}
\tag{140}
$$

This inequality implies

$$
r^{\alpha,\beta}_{\omega,V} > S_{\omega,V}(u(t)) > \frac{(p-1)\alpha-2\beta}{2\{(p+1)\alpha-d\beta\}}J_{\omega,V}(u(t)) \sim \|u(t)\|^2_{H^1}. \tag{141}
$$

Therefore, we see that u exists globally in time. □

Acknowledgments The authors would like to express deep appreciation to Professor Takahisa Inui and Professor Masahito Ohta. Professor Masahito Ohta introduced this problem to the authors and Professor Takahisa Inui gave a lot of useful comments. The first author is supported by the Grant-in-Aid for Japan Society for the Promotion of Science (No.19J13300). The second author is supported by the Grant-in-Aid for Scientific Research (B) (No.18H01132), Young Scientists Research (No.19K14581), Japan Society for the Promotion of Science and JST CREST (JPMJCR1913), Japan.

References

1. Akahori, T., Nawa, H.: Blowup and scattering problems for the nonlinear Schrödinger equations. Kyoto J. Math. **53**(3), 629–672 (2013). MR3102564
2. Brézis, H., Lieb, E.: A relation between pointwise convergence of functions and convergence of functionals. Proc. Amer. Math. Soc. **88**(3), 486–490 (1983). MR0699419
3. Cazenave, T.: Semilinear Schrödinger Equations. Courant Lecture Notes in Mathematics, vol. 10, xiv+323 pp. New York University, Courant Institute of Mathematical Sciences, New York; American Mathematical Society, Providence (2003). MR2002047
4. Dodson, B., Murphy, J.: A new proof of scattering below the ground state for the 3D radial focusing cubic NLS. Proc. Amer. Math. Soc. **145**(11), 4859–4867 (2017)
5. Du, D., Wu, Y., Zhang, K.: On blow-up criterion for the nonlinear Schrödinger equation. Discrete Contin. Dyn. Syst. **36**(7), 3639–3650 (2016)
6. Duyckaerts, T., Holmer, J., Roudenko, S.: Scattering for the non-radial 3D cubic nonlinear Schrödinger equation. Math. Res. Lett. **15**(6), 1233–1250 (2008)
7. Fang, D., Xie, J., Cazenave, T.: Scattering for the focusing energy-subcritical nonlinear Schrödinger equation. Sci. China Math. **54**(10), 2037–2062 (2011)
8. Foschi, D.: Inhomogeneous Strichartz estimates. J. Hyperbolic Differ. Equ. **2**(1), 1–24 (2005)
9. Hamano, M., Ikeda, M.: Global dynamics below the ground state for the focusing Schrödinger equation with a potential. J. Evol. Equ. **20**(3), 1131–1172 (2020)

10. Hamano, M., Ikeda, M.: Global well-posedness below the ground state for the nonlinear Schrödinger equation with a linear potential, Proc. Amer. Math. Soc. To appear
11. Holmer, J., Roudeko, S.: A sharp condition for scattering of the radial 3D cubic nonlinear Schrödinger equation. Comm. Math. Phys. **282**(2), 435–467 (2008)
12. Hong, Y.: Scattering for a nonlinear Schrödinger equation with a potential. Commun. Pure Appl. Anal. **15**(5), 1571–1601 (2016)
13. Ibrahim, S., Masmoudi, N., Nakanishi, K.: Scattering threshold for the focusing nonlinear Klein-Gordon equation. Anal. PDE **4**(3), 405–460 (2011)
14. Ikeda, M., Inui, T.: Global dynamics below the standing waves for the focusing semilinear Schrödinger equation with a repulsive Dirac delta potential. Anal. PDE **10**(2), 481–512 (2017)
15. Keel, M., Tao, T.: Endpoint Strichartz estimates. Amer. J. Math. **120**(5), 955–980 (1998)
16. Kenig, C.E., Merle, F.: Global well-posedness, scattering and blow-up for the energy-critical, focusing, non-linear Schrödinger equation in the radial case. Invent. Math. **166**(3), 645–675 (2006)
17. Killip, R., Masaki, S., Murphy, J., Visan, M.: Large data mass-subcritical NLS: critical weighted bounds imply scattering. NoDEA Nonlin. Differ. Equ. Appl. **24**(4), 38 (2017)
18. Killip, R., Murphy, J., Visan, M., Zheng, J.: The focusing cubic NLS with inverse-square potential in three space dimensions. Differ. Integ. Equ. **30**(3–4), 161–206 (2017). MR3611498
19. Mizutani, H.: Wave operators on Sobolev spaces. Proc. Amer. Math. Soc. **148**(4), 1645–1652 (2020)
20. Rose, H.A., Weinstein, M.I.: On the bound states of the nonlinear Schrödinger equation with a linear potential. Phys. D **30**(1–2), 207–218 (1988)

Well-Posedness for a Generalized Klein-Gordon-Schrödinger Equations

Jishan Fan and Tohru Ozawa

Abstract We first prove uniform-in-ϵ regularity estimates of strong solutions to a generalized Klein-Gordon-Schrödinger equations in $\mathbb{R}^n$ $(n \geq 1)$. Here ϵ is the dispersion coefficient. Then we prove the global well-posedness of strong solutions to the limit problem $(\epsilon \to 0)$ when $n \leq 3$.

Keywords Klein-Gordon-Schrödinger · Well-posedness · Uniform estimates

Mathematics Subject Classifications (2010) 35Q55, 35L70

1 Introduction

In this paper, we consider the following generalized Klein-Gordon-Schrödinger system [1]:

$$i\partial_t\psi + \epsilon\Delta\psi = \phi^2\psi - \phi\psi + f(x,t), \tag{1}$$

$$(\Box + 1)\phi = -2|\psi|^2\phi + |\psi|^2 \quad \text{in } \mathbb{R}^n \times (0,\infty), \tag{2}$$

$$(\psi, \phi, \partial_t\phi)(\cdot, 0) = (\psi_0, \phi_0, \phi_1)(\cdot) \quad \text{in } \mathbb{R}^n, \tag{3}$$

where $n \geq 1$, ψ is a complex nucleon field, ϕ is a real meson field, $\epsilon \in (0, 1]$ is the dispersion coefficient, f is a complex smooth function, and $\Box := \partial_t^2 - \Delta$.

J. Fan (✉)
Department of Applied Mathematics, Nanjing Forestry University, Nanjing, P.R. China
e-mail: fanjishan@njfu.com.cn

T. Ozawa
Department of Applied Physics, Waseda University, Tokyo, Japan
e-mail: txozawa@waseda.jp

© Springer Nature Switzerland AG 2020

V. Georgiev et al. (eds.), *Advances in Harmonic Analysis and Partial Differential Equations*, Trends in Mathematics, https://doi.org/10.1007/978-3-030-58215-9_13

The classical Klein-Gordon-Schrödinger system without the cubic nonlinearities was considered extensively in the literature [2–16]. Guo and Li [10] showed existence of global attractor when $n = 3$. Pecher [11, 12] proved well-posedness in Bourgain-Klainerman-Machedon type spaces when $n = 2$ and $n = 3$.

For the generalized Klein-Gordon-Schrödinger system, much less is known. Tang and Ding [17] considered modulational instability and exact solutions. Shi et al. [18] proved global well-posedness in the energy space using Kato type smoothing estimates for the nonlinear Schrödinger equation. Very recently, Pecher [1] proves local existence, uniqueness and continuous dependence on the data in Bourgain-Klainerman-Machedon spaces for low regularity data, which improves earlier results by Shi et al. [19]. Global well-posedness in the energy space is also obtained as a special case.

All the above results dealt with the case $\epsilon = 1$. The aim of this paper is to study uniform regularity independent of $\epsilon > 0$ and consider the limit problem when $\epsilon \to 0$ and $n \le 3$. We will prove

Theorem 1.1 *Let $n \ge 1, s \in [1, \infty) \cap \left(\frac{n}{2}, \infty\right)$, and $\epsilon \in (0, 1]$. Let $\psi_0, \phi_0 \in H^s$, and $\phi_1 \in H^{s-1}$. Then there exists $T_0 > 0$ independent of $\epsilon > 0$ such that for some $T \in [T_0, \infty)$ there exist C_T independent of $\epsilon > 0$ such that the problem* (1)–(3) *has a unique solution $(\psi^\epsilon, \phi^\epsilon)$ satisfying*

$$\|(\psi^\epsilon, \phi^\epsilon)\|_{L^\infty(0,T;H^s)} + \|\partial_t \phi^\epsilon\|_{L^\infty(0,T;H^{s-1})} \le C_T \tag{4}$$

Theorem 1.2 *Let $\epsilon = 0$ and $n \le 3$. Let $\phi_0, \psi_0 \in H^2$, and $\phi_1 \in H^1$. Then the limit problem (when $\epsilon = 0$) has a unique global strong solution (ψ^0, ϕ^0) satisfying* (4) *with $s = 2$ for some $T > 0$.*

Theorem 1.3 *Let $n \le 3$. Let $(\psi^\epsilon, \phi^\epsilon)$ and (ψ^0, ϕ^0) be H^2-solutions given respectively by Theorems 1.1 and 1.2. Then for some $T > 0$*

$$\|(\psi^\epsilon - \psi^0, \phi^\epsilon - \phi^0)\|_{L^\infty(0,T;H^2)} \to 0 \tag{5}$$

as $\epsilon \to 0$.

In the following proofs, we will use the bilinear product estimates due to Kato and Ponce [20]:

$$\|\Lambda^s(fg)\|_{L^p} \le C(\|f\|_{L^{p_1}}\|\Lambda^s g\|_{L^{q_1}} + \|\Lambda^s f\|_{L^{p_2}}\|g\|_{L^{q_2}}) \tag{6}$$

with $s > 0$, $\Lambda := (-\Delta)^{\frac{1}{2}}$ and $\dfrac{1}{p} = \dfrac{1}{p_1} + \dfrac{1}{q_1} = \dfrac{1}{p_2} + \dfrac{1}{q_2}$.

2 Proof of Theorem 1.1

Since the local existence of smooth solutions is well-known, we only need to show (4) for smooth solutions.

Proof ***(Theorem 1.1)*** Equation (1) can be written as

$$\partial_t \psi = i\epsilon\Delta\psi - i\phi^2\psi + i\phi\psi - if. \tag{7}$$

Testing (7) by $\overline{\psi}$ and taking the real parts, we see that

$$\frac{1}{2}\frac{\mathrm{d}}{\mathrm{d}t}\int |\psi|^2 \mathrm{d}x = -\mathrm{Re}\, i \int f\overline{\psi}\mathrm{d}x \le \|f\|_{L^2}\|\psi\|_{L^2}, \tag{8}$$

and thus

$$\|\psi\|_{L^2} \le \|\psi_0\|_{L^2} + \int_0^t \|f(s)\|_{L^2}\mathrm{d}s. \tag{9}$$

Applying Λ^s to (7), testing by $\Lambda^s\overline{\psi}$, and taking the real parts, and using (6), we find that

$$\begin{aligned}
&\frac{1}{2}\frac{\mathrm{d}}{\mathrm{d}t}\int |\Lambda^s\psi|^2\mathrm{d}x \\
&= -\mathrm{Re}\, i\int \Lambda^s(\phi^2\psi)\Lambda^s\overline{\psi}\mathrm{d}x + \mathrm{Re}\, i\int \Lambda^s(\phi\psi)\Lambda^s\overline{\psi}\mathrm{d}x - \mathrm{Re}\, i\int \Lambda^s f\Lambda^s\overline{\psi}\mathrm{d}x \\
&\le C\|\Lambda^s(\phi^2\psi)\|_{L^2}\|\Lambda^s\psi\|_{L^2} + C\|\Lambda^s(\phi\psi)\|_{L^2}\|\Lambda^s\psi\|_{L^2} + C\|\Lambda^s f\|_{L^2}\|\Lambda^s\psi\|_{L^2} \\
&\le C(\|\phi\|_{L^\infty}\|\Lambda^s(\phi\psi)\|_{L^2} + \|\Lambda^s\phi\|_{L^2}\|\phi\psi\|_{L^\infty})\|\Lambda^s\psi\|_{L^2} \\
&\quad + C\|\Lambda^s(\phi\psi)\|_{L^2}\|\Lambda^s\psi\|_{L^2} + C\|\Lambda^s\psi\|_{L^2} \\
&\le C(\|\phi\|_{L^\infty}^2\|\Lambda^s\psi\|_{L^2} + \|\Lambda^s\phi\|_{L^2}\|\phi\|_{L^\infty}\|\psi\|_{L^\infty})\|\Lambda^s\psi\|_{L^2} \\
&\quad + C(\|\phi\|_{L^\infty}\|\Lambda^s\psi\|_{L^2} + \|\psi\|_{L^\infty}\|\Lambda^s\phi\|_{L^2})\|\Lambda^s\psi\|_{L^2} + C\|\Lambda^s\psi\|_{L^2} \\
&\le C\|\phi\|_{H^s}^4 + C\|\psi\|_{H^s}^4 + C.
\end{aligned} \tag{10}$$

Testing (2) by $\partial_t \phi$, we get

$$\begin{aligned}
&\frac{1}{2}\frac{\mathrm{d}}{\mathrm{d}t}\int((\partial_t\phi)^2 + |\nabla\phi|^2 + \phi^2)\mathrm{d}x \\
&\quad = \int(-2|\psi|^2\phi + |\psi|^2)\partial_t\phi \mathrm{d}x \\
&\quad \le (2\|\psi\|_{L^\infty}^2\|\phi\|_{L^2} + \|\psi\|_{L^4}^2)\|\partial_t\phi\|_{L^2} \\
&\quad \le C\|\psi\|_{H^s}^4 + C\|\phi\|_{L^2}^4 + C\|\partial_t\phi\|_{L^2}^4 + C.
\end{aligned} \tag{11}$$

Applying Λ^{s-1} to (2), testing by $\Lambda^{s-1}\partial_t\phi$ and using (6), we have

$$\begin{aligned}
&\frac{1}{2}\frac{\mathrm{d}}{\mathrm{d}t}\int((\Lambda^{s-1}\partial_t\phi)^2 + (\Lambda^s\phi)^2 + (\Lambda^{s-1}\phi)^2)\mathrm{d}x \\
&\quad = \int(\Lambda^{s-1}(-2|\psi|^2\phi) + \Lambda^{s-1}(|\psi|^2))\Lambda^{s-1}\partial_t\phi \mathrm{d}x \\
&\quad \le C(\|\psi\|_{H^s}^2\|\phi\|_{H^s} + \|\psi\|_{H^s}^2)\|\Lambda^{s-1}\partial_t\phi\|_{L^2} \\
&\quad \le C\|\Lambda^{s-1}\partial_t\phi\|_{L^2}^4 + C\|\psi\|_{H^s}^4 + C\|\phi\|_{H^s}^4 + C.
\end{aligned} \tag{12}$$

Summing up (9), (10), (11) and (12), we easily get $\frac{\mathrm{d}y}{\mathrm{d}t} \le cy^2$ for

$$y := 1 + \|\psi\|_{H^s}^2 + \|\phi\|_{H^s}^2 + \|\partial_t\phi\|_{H^{s-1}}^2, \tag{13}$$

which shows that

$$\frac{\mathrm{d}y}{y^2} \le c\mathrm{d}t, \tag{14}$$

and therefore

$$y \le \frac{y_0}{1 - cy_0 t} \quad \left(t \le T < \frac{1}{cy_0}\right). \tag{15}$$

The proof is complete. □

3 Proof of Theorem 1.2

***Proof** (Theorem 1.2)* We only deal with the case $n = 3$. First, we have

$$\partial_t\psi = -i\phi^2\psi + i\phi\psi - if. \tag{16}$$

We also have (9). It is clear that

$$\frac{1}{2}\frac{\mathrm{d}}{\mathrm{d}t}|\psi|^2 = \mathrm{Im}(\overline{\psi} f), \tag{17}$$

and therefore

$$|\psi|^2 \le |\psi_0|^2 + \int_0^t 2|\psi f|\mathrm{d}\tau \le |\psi_0|^2 + \int_0^t (|\psi|^2 + |f^2|)\mathrm{d}\tau, \tag{18}$$

which gives

$$\|\psi\|_{L^\infty} \le C. \tag{19}$$

It follows from (11) that

$$\begin{aligned}\frac{1}{2}\frac{\mathrm{d}}{\mathrm{d}t}&\int((\partial_t\phi)^2 + |\nabla\phi|^2 + \phi^2)\mathrm{d}x\\ &= \int(-2|\psi|^2\phi + |\psi|^2)\partial_t\phi\mathrm{d}x\\ &\le (2\|\psi\|_{L^\infty}^2\|\phi\|_{L^2} + \|\psi\|_{L^4}^2)\|\partial_t\phi\|_{L^2}\\ &\le C(\|\phi\|_{L^2} + 1)\|\partial_t\phi\|_{L^2},\end{aligned} \tag{20}$$

and hence

$$\int(\phi^2 + |\nabla\phi|^2 + (\partial_t\phi)^2)\mathrm{d}x \le C. \tag{21}$$

It follows from (16), (19) and (21) that

$$\int|\partial_t\psi|^2\mathrm{d}x + \int|\partial_t\psi|^3\mathrm{d}x \le C. \tag{22}$$

Applying ∂_t to (2), testing by $\partial_t^2\phi$, using (19), (21) and (22), we obtain

$$\begin{aligned}&\frac{1}{2}\frac{\mathrm{d}}{\mathrm{d}t}\int((\partial_t\phi)^2 + |\nabla(\partial_t\phi)|^2 + (\partial_t^2\psi)^2)\mathrm{d}x\\ &= \int\partial_t(-2|\psi|^2\phi + |\psi|^2)(\partial_t^2\phi)\mathrm{d}x\\ &\le C\int(|\partial_t\psi||\phi| + |\partial_t\phi| + |\partial_t\psi|)|\partial_t^2\phi|\mathrm{d}x\\ &\le C(\|\partial_t\psi\|_{L^3}\|\phi\|_{L^6} + \|\partial_t\phi\|_{L^2} + \|\partial_t\psi\|_{L^2})\|\partial_t^2\phi\|_{L^2}\\ &\le C\|\partial_t^2\phi\|_{L^2},\end{aligned} \tag{23}$$

which gives

$$\int ((\partial_t\phi)^2 + |\nabla \partial_t\phi|^2 + (\partial_t^2\phi)^2)\mathrm{d}x \le C. \tag{24}$$

Equation (2) can be written as

$$-\Delta\phi + \phi = -\partial_t^2\phi - 2|\psi|^2\phi + |\psi|^2 \in L^\infty(0, T; L^2). \tag{25}$$

Using the H^2-estimates of Poisson equation, we have

$$\|\phi(\cdot, t)\|_{H^2} \le C. \tag{26}$$

Applying Δ to (16), testing by $\Delta\overline{\psi}$ and taking the real parts, and using (19) and (26), we arrive at

$$\begin{aligned}\frac{1}{2}\frac{\mathrm{d}}{\mathrm{d}t}\int |\Delta\psi|^2\mathrm{d}x &\le C\|\phi\|_{H^2}^2\|\psi\|_{H^2}^2 + C\|\phi\|_{H^2}\|\psi\|_{H^2}^2 + C\|\Delta f\|_{L^2}\|\Delta\psi\|_{L^2}\\ &\le C\|\psi\|_{H^2}^2 + C,\end{aligned} \tag{27}$$

which gives

$$\|\psi(\cdot, t)\|_{H^2} \le C. \tag{28}$$

This proof is complete. □

4 Proof of Theorem 1.3

This section is devoted to the proof of Theorem 1.3.

Proof (Theorem 1.3) The solutions $(\psi^\epsilon, \phi^\epsilon)$ and (ψ^0, ϕ^0) satisfy the following integral equations

$$\begin{aligned}\psi^\epsilon(t) &= U(\epsilon t)\psi_0 - i\int_0^t U(\epsilon(t-s))((\phi^\epsilon)^2\psi^\epsilon - \phi^\epsilon\psi^\epsilon + f)(s)\mathrm{d}s,\\ \phi^\epsilon(t) &= \dot{K}(t)\phi_0 + K(t)\phi_1 + \int_0^t K(t-s)(-2|\phi^\epsilon|^2\phi^\epsilon + |\psi^\epsilon|^2)(s)\mathrm{d}s,\\ \psi^0(t) &= \psi_0 - i\int_0^t ((\phi^0)^2\psi^0 - \phi^0\psi^0 + f)(s)\mathrm{d}s,\\ \phi^0(t) &= \dot{K}(t)\phi_0 + K(t)\phi_1 + \int_0^t K(t-s)(-2|\psi^0|^2\phi^0 + |\psi^0|^2)(s)\mathrm{d}s,\end{aligned} \tag{29}$$

where $U(\epsilon t) = \exp(i\epsilon t\Delta)$, $K(t) = (I-\Delta)^{-\frac{1}{2}}\sin(t(I-\Delta)^{\frac{1}{2}})$, and $\dot{K}(t) = \cos(t(I-\Delta)^{\frac{1}{2}})$. We write the difference $(\psi^\epsilon - \psi^0, \phi^\epsilon - \phi^0)$ as

$$\begin{aligned}
&\psi^\epsilon(t) - \psi^0(t)\\
&\quad = (U(\epsilon t) - I)\psi_0\\
&\quad\quad - i\int_0^t U(\epsilon(t-s))((\phi^\epsilon + \phi^0)(\phi^\epsilon - \phi^0)\psi^\epsilon + (\phi^0)^2(\psi^\epsilon - \psi^0)\\
&\quad\quad - (\phi^\epsilon - \phi^0)\psi^\epsilon - \phi^0(\psi^\epsilon - \psi^0))(s)\mathrm{d}s\\
&\quad\quad - i\int_0^t (U(\epsilon(t-s)) - I)((\phi^0)^2\psi^0 - \phi^0\psi^0 + f)(s)\mathrm{d}s,\\
&\phi^\epsilon(t) - \phi^0(t)\\
&\quad = \int_0^t K(t-s)(-2\mathrm{Re}(\overline{(\psi^\epsilon + \psi^0)}(\psi^\epsilon - \psi^0))\phi^\epsilon + 2|\psi^0|^2(\phi^\epsilon - \phi^0)\\
&\quad\quad + \mathrm{Re}(\overline{(\psi^\epsilon + \psi^0)}(\psi^\epsilon - \psi^0)))(s)\mathrm{d}s.
\end{aligned} \tag{30}$$

Since $U(\epsilon t)$ is unitary in H^2, $K(t)$ is bounded in H^2 with operator norm bounded by t, and H^2 forms an algebra, it follows that

$$\begin{aligned}
&\|\psi^\epsilon(t) - \psi^0(t)\|_{H^2} + \|\phi^\epsilon(t) - \phi^0(t)\|_{H^2}\\
&\quad \le \|(U(\epsilon t) - I)\psi_0\|_{H^2}\\
&\quad\quad + \int_0^t \|(U(\epsilon(t-s)) - I)((\phi^0)^2\psi^0 - \phi^0\psi^0 + f)(s)\|_{H^2}\mathrm{d}t\\
&\quad\quad + CT(1 + \sup_{\varepsilon\in[0,1]} \|\phi_\epsilon\|^2_{L^\infty(0,T;H^2)} + \sup_{\varepsilon\in[0,1]} \|\psi_\epsilon\|_{L^\infty(0,T;H^2)})\\
&\quad\quad\quad \times \int_0^t (\|\psi^\epsilon(s) - \psi^0(s)\|_{H^2} + \|\phi^\epsilon(s) - \phi^0(s)\|_{H^2})\mathrm{d}s.
\end{aligned} \tag{31}$$

By Gronwall inequality, we have

$$\begin{aligned}
&\|\psi^\epsilon(t) - \psi^0(t)\|_{H^2} + \|\phi^\epsilon(s) - \phi^0(s)\|_{H^2}\\
&\quad\quad \le L_\epsilon(t) + \int_0^t e^{c(t-s)}(cL_\epsilon(s) + M_\epsilon(s))\mathrm{d}s,
\end{aligned} \tag{32}$$

where

$$
\begin{aligned}
L_\epsilon(t) &= \|(U(\epsilon t) - I)\psi_0\|_{H^2}, \\
M_\epsilon(s) &= \|(U(\epsilon(t-s)) - I)((\phi^0)^2\psi^0 - \phi^0\psi^0 + f)(s)\|_{H^2}, \\
c &= CT(1 + \sup_{\varepsilon\in[0,1]} \|\phi_\epsilon\|^2_{L^\infty(0,T;H^2)} + \sup_{\varepsilon\in[0,1]} \|\psi_\epsilon\|^2_{L^\infty(0,T;H^2)}).
\end{aligned} \tag{33}
$$

The RHS of the last inequality tends to zero uniformly on $[0, T]$ as $\epsilon \to 0$. This completes the proof. □

Acknowledgement Jishan Fan is partially supported by NSFC (No. 11971234).

References

1. Pecher, H.: Well-posedness results for a generalized Klein-Gordon-Schrödinger system (2019). arXiv: 1910.06612 v1 [math.AP] 15 Oct 2019
2. Colliander, J., Holmer, J., Tzirakis, N.: Low regularity global well-posedness for the Zakharov and Klein-Gordon-Schrödinger systems. Trans. Amer. Math. Soc. **360**, 4619–4638 (2008)
3. Fukuda, I., Tsutsumi, M.: On coupled Klein-Gordon-Schrödinger equations II. J. Math. Anal. Appl. **66**, 358–378 (1978)
4. Hayashi, N., von Wahl, W.: On the global strong solution of coupled Klein-Gordon-Schrödinger equations. J. Math. Soc. Japan **39**, 489–497 (1987)
5. Hayashi, N.: Global strong solutions of coupled Klein-Gordon-Schrödinger equations. Funkcial. Ekvac. **29**, 299–307 (1986)
6. Kikuchi, H.: Orbital stability of semitrivial standing waves for the Klein-Gordon-Schrödinger system. Ann. Inst. H. Poincaré Anal. Non Linéaire **28**(2), 315–323 (2011)
7. Kikuchi, H., Ohta, M.: Stability of standing waves for the Klein-Gordon-Schrödinger system. J. Math. Anal. Appl. **365**(1), 109–114 (2010)
8. Ozawa, T., Tsutsumi, Y.: Asymptotic behavior of solutions for the coupled Klein-Gordon-Schrödinger equations. Adv. Stud. Pure Math. **23**, 295–305 (1993)
9. Tzirakis, N.: The Cauchy problem for the Klein-Gordon-Schrödinger system in low dimensions below the energy space. Commun. Partial Differential Equations **30**, 605–641 (2005)
10. Guo, B., Li, Y.: Attractor for dissipative Klein-Gordon-Schrödinger equations in $\mathbb{R}^3$. J. Differential Equations **136**, 356–377 (1997)
11. Pecher, H.: Some new well-posedness results for the Klein-Gordon-Schrödinger system. Differential Integral Equations **25**, 117–142 (2012)
12. Pecher, H.: Low regularity well-posedness for the 3D Klein-Gordon-Schrödinger system. Commun. Pure Appl. Anal. **11**, 1081–1096 (2012)
13. Shi, Q., Wang, S., Li, Y.: Existence and uniqueness of energy solution to Klein-Gordon-Schrödinger equations. J. Differential Equations **252**(1), 168–180 (2012)
14. Shi, Q., Li, W., Wang, S.: Well-posedness in energy space for the nonlinear Klein-Gordon-Schrödinger system. Appl. Math. Comput. **251**, 55–64 (2015)
15. Banquet, C., Ferreira, L., Villamizar-Roa, E.: On existence and scattering theory for the Klein-Gordon-Schrödinger system in an infinite L^2-norm setting. Ann. Mat. Pura Appl. (4) **194**(3), 781–804 (2015)
16. Compaan, E.: Smoothing for the Zakharov and Klein-Gordon-Schrödinger systems on Euclidean spaces. SIAM J. Math. Anal. **49**(5), 4206–4231 (2017)

17. Tang, X.-Y., Ding, W.: The general Klein-Gordon-Schrödinger system: modulational instability and exact solutions. Phys. Scr. **77**, 1–8 (2008)
18. Shi, Q., Li, W., Wang, S.: Kato type for NLS equation in a scalar field and unique solvability of NKGS system in energy space. J. Differential Equations **256**, 3440–3462 (2014)
19. Shi, Q., Wang, S., Li, Y., Wang, C.: Well-posedness for the nonlinear Klein-Gordon-Schrödinger equations with heterointeractions. J. Math. Phys. **51**, 032102 (2010)
20. Kato, T., Ponce, G.: Commutator estimates and the Euler and Navier-Stokes equations. Commun. Pure Appl. Math. **41**, 891–907 (1988)

Printed by Printforce, the Netherlands